Mineral Engineering Conference (MEC 2017)

E3S Web of Conferences Volume 18 (2017)

Wisla, Poland
20 - 23 September 2017

Editors:

Andrzej Chmielarz
Andrzej Wieniewski

ISBN: 978-1-5108-4972-3

Printed from e-media with permission by:

Curran Associates, Inc.
57 Morehouse Lane
Red Hook, NY 12571

proceedings.com

Some format issues inherent in the e-media version may also appear in this print version.

Printed by Curran Associates, Inc. (2017)

For additional information, please contact EDP Sciences – Web of Conferences at the address below.

EDP Sciences – Web of Conferences
17, Avenue du Hoggar
Parc d'Activité de Courtabœuf
BP 112
F-91944 Les Ulis Cedex A
France

Phone: +33 (0) 1 69 18 75 75
Fax: +33 (0) 1 69 28 84 91

contact-edps@webofconferences.org

Additional copies of this publication are available from:

Curran Associates, Inc.
57 Morehouse Lane
Red Hook, NY 12571 USA
Phone: 845-758-0400
Fax: 845-758-2633
Email: curran@proceedings.com
Web: www.proceedings.com

TABLE OF CONTENTS

ADVANCED SEPARATION TECHNOLOGIES FOR PRE-CONCENTRATION OF METAL ORES AND THE ADDITIONAL PROCESS CONTROL 1
Kolacz Jacek

THE INFLUENCE OF LEAD ON CONCENTRATES QUALITY PRODUCED IN KGHM POLSKA MIEDZ CONCENTRATOR PLANTS 5
Pawlos Witold, Krzeminska Malgorzata

THE IMPORTANCE OF MIDDLINGS IN THE BENEFICIATION TECHNOLOGY OF POLISH COPPER ORE AND NEW POSSIBILITIES OF THEIR PROCESSING 8
Skorupska Bozena, Wieniewski Andrzej, Kubacz Norbert

RESEARCH ON THE POSSIBILITY OF SORTING APPLICATION FOR SEPARATION OF SHALE AND/OR GANGUE FROM THE FEED OF RUDNA CONCENTRATOR 15
Grotowski Andrzej, Witecki Kajetan

A MULTIDIMENSIONAL ANALYSIS AND MODELLING OF FLOTATION PROCESS FOR SELECTED POLISH LITHOLOGICAL COPPER ORE TYPES 21
Niedoba Tomasz, Surowiak Agnieszka, Pieta Paulina

EVALUATION OF FLOTATION PROCESS COURSE ON THE EXAMPLE OF SULPHIDE ORES 27
Oleksik Konrad, Saramak Daniel, MLynarczykowska Anna

THE EFFECT OF PROCESS WATER SALINITY ON FLOTATION OF COPPER ORE FROM LUBIN MINING REGION (SW POLAND) 31
Bakalarz Alicja, Duchnowska Magdalena, Luszczkiewicz Andrzej

POSSIBILITIES FOR REUSING THE WASTE FROM THE PROCESS OF ZN-PB ORE BENEFICIATION 37
Cichy Krystian, Tobiczyk Kornel, Manka Arkadiusz, Pajor Grazyna, ReguLa Cezary

EVALUATION OF FROTHERS BASED ON PROPERTIES OF THE SOLIDS/LIQUID/GAS SYSTEM 43
Drzymala Jan

DYNAMIC ASPECTS OF DESIGN AND MAINTENANCE OF THE ROTATING MACHINERY APPLIED IN THE MINING INDUSTRY 47
Szolc Tomasz, Konowrocki Robert, Pochanke Andrzej, MichajLow Maciej

TWO-STAGE METHOD FOR THE SIMULATION OF THE COMMINUTION PROCESS IN A HIGH-SPEED BEATER MILL 57
Marijnissen Michal J., Graczykowski Cezary, Rojek Jerzy

ANALYSIS OF A GRINDING EFFICIENCY IN THE ELECTROMAGNETIC MILL FOR VARIABLE PROCESS AND FEED PARAMETERS 64
Wolosiewicz-Glab Marta, Foszcz Dariusz, Saramak Daniel, Gawenda Tomasz, Krawczykowski Damian

EFFECT OF SOME OPERATING VARIABLES ON THE PERFORMANCE OF A 150 MM HEAVY MEDIUM CYCLONE TREATING HIGH ASH INDIAN COKING COAL 70
Charan Thonangi Gouri, Jha G.S., Sinha K.M.K., Chattopadhyay U.S., Sen Kalyan

REUSE OF CEMENT KILN DUST FOR BACKFILLING AND CO_2 CARBONATION 73
Lutynski Marcin, Pierzyna Piotr

AN UPDATE ON POSSIBILITIES OF METALS RECOVERY FROM POLISH COPPER ORES BY BIOTECHNOLOGY. PROJECT ECOMETALS 78
Szubert Agnieszka, Guezennec Anne-Gwénaëlle, Bodénan Françoise, Dirlich Stefan, Pawlowska Agnieszka, Grotowski Andrzej, Sadowski Zygmunt, Witecki Kajetan

STUDIES OF THE AROMATIC ß-DIKETONES AS EXTRACTANT OF COPPER IONS 84
Witt Katarzyna, Radzyminska-Lenarcik Elzbieta

STUDIES ON THE SEPARATION OF SOME TRANSITION METALS USING TRIALKYLIMIDAZOLE AS SELECTIVE EXTRACTANT 89
Radzyminska-Lenarcik Elzbieta, Witt Katarzyna

THE UPTAKE OF SILVER(I) FROM CHLORIDE SOLUTIONS BY AMINE EXTRACTANTS 94
Wejman Katarzyna, Pilsniak-Rabiega Magdalena

THE SORPTION OF METAL IONS ON NANOSCALE ZERO-VALENT IRON 99
Suponik Tomasz, Popczyk Marcin, Pierzyna Piotr

CONSIDERATIONS ABOUT RECOVERY OF CRITICAL METALS USING BIO-METALLURGY 104
Gherghe Sandica Liliana

ENVIRONMENTALLY FRIENDLY METHOD FOR REGENERATION OF COPPER CHLORIDE ACIDIC SOLUTIONS USED IN ETCHING OF PRINTED CIRCUITS 109
Kopyto Dorota, Baranek Wit, Myczkowski Zbigniew, Leszczynska-Sejda Katarzyna, Drzazga Michal, Matusewicz Maciej, Matusewicz Tadeusz, Matusewicz Wojciech

ROMANIAN BENTONITE AND FLY ASH CHARACTERIZATION AND THEIR USE IN HEAVY METAL "IN-SITU" IMMOBILIZATION IN POLLUTED SOILS 114
Gherghe Sandica Liliana, Anger Ildiko, Moise Georgiana, Trusca Roxana

SULFUR REDUCTION IN SANGAN IRON ORE BY FLOTATION 121
Rezaee Bahram, Sarvi Atefe, Eslamian Atiyeh, MehdiJebraeeli Seyed, Zabihi Abolfazl

CIRCULAR ECONOMY IN MINERAL PROCESSING 125
Pomykala Radoslaw, Tora Barbara

INFLUENCE OF LIBERATION OF SULPHIDE MINERALS ON FLOTATION OF SEDIMENTARY COPPER ORE 131
Bakalarz Alicja, Duchnowska Magdalena, Luszczkiewicz Andrzej

SOLIDIFICATION OF HAZARDOUS WASTE AS A PART OF THE RAW MATERIAL RECOVERY PROCESS 139
Polek Daria

SILVER RECOVERY FROM ZINC METALLURGICAL SLUDGE – ANALYSIS OF SOLUTIONS 143
Pietrzyk Stanislaw, Palimaka Piotr, Stepien Michal, Krawczykowska Aldona, Tora Barbara, Fatyga Miroslaw, Pajor Grazyna, Manka Arkadiusz

FLOTATION OF COPPER-BEARING SHALE IN SOLUTIONS OF INORGANIC SALTS AND ORGANIC REAGENTS 149
Ratajczak Tomasz

EFFECT OF FLY AND BOTTOM ASH MIXTURE FROM COMBUSTION OF BIOMASS ON STRENGTH OF CEMENT MORTAR 152
Ulewicz Malgorzata, Jura Jakub

EVALUATION OF THE RESULTS OF COAL JIGGING PROCESS 155
Surowiak Agnieszka

PHYSICAL CHEMICAL CHARACTERIZATION OF HISTORICAL MINING WASTE AND ARD PREDICTION TESTS 161
Filcenco-Olteanu Antoneta, Alakangas Lena, Fiuza Antoio, Zlagnean Marius, Tomus Nicolae

POSSIBILITIES OF USE OF WASTE SULFUR FOR THE PRODUCTION OF TECHNICAL CONCRETE 169
Grabowski Lukasz, Gliniak Maciej, Polek Daria

TESTING THE POSSIBILITY OF LEACHING SALT DEBRIS OBTAINED FROM UNDERGROUND EXCAVATIONS 171
Pierzyna Piotr, Popczyk Marcin, Suponik Tomasz

THE INFLUENCE OF ULTRASOUNDS ON THE PROCESS OF URANIUM DESORPTION 177
Panturu Eugenia, Filcenco-Olteanu Antoneta, Radu Aura Daniela, Zlagnean Marius

IMPACT OF SPIRAL SEPARATOR GEOMETRICAL PARAMETERS ON THE DENSITY SEPARATION OF VARIOUS FINE-GRAINED MATERIALS 188
Szpyrka Jan, Suponik Tomasz, Lutynski Marcin

Author Index

Mineral Engineering Council (MEC Council)

The MEC Council is a body elected from the organizing and supporting Mineral Engineering Conference institutions, and is responsible for choosing the date and venue of the MEC conference and ensuring a high standard of the conference. The Members of the MEC Council are:

Dariusz Foszcz – AGH University of Science and Technology, Kraków
Przemyslaw B. Kowalczuk – Wroclaw University of Technology
Marcin Lutyński – Silesian University of Technology, Gliwice
Witold Pawlos – KGHM Polska Miedź S.A., Lubin
Andrzej Grotowski – KGHM Cuprum Research & Development Centre, Wrocław
Andrzej Wieniewski – Institute of Non-Ferrous Metals, Gliwice

Conference Chairman

Zbigniew Śmieszek – Institute of Non-Ferrous Metals

Honorary Committee of MEC2017

Honorary Chairmen
Kazimierz Sztaba – Honorary Professor AGH University of Science and Technology
Janusz Laskowski – Profesor University of British Columbia, Canada

Radosław Domagalski-Łabędzki - President of the Board of KGHM Polska Miedź S.A.
Piotr Dytko - President of the Board of KGHM CUPRUM sp. z o.o.,CBR
Mirosław Kidoń - Director of O/ZWR KGHM „Polska Miedz" S.A.
Cezary Madryas - Rector of Wroclaw University of Science and Technology
Arkadiusz Mężyk - Rector of Silesian University of Technology
Bogusław Ochab - President of Board ZGH „Bolesław" S.A
Tadeusz Słomka - Rector of AGH University of Science and Technology

Programme Council of MEC2017

Chairman– Andrzej Wieniewski
Vice Chairman– Andrzej Chmielarz
Tomasz Chmielewski
Jan Drzymała
Andrzej Grotowski
Ewelina Kasińska-Pilut
Przemysław B. Kowalczuk

Marcin Lutyński
Tomasz Niedoba
Bożena Skorupska
Barbara Tora

Scientific Committee of the Conference

Ireneusz Baic, Instytut Mechanizacji Budownictwa i Górnictwa Skalnego, Katowice
Wiesław Blaschke, Polska Akademia Nauk, Kraków
Stanisław Chibowski, Uniwersytet Marie Curie-Skłodowskiej, Lublin
Andrzej Chmielarz , Instytut Metali Nieżelaznych, Gliwice
Tomasz Chmielewski, Politechnika Wrocławska, Wrocław
Jan Drzymała, Politechnika Wrocławska, Wrocław
Dariusz Foszcz, Akademia Górniczo-Hutnicza, Kraków
Tomasz Gawenda, Akademia Górniczo-Hutnicza, Kraków
Janusz Girczys, Politechnika Częstochowska, Częstochowa
Andrzej Grotowski, KGHM Cuprum Badawczo-Rozwojowe, Wrocław
Jan Hupka, Politechnika Gdańska, Gdańsk
Teofil Jesionowski, Politechnika Poznańska, Poznań
Andrzej Konieczny, KGHM Polska Miedź S.A., Lubin
Przemysław B. Kowalczuk, Politechnika Wrocławska, Wrocław
Aleksander Lutyński, Instytut Techniki Górniczej KOMAG, Gliwice
Marcin Lutyński, Politechnika Śląska, Gliwice
Andrzej Łuszczkiewicz, Politechnika Wrocławska, Wrocław
Kazimierz Małysa, Polska Akademia Nauk, Kraków
Tomasz Niedoba, Akademia Górniczo-Hutnicza, Kraków
Paweł Nowak, Polska Akademia Nauk, Kraków
Witold Pawlos, KGHM Polska Miedz S.A., Lubin
Joachim Pielot, Politechnika Śląska, Gliwice
Aleksandra Potulska, KGHM Cuprum Badawczo-Rozwojowe, Wrocław
Ireneusz Pyka, Główny Instytut Górnictwa
Tomasz Ratajczak, Politechnika Wrocławska, Wrocław
Zygmunt Sadowski, Politechnika Wrocławska, Wrocław
Daniel Saramak, Akademia Górniczo-Hutnicza, Kraków
Bożena Skorupska, Instytut Metali Nieżelaznych, Gliwice
Tomasz Suponik, Politechnika Śląska, Gliwice
Agnieszka Surowiak, Akademia Górniczo-Hutnicza, Kraków
Barbara Tora, Akademia Górniczo-Hutnicza, Kraków
Andrzej Wieniewski, Instytut Metali Nieżelaznych, Gliwice
Piotr Wodziński, Politechnika Częstochowska, Częstochowa

International Scientific Committee MEC2017

El-Sayed Abdel-Aal, Central Metallurgical R&D Institute, Egypt
Emre Naci Altun, Middle East Technical University, Turkey
Corby Anderson, Colorado School of Mines, USA
Jari Aromaa, Aalto University, Finland
Vedat Deniz, Hitit University, Corum, Turkey
Olof Forsen, Aalto University, Finland
Hylke J Glass, Camborne School of Mines, Univ. Exeter, UK
Shekwonyadu Iyakwari, Nasarawa State University, Nigeria
Janusz Laskowski, University of British Columbia, Canada
Dariusz Lelinski, FLSmidth, USA
Mari Lundstrom, Aalto University, Finland
Jan D. Miller, University of Utah, USA
Bahri Oteyaka, Osmangazi University, Turkey
Pradip, Tata R&D Design Centre, India
Fereshteh Rashchi, University of Tehran, Iran
Oktay Sahbaz, Dumlupinar University, Turkey
Yakup Umucu, Suleyman Demirel University, Turkey

Organizing Committee MEC2017

Chairman – Andrzej Wieniewski
Vice Chairman – Bożena Skorupska

Maria Grzesik
Anna Królikowska
Adam Mańka
Alicja Skotnicka
Łukasz Skrzypiec
Anna Suruło
Kornel Tobiczyk

edp sciences

E3S Web of Conferences

Statement of Peer review

In submitting conference proceedings to E3S Web of Conferences, I certify to the Publisher that I adhere to the Policy on Publishing Integrity of the journal in order to safeguard good scientific practice in publishing.

1. All articles have been subjected to peer review administered by the proceedings editors.
2. Reviews have been conducted by expert referees, who have been requested to provide unbiased and constructive comments aimed, whenever possible, at improving the work.
3. Proceedings editors have taken all reasonable steps to ensure the quality of the materials they publish and their decision to accept or reject a paper for publication has been based only on the merits of the work and the relevance to the journal.

Title, date and place of the conference

MEC2017 - Mineral Engineering Conference, 20-23.09.2017, Wisla, Poland

Proceedings editor(s):

1. Andrzej CHMIELARZ PhD, prof.IMN, Institute of Non-Ferrous Metals Gliwice, Poland e-mail: andrzejch@imn.gliwice.pl
2. Andrzej WIENIEWSKI, PhD, prof.IMN, Institute of Non-Ferrous Metals, Gliwice, Poland, e-mail: andrzejw@imn.gliwice.pl

Date and editor's signature 24.08.2017

Advanced separation technologies for pre-concentration of metal ores and the additional process control

Jacek Kolacz[1,*]

[1]Comex AS, Box 53, 1309 Rud, Norway

Abstract. A new sensor based sorting system has been developed at Comex, based on complex image analysis. The sorting system employs different images generated by visible light and X-ray to provide maximal information about processed particles. Application of X-ray makes it possible to analyse the internal structure of the particles, which provides a 2D image with the valuable information. The advanced system for the texture and the pattern recognition is applied to analyse these images simultaneously, to provide the required identification of the particles for efficient separation. The paper describes a number of tests during sorting of Cu-Zn-Sn ores with high separation efficiency. As an example, the Cu content is enriched from 0.4% to 1.25% and the Zn content is increased from 0.83% to 2.24%. In addition, the performance of the X-ray sorting system is described for enrichment of Zn-Pb ore and compared with the heavy media separation technology. Simplifying the pre-concentration through the use of the sorting technology can significantly reduce the cost of these operations. Finally, the operating data from the sorting system have been analysed regarding a potential use for controlling the plant operation. This can bring the new tools for control strategies for the complex processing plants.

1 Current technology

Processing of metal ores including zinc and lead, is mainly based on gravity separation equipment. It often includes applying dense media separation systems, especially for large particles. Such equipment provides an efficient separation, however, it has a number of disadvantages and limitations. The most important of them can are listed below:

- It is necessary to apply complicated media conditioning equipment to achieve a stable liquid density, like pumps, dosage systems, mixers, density sensors, etc.
- It requires applying heavy density materials like ferrosilicon, which are expensive.
- The high density materials must be washed thoroughly from the particles to avoid loses and limit product contamination.
- The system requires use of water, which must be purified and recycled. It also requires heating in low temperature environments.
- The washing system is mechanically complicated with many moving parts, which require servicing and maintenance.
- The complete system requires a lot of space in form of both height and foot print area.

All these disadvantages provide a high cost of operation [1] especially for the low production capacity. It is very often not feasible to apply such systems for low or medium production scale. Consequently, it is necessary to build the central processing plants, which require more transportation lines and complicated logistics.

Finally, the dense media separation process provides a relatively low separation sharpness. The particles, which have very high or very low density, when compared to the suspension density, will very quickly sink or float, thus providing a good separation effect. However, when the separated material has a density closer to the suspension, the sinking or floating effect will require more time and this can create more chances for misplaced particles during this process. Such effect is not taking place when sensor based sorting is applied.

2 X-ray sorting technology

Application of the sensor based sorting systems, without any doubt, brings a lot of possible advantages in many mineral operations. The main advantage of the new system is related to its universality and at the same time very sophisticated image processing functions, which can be carried out in the same processing unit [2]. Figure 1 shows the CXR-1000 system from Comex with 1 m belt width, in the configuration where many different analysed parameters can be used to provide particle separation. The image analysis system includes a camera installed either over the transport belt conveyor or at its discharge end. The system includes the X-ray attenuation analysis realized by the XRT system in the central part of the conveyor belt. The sorting system can be used with both

* Corresponding author: jacek.kolacz@comex-group.com

optical and XRT analysis or separately depending on an application.

Particle recognition used to separate different materials is based on a complex shape and colour analysis where the particles can also be identified by over 20 parameters used for shape description. Some of them are: diameter in different orientations, perimeter, centre of mass, moment of inertia, particle elongation factor, edge sharpness, etc. Additional combinations of these parameters can also be used for distinguishing particles of interest. The surface of particles where different colours or contours vary in intensity and frequency can be analysed by FFT filtration (Fast Fourier Transformation) to recognize differences in texture and structure of the processed particles. This analysis brings much more complex information about the analysed particles rather than colour recognition alone. Finally, the XRT picture is integrated into the optical analysis, which provides a lot more information about the particle surface properties and its internal structure. All these sophisticated analysing functions require a lot of computation power and they have to be optimized to allow high capacity sorting. This is done by special program architecture and algorithm solutions allowing efficient management of the calculation routines and sorting priorities. This allows achieving still high separation capacity and extremely high efficiency [2], where the product purity can reach even 99.9 %. Such results can be achieved in different production scale depending on the process requirements [3].

The sorting system as shown above can be configured with different sensors depending on the particular application. In the test work as described in this paper, only the X-ray sensor was applied since the optical properties did not show any possibility for differentiation of particles of interests. The X-ray sensor provided possibility to calculate a density of separated particles. Therefore, it was possible to compare this new sorting technology with dense media separation as the traditional processing method in this process. Such sensor based sorting systems are already successfully tested and applied in the metal ore processing like iron ore [2] including Zn/Pb ore [4] and coal [5].

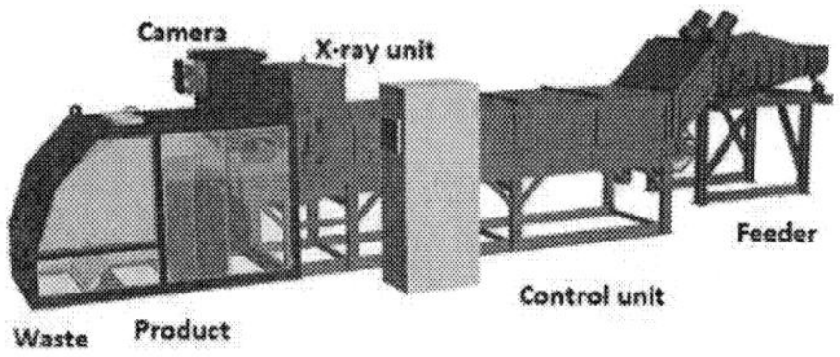

Fig. 1. Comex CXR sorting system – configuration with the X-ray and the optical analysis.

Application of the sensor based sorting is very attractive when compared to other separation technologies. It provides the following advantages, when compared to the DMS systems:

- Possibility to apply in small capacity production lines
- Small space requirement
- No use of water (it is completely dry technology)
- No complicated additional equipment like, pumps, mixers, conditioners, recuperation devices, instrumentation, etc.
- No requirement of high density powder (ferrosilicon)
- Lower investment cost (about 25-50%)
- Very low operating cost (about 15-20 time lower)

It was therefore very interesting to compare the separation efficiency using the similar input material.

3 Comparison of the DMS and X-ray separation results

The DMS and the X-ray sorting system have been compared on the base of the same feed material. The test work was performed in the zinc and lead mine DPM „Olkusz – Pomorzany" in Poland, where the DMS system was applied together with flotation as the main enrichment process [4]. The DMS system was employing the Wemco unit operating in 20-60 mm size range. It was possible to sample the input and outlet fractions and carry out the necessary chemical analysis and gravity separation tests. The samples included few hundred kg material fractions, taken under the stable operating conditions of the DMS system. The results of the detailed analysis of the feed, sink and float fractions are shown in Table 1.

Table 1. Separation results in the Wemco DMS system.

Fraction	Rec. [%]	Grade [%]			Recovery [%]		
		Zn	Pb	Fe	Zn	Pb	Fe
Feed	100	1.80	0.80	4.17	100	100	100
Sink (heavy)	19.13	7.63	3.93	16.80	81.31	93.77	76.77
Float (light)	80.87	0.41	0.06	1.20	18.69	6.23	23.23

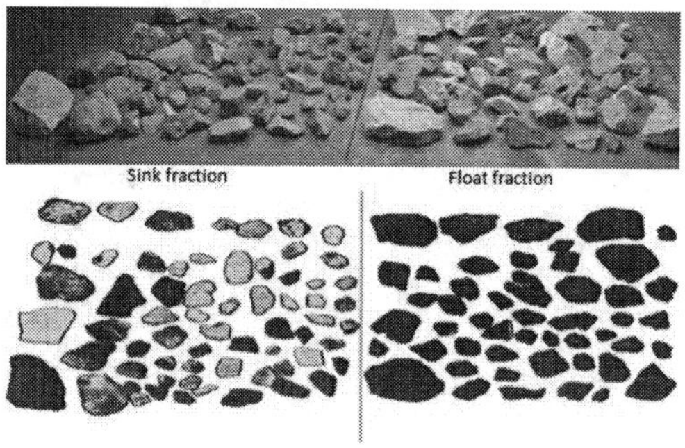

Fig. 2. Picture of the DMS outlet fractions together with the X-ray image of the same particles.

In addition, the sink and float fractions, were analysed by the Comex CXR sorting unit to investigate the content of high and low density particles. Figure 2 shows these fractions in form of the optical picture and the X-ray image. The blue colure shows the low density particles and the red and yellow colours indicate the high density material. It can be noticed that the sink fraction still contains the low density particles which can be recovered by the sorting unit and separated to the waste fraction. It was therefore concluded that the separation process in the

DOI:10.1051/e3sconf/20171801001

X-ray sorting unit might provide better result in form of increased recovery or grade values.

The Comex CXR-1000 X-ray sorting unit with 1 m belt width was used for testing. It was fed with the same material as the feed sample taken from the DMS unit. It was therefore possible to compare the separation effect from both separating units using the identical feed material composition. The separation results are shown in Table 2.

Table 2. Separation results in the CXR sorting system.

Frac.	Rec. [%]	Grade [%]			Recovery [%]		
		Zn	Pb	Fe	Zn	Pb	Fe
Feed	100	1.80	0.80	4.17	100	100	100
High dens.	10.96	5.42	15.36	79.16	87.12	53.40	10.96
Low dens.	87.00	0.43	0.12	2.00	20.84	12.88	46.60

When comparing the results in Table 1 and 2, it can already be noticed that the a different enrichment level of investigated minerals was registered for various recoveries. For better illustration, the rade and the recovery points were shown on Figure 3.

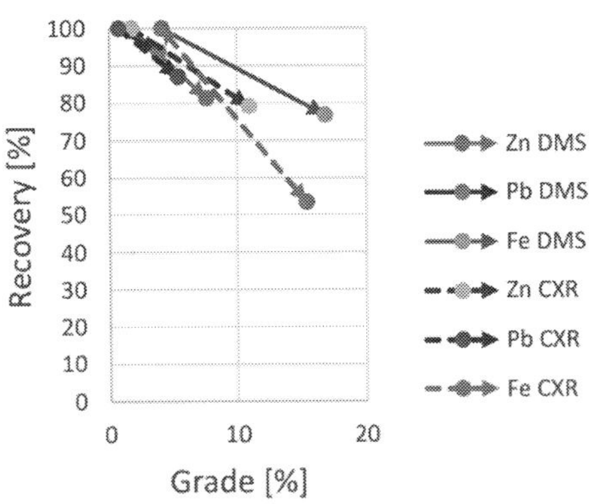

Fig. 3. Grade-recovery illustration for the DMS and X-ray sorting processes.

For the DMS process, the Pb content was with lower grade but higher recovery, when compared to the X-ray sorting. However, this can be adjusted by setting of the sorter to achieve a similar level of both parameters. Regarding the Zn separation, there is a significant difference in grade for the similar recovery level, where the X-ray sorting provides almost 11% grade, while DMS only 7.6%, for the similar recovery of about 80%. Finally, the recovery of Fe is much lower for and the grade is reduced, which is beneficial for this particular process of Zn-Pb production.

4 Other sorting applications

Another application of the CXR sorting unit is shown for the Cu-Zn-Sn ore in one of the mines in China. In this case the ore was investigated in two main material streams where the Cu concentration was the main component in the first case, and the second stream contained mainly Zn and Sn. In this case the material was processed by the external processing plant where the large quantities of the ore had to be transported with significant distances. The sorting solution with the CXR system was suggested to provide improvements in the overall process. The separation results are shown in Table 3 for the Cu ore. In this case, the feed material containing 0.404% of Cu was enriched to 1.258% with 85.65% recovery of Cu. At the same time the concentrate stream after sorting system, corresponded to only 27.52% of the incoming mass. It means the pre-concentrated material was significantly reduced in quantity before further processing.

Table 3. Cu ore separated in the CXR sorting system.

Fraction	Yield [%]	Density [kg/l]	Test Elements [%]		
			Cu	Zn	Sn
Cu Conc.	27.52	3.09	1.258	0.989	0.299
Cu Tailing	72.48	1.92	0.080	0.151	0.096
Cu Feed	100	2.24	0.404	0.382	0.152
Fraction	**Yield [%]**	**Density [kg/l]**	**Recovery [%]**		
			Cu	**Zn**	**Sn**
Cu Conc.	27.52	3.09	85.65	71.32	54.18
Cu Tailing	72.48	1.92	14.35	28.68	45.82
Cu Feed	100	2.24	100	100	100

Table 4. Zn/Sn ore separation in the CXR sorting system.

Description	Yield [%]	Density [kg/l]	Test Elements [%]		
			Cu	Zn	Sn
Zn/Sn Conc.	28.30	2.87	0.145	2.249	0.655
Zn/Sn Tailing	71.70	2.65	0.023	0.273	0.097
Zn/Sn Feed	100	2.71	0.058	0.832	0.255
Description	**Yield [%]**	**Density [kg/l]**	**Recovery [%]**		
			Cu	**Zn**	**Sn**
Zn/Sn Conc.	28.30	2.87	71.33	76.48	72.72
Zn/Sn Tailing	71.70	2.65	28.67	23.52	27.28
Zn/Sn Feed	100	2.71	100	100	100

The similar results were achieved for the Zn-Sn ore as shown in Table 4. In this case, the feed contained 0.832% of Zn and 0.255% of Sn. After sorting, the metal content was increased to 2.249% of Zn and 0.655% of Sn with the recoveries of 76.48% and 72.72% correspondingly. At the same time, the material quantity was reduced to 28.3% of the total incoming mass to the sorting process. Again, this provided huge savings in terms of the further processing cost.

5 Operating cost improvements

Sorting systems have the important limitation related to the capacity of the single processing unit. This is related to the operating principle where the sorting process is carried out in the mono-layer of the particles introduced to the sorting unit. It means that higher throughputs must be provided by the parallel configuration of the sorting units. Dry X-ray separation equipment provided by Comex fulfils this main cost requirement for multiple stage separation keeping the total cost at a competitive level. The resent development and progress in microelectronics provided quite a number of new possible solutions within automation and separation techniques. Comparison with the traditional processing methods like DMS, indicate very clearly that the dry X-ray sorting systems can be very competitive regarding investment cost, operating cost and other environmental aspects like no need of water. Simple comparison of the DMS washing plant and the sorting plant based on CXR Comex sorters, is shown in Table 5.

Table 5. Cost comparison between the traditional washing plant and the Comex dry beneficiation plant.

Option		DMS plant	Comex dry sorting plant
Processing capacity	[MTPA]	4.5-5	4.5-5
Water consumption	[l/t]	1-2	0.01
Investment cost	[USD]	15 mln	10-12 mln
Operating cost	[USD/t]	**1-1.5**	**0.05-0.1**

The traditional DMS plant is characterised by a simple operating principle, however, it requires a lot of moving parts, motors, belts, pumps, magnetic recuperation as well as access to water and heavy media like ground magnetite or ferrosilicon. This result in high both investment and especially operating cost. Based on the average plant size providing 5 MTPA the investment cost is about 15 mln USD and the operating cost ranges about 1-1.5 USD per ton. In addition there is a cost related to water supply which is very difficult to calculate especially for hot climate countries.

For the alternative process where the CXR sorters are employed, the separation can be done in combination of parallel-serial connections of the sorting units and the necessary screening equipment. The investment cost ranges between 10-12 mln USD and the operating cost is about 0.05-0.1 USD per ton. It means the operating cost of the X-ray sorting plant is about 15-20 times lower then the corresponding process employing DMS. The benefit of using the dry X-ray sorting plant is growing rapidly when the processing capacity is lower. For smaller mines, it is often not feasible to build the washing plant while the dry sorting system having smaller capacity, can easily be applied.

6 Data processing potential

The sorting system is usually installed at the high throughput streams, which are at the early stage of processing. It means many particles are passing the sensors of the sorting equipment and these particles very often correspond to the ROM material or the main material introduced for the enrichment process. By collecting this data on-line, it is possible to control the rest of the process by adjusting its parameters to obtain the optimal conditions and adjusting the flowsheets streams. The control system can adjust e.g. the reagents concentration used in floatation based on the information about the metal or rejects concentration in the ore. It is also possible to use a fuzzy logic control to adjust the processing paths in the flowsheet to get the best performance of the plant. It brings new possibilities for optimising the complex processing plants. This knowledge has never been used before and it will be further investigated in the future work.

7 Conclusions

The CXR X-ray sorting unit can provide a similar or even better separation effect than DMS system, during pre-concentration of the Zn-Pb ore before further processing. The similar separation process can be realised by X-ray sorting as a much simpler system, regarding lower number of moving parts, lower complication level, as well as reduced maintenance and servicing. No need of water and the high density weight material, can be important for more flexible and more environmentally friendly operation. Finally, in terms of the reduced investment and operating cost (by 15-20 times), it can dramatically improve the mine operation regarding economic aspects. Finally, the sorting results can be transferred to the valuable data describing the input material. Such information can be used for controlling the rest of the processing plant by adjusting the control models for flotation regents and different flowsheet configurations.

References

1. Wills B.A., *Wills' Mineral Processing Technology: An Introduction to the Practical Aspects of Ore Treatment and Mineral Recovery* (2006)
2. Kolacz J. Sensor based sorting with signal pattern recognition: the new powerful tool in mineral processing (XXVII International Mineral Processing Congress, Santiago-Chile,106-115, 2014)
3. Kolacz J. Efficient scale up of sensor based sorting systems. XXVIII International Mineral Processing Congress, Quebec, Canada, 2016)
4. Wieniewski at al., A. Wieniewski , E. Szczerba, A. Nad, R. Luczak, J. Kolacz, Szewczuk, Evaluation of the application possibilities of modern separation techniques for pre-concentration of the Zn-Pb ore, (XI International Non-Ferrous Metals Conference ICNOP, Trzebieszowice, Poland, 2015)
5. Kolacz J. Dry coal beneficiation by sensor based sorting (Mineral Engineering Conference, Szczawnica, Poland, 2015)

E3S Web of Conferences **18**, 01002 (2017) DOI:10.1051/e3sconf/20171801002
MEC2017

The influence of lead on concentrates quality produced in KGHM Polska Miedz concentrator plants

Witold Pawlos[1,*], and *Małgorzata* Krzemińska[1]

[1]KGHM Polska Miedź S.A., Oddział Zakłady Wzbogacania Rud, ul. Kopalniana 1, 59-101 Polkowice, Polska

Abstract. The behaviour of lead compounds present in feed of beneficiation process has been discussed. Changeability of copper and lead content in feed during last decades has been described. Galena was pointed as a major source of lead and the beneficiation kinetic of this mineral was presented comparing with other sulphides, including copper sulphides. Significant amount of galena together with its very good concentrating properties crucially effect on quality of concentrates produced in Lubin concentrator plant.

1 Introduction

One of the most important issue of ore processing in Lubin concentrator plant is occurrence accompanied metals with crucial position of lead. The presence of lead in feed significantly influences on grade and quality produced concentrates. Lead is strictly connected with copper orebody or occurs in adjacent roof rocks formations. It makes selective excavation of copper ore difficult or even impossible. Attempts of elimination of lead during flotation process were also unsuccessful. The gradually increase of lead content in feed and consistently raised lead amounts in concentrates effects on achieved ore processing indexes, what was discussed in this paper.

2 The lead content in ore

During last years the increase of lead content with simultaneous decrease of copper amounts in feed is being observed at Lubin concentrator plant. The graph showing these phenomena is presented on Fig. 1 where changeability curves of Cu and Pb content in ore in 1999-2015 are plotted.

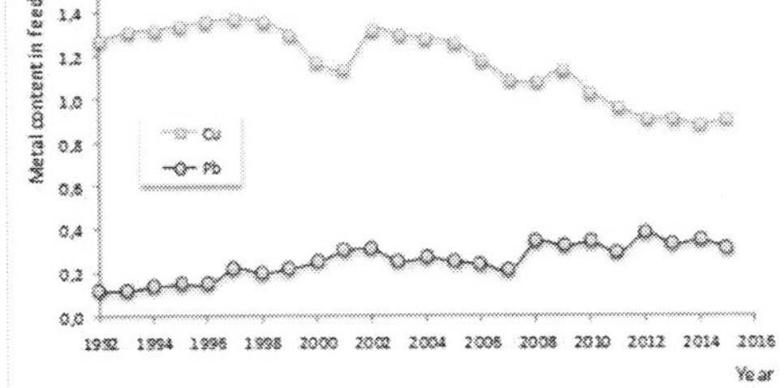

Fig. 1. The changebility of copper and lead in ore processed at Lubin concentrators in 1992-2015 (O/ZWR own materials).

The changes in copper and lead contents observed during last years allow us to make a conclusion that feed to process realized in Lubin concentrator is copper-lead ore (the copper to lead ratio is 2:1). Lowering the Cu;Pb ratio is the reason of emerging technological issues, especially with achieving the proper quality of final copper concentrates.

In the past many efforts to control the lead content in produced concentrates have been undergone [1]. During flotation research some lead depressors were examined, which are well known and widely used in froth flotation of copper and lead sulphide minerals. Among others the addition starch and dextrines was applied [2-5]. I spite of broad attempts facing the problem of high leads levels in concentrates, no proven solution has been found. Getting higher lead contents effects negatively on concentrate grade at acceptable copper recovery values.

3 The mineral formations of lead

In polymetallic copper ore deposit recognised as Legnica-Glogow Copper Basin, lead is present exclusively in a form of galena (PbS), and apart from it there is no other significant lead sources.

Galena creates grains in size from 10 μm to some millimetres, mainly in range 30-200 μm. The grains are dispersed in matrix made of clay-dolomite and carbonates rock formations. Galena can occurs also as aggregates of irregular shapes. In some small amounts, not more than 15-20% galena is present as associations, mainly with sphalerite, but also with pyrite, chalcopyrite, rarely with bornite [6].

Mineral composition of ore processed in Lubin concentrator plant varied significantly for last decade (Fig. 2). The considerable increase of galena content among other sulphides is noticeable (starting at 2% level to 20% of all sulphides) while copper bearing sulphides decrease from 80% to 50% of all sulphides. It is confirmed in chemic al assay what is presented on Fig. 1.

* Corresponding author: witold.pawlos@kghm.com

DOI:10.1051/e3sconf/20171801002

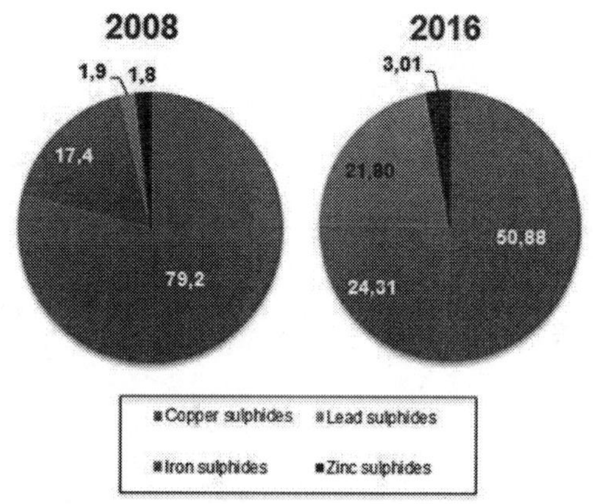

Fig. 2. The mineral composition of feed to Lubin concentrator plant [7,8].

4 The mineral formations of lead

During froth flotation process lead has worse upgrading properties than copper [9], see Fig. 3.

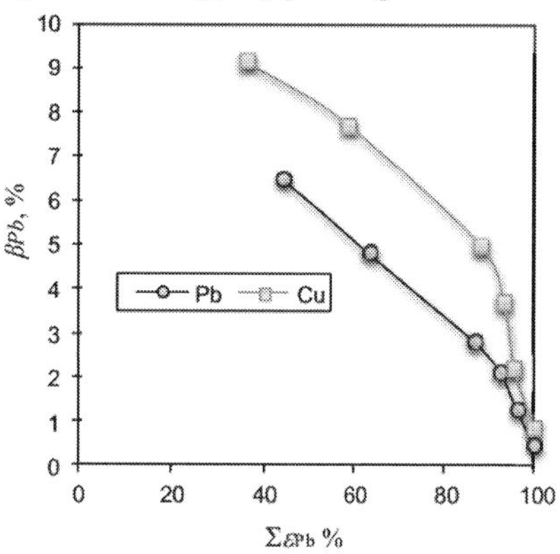

Fig. 3. Halbich upgrading curves for copper and lead at Lubin concentrator plant.

According to bibliography sources [10] galena is classified as a naturally slight hydrophobic mineral and collector-less floating sulphide. Insignificant hydrophobic properties of galena was also confirmed by Drzymala [11] in his research. Some investigations undergone in KGHM Division of Concentrators proved that galena is easily upgraded mineral (see Fig. 4).

High upgrading properties of galena effects elevated amounts lead in concentrates accompanied with high recoveries of this metal during ore beneficiation process realized in Lubin plant. Increasing galena content with simultaneously decreasing content of copper bearing sulphides is responsible for changes in Cu/Pb content in ore processed. It affects directly on concentrates grade. Hence the influence of lead present in concentrate on lead content in concentrate in period 2005-2015 is shown on Fig. 5.

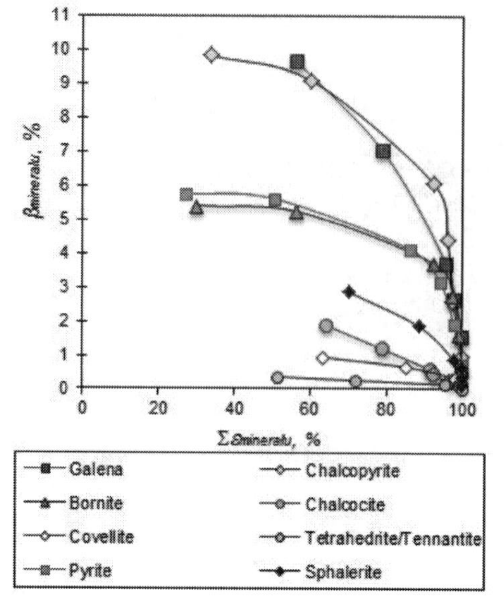

Fig. 4. Halbich upgrading curves for main sulphide minerals present in feed to Lubin concentrator plant.

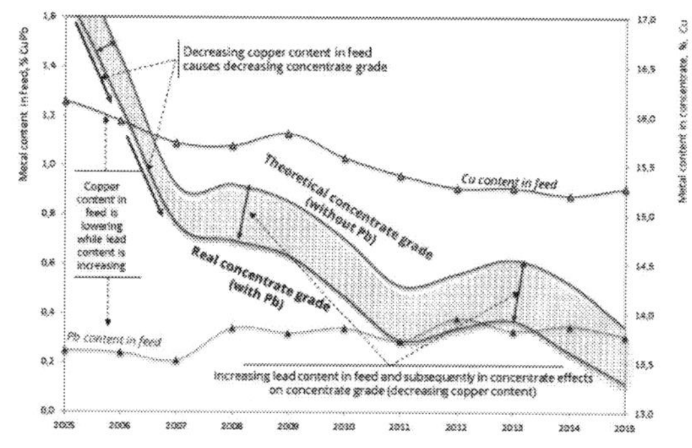

Fig. 4. Influence of Pb content on concentrate grade in 2005-2015.

4 Summary

The technology applied in KGHM Polska Miedz S.A. Division of Concentrators is based on froth flotation beneficiation process focused on production of bulk concentrates collecting mainly copper, silver and other metals sulphides. The concentrate as the final product of ore processing contains copper but also significant amounts of other components including lead. This is one of the reason of decreasing copper concentrate grade. It is concluded that concentrate produced in Lubin plant is rather copper-lead than "pure copper" concentrate. It should be consider to apply solutions to allow separate copper sulphide apart of other sulphides and waste minerals responsible for decreasing ore beneficiation indexes. However, it must be emphasized that effective lead recovery during pyrometallurgical process in copper smelters can provide benefits as it is one of commercial product of KGHM Polska Miedz S.A.

References

1. R. Bortel, N. Kubacz, Zeszyty Naukowe AGH, Górnictwo z. **132**, (1987)
2. A.M. Gaudin A.M.: *Flotacja*. (Wydawnictwo Śląsk, Katowice, 1963)
3. Lin K.F., Budrick C., L.: *Polymeric Depressants. Reagents in Mineral Technology*, (1988)
4. Q. Liu, J.S. Laskowski, Int. J. Miner. Process. **27**, (1989)
5. A.V. Lopez, A.A. Sanchez, S. Song, H.A. Martinez Garcia, Canadian J. Metall. Mater. Sci. vol. **46**, (2003)
6. P. Kijewski, Physicochem. Probl. Miner. Process., **20**, 89-96 (1988)
7. K. Trybalski, A. Łuszczkiewicz, J. Drzymała, T. Tumidajski, D. Foszcz, A. Muszer, T. Niedoba, T. Henc, *Określenie wpływu wzbogacalności przerabianych rud na jakość koncentratów miedziowych dla potrzeb optymalizacji górniczo-hutniczego procesu wytwarzania miedzi, Etap II: Badania wzbogacalności urobku planowanego do wydobycia*, Akademia Górniczo-Hutnicza, (raport z badań), Kraków (2008)
8. A. Łuszczkiewicz, M. Duchnowska, A. Bakalarz, A. Muszer, J. Drzymała, M. Komorowska, *Ocena i charakterystyka krajowych rud miedzi, produktów i półproduktów obecnych i badanych technologii pod kątem innowacyjnych rozwiązań technologicznych ich wzbogacania*, Raport nr W-6/S-010/2016. (2016)
9. A. Łuszczkiewicz, D. Foszcz, J. Drzymała, M. Duchnowska, A. Bakalarz, D. Szyszka, T. Tumidajski, Ż. Konopacka, T. Niedoba, P.B. Kowalczuk, J. Hupka, M. Niewiadomski, P. Karwowski, K. Księżniak, A. Rogala, *Opracowanie metodyki badania odczynników flotacyjnych pod kątem ich własności użytkowych,* Raport Instytutu Górnictwa Politechniki Wrocławskiej, nr I-11/2013/S-30, Wrocław (2013)
10. G.W. Heyes, W.J. Trahar, Int. J. Miner. Process. **4**, 317-344 (1977)
11. J. Drzymała, *Podstawy mineralurgii,* (Oficyna Wydawnicza Politechniki Wrocławskiej, Wrocław, 2001)

E3S Web of Conferences **18**, 01003 (2017)
MEC2017
DOI:10.1051/e3sconf/20171801003

The importance of middlings in the beneficiation technology of Polish copper ore and new possibilities of their processing

Bozena Skorupska[1,*], *Andrzej* Wieniewski[1], and *Norbert* Kubacz[1]

[1]Institute of Non-ferrous Metals, ul. Sowinskiego 5, 44-100 Gliwice, Poland

Abstract. In the paper, the middlings (tailings of I cleaner flotation) were characterized, and new possibilities of their processing were presented. The advantages of introducing a separate circuit for middlings processing were emphasized. This circuit should be adapted to the properties of the processed middlings and characterized by relative simplicity. An important element of the separate processing of middlings is grinding. The paper presents the advantages of the grinding process in the magnetic mill (MM), which has not been used until now in the enrichment of mineral raw materials. The importance of introducing for the difficult to treat middlings a separate flotation in a flotation machine characterized by a high dynamics of the process was emphasized.

1 Introduction

One of the reasons for technological problems of copper ore enrichment at KGHM PM SA are circulating in the system so called middlings. They arise as a result of cleaning the concentrates, as tailings of I cleaning operation. The diagram of middlings formation is shown in Figure 1. The scale of the problems posed by the middlings is high due to their poor enrichability. The high content of copper in the middlings does not allow them to be sent to tailings, but it is too low to be a component of the final concentrate. As a result, how much of copper obtained from the middlings will be directed to the concentrate and how much to tailings depends on the technology of middlings processing used and the quality requirements for the final concentrate.

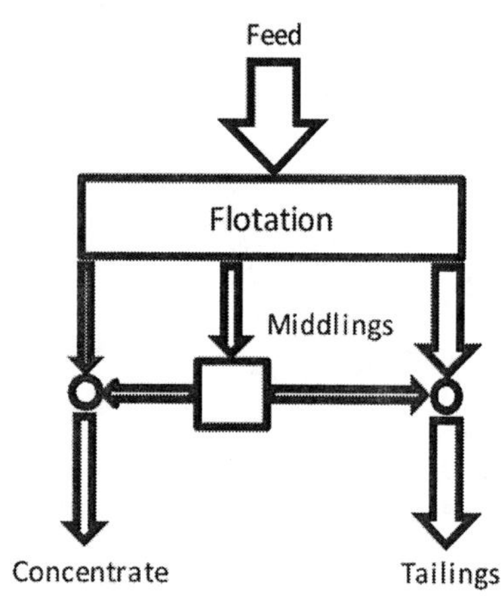

Fig. 1. The basic diagram for the division of the middlings into concentrate and tailings.

In recent years, in all enrichment plants, despite the decrease in copper content in the ore directed to the enrichment, the copper content in the middlings increased. This is due to a change in the enrichability of the processed ore resulting from an increased share of the difficult to treat shale fraction in the extracted ore. This fraction is a carrier of large quantity of organic matter which influences negatively on the flotation process. The increased amount of shale in the processed ore results in yield increase of froth product in the rougher flotation, thereby increasing the circulating amount of middlings. It negatively affects the enrichment process and causes an increase of metal losses in tailings.

At present (2015), it is estimated that in the middlings circulates around 70 to over 80% of mass of ore carrying about 180 to over 200% of copper, which occurs mainly in the form of unliberated copper sulphides. An additional problem is the accumulation of organic carbon – the middlings contain it between tens and hundreds of percent.

Efforts should be made to reduce the amount of middlings by improving their floatability. This task is not easy and has not yet been resolved in a satisfactory way. It should be emphasized that the kind of ore determines the scale of problems posed by middlings.

Therefore, in most enrichment plants in the world, in the case of processing the easy to treat porphyry ores, the problem of middlings is of less importance and the existing method for their processing is satisfactory.

The problem of middlings is usually solved in two ways. One is to return the middlings to the front of the rougher flotation, where together with the new batch of ore is subjected to flotation. In some cases, the middlings are subjected, before they are directed to the front of flotation, to classification and regrinding. The second solution is to regrind the entire middlings with the

* Corresponding author: bozenas@imn.gliwice.pl

subsequent flotation until the final tailings are obtained. Each method has its own positive and negative sides.

In the case of the easy to treat porphyry ores, it is preferable to return the middlings to the front of flotation. A necessary condition is a low yield of froth product so that the amount of circulating middlings is low.

This condition is not met for ores with clay content and/or containing gangue components of good floatability (e.g. organic matter). In this case, the amount of circulating middlings is excessively increasing, and what is more, the unwanted ore components accumulate in them in large quantities.

Therefore, in the case of processing the domestic copper ores, the quantity and quality of the circulating middlings justify their separate processing with the separation of the second component of the final tailings. The separation of the middlings stream will facilitate the selection of the most advantageous conditions for their enrichment and the maintenance of the technological regime of processing.

The additional benefit of separate processing of middlings would also be the improvement of operating conditions of the rougher flotation, which feed would be deprived of the difficult to treat fraction derived from middlings.

2 Description of middlings and their current processing

Middlings are characterized by significantly poorer enrichability than raw ore. The main cause is the fine graining of useful minerals and the presence of organic matter that promotes good floatability of the waste rock. The fine particle size distribution of copper sulphide minerals causes that most of the mineral grains are present in unliberated form.

The chemical characteristics of middlings produced in all enrichment plants are shown below, while Table 1 shows the basic granulometric and mineralogical characteristics of the middlings originating from the Rudna Concentrator (2015).

Rudna Concentrator

• sandstone middlings – Cu about 13%; C_{org} about 5%

• carbonate middlings – Cu about 6%; C_{org} about 6%

Polkowice Concentrator

• middlings from II technological line – Cu about 6%; C_{org} about 8%

Lubin Concentrator

• middlings from I technological line – Cu 2.3%; C_{org} 9.8%

Current processing of middlings consists in their regrinding in mills with cylpebs as grinding media, then classification in hydrocyclones and return to the rougher flotation. Recirculation of middlings causes the overgrinding of copper sulphides and their accumulation in circuits. This also applies to other unfavourable ore constituents, including organic carbon, clay minerals, etc. This negatively affects the operation of the rougher flotation and the achieved final indexes of the enrichment process.

Another solution for the processing of middlings was applied at the Polkowice Concentrator. It consists in changing the enrichability of the floated middlings by chemical treatment - partial leaching of the gangue components from the intergrowths (chemical grinding). This process involves the liberation of fine copper sulphides, [1, 6].

As a result of the chemical modification of the middlings from the enrichment of carbonate copper ores process at the Polkowice Concentrator, the recovery of copper in the concentrate was increased by 2.60% and Cu content by about 1.50%, [2].

3 Technology of separate processing of middlings

In the case of domestic enrichment plants, both the mass and the amount of circulating copper justify a separate middlings processing circuit. On the one hand, its task would be to improve the enrichability of the middlings (e.g. through better liberation of useful minerals) and, on the other, an effective flotation. The middlings processing circuit should be tailored to the properties of the processed middlings and characterized by relative simplicity.

Table 1. Particle size distribution, chemical composition and the degree of liberation of sulphide minerals in the middlings at the Rudna Concentrator (QEMSCAN® analysis).

Grain class [mm]	Sandstone fraction			Carbonate fraction		
	Yield [%]	Cu [%]	Liberation degree*	Yield [%]	Cu [%]	Liberation degree*
0.2-0.1	35.51	16.41	18.94	22.93	5.42	4.40
0.1-0.075	18.43	16.30	9.50	14.14	4.51	6.05
0.075-0.045	17.80	12.69	6.49	20.09	3.24	8.71
0.045-0.036	3.97	13.76	1.80	5.06	3.51	2.61
0-0.036	24.29	9.53	8.51	37.78	2.89	15.80
Sum	100.00	13.6	45.25	100.00	6.20	37.55

* - degree of liberation of copper sulphide minerals 80-100%

Essential components of middlings processing (Fig. 2):
• regrinding in mills of special construction (Vertimill®, Isamill™ or other grinding mills which ensure fine grinding such as magnetic mills)
• separate flotation in a flotation machine with a limited chamber size (not more than 10-15m^3) operating in dynamic conditions; flotation is preferably preceded by intensive conditioning of the middlings (HIC), [3].
• maintaining an optimal technological regime (set of reagents, flotation density, etc.)
• directing the concentrate from the middlings processing circuit to the cleaner flotation and the tailings to the final tailings.

Due to the fact that significant amounts of copper sulphides in middlings are present in unliberated form, effective regrinding is an important issue.

This task is well met, for instance, by mills of special construction (Vertimill® or IsaMill™), which allow to obtain fine grain composition, usually below 75μm, with less energy consumption by about 30% than in conventional tumbling mills, [4]. In recent years on Polish market has appeared innovative technology of fine grinding in magnetic mill, [5].

The Institute of Non-ferrous Metals has a laboratory stand equipped with a magnetic flow mill. The mill's operating chamber is stationary, while the grinding operation is performed by very small grinding media that move with great intensity under variable magnetic field. The field is generated by a coil winding on the body of the mill chamber. This mill has successfully passed the laboratory tests in the range of grinding of middlings.

The mill is currently under construction on a semi-technical scale. A pilot plant for the flotation in a closed circuit will be equipped with this mill.

The parameters of the mill on semi-technical scale (Figure 3):
• dimensions of the working chamber 100x150 mm
• induction 90-120 mT
• active power 5 kW
• power supply 3x12 A
• frequency 50 Hz
• size of ferromagnetic grinding media 0.1x1 mm
• number of grinding media approx. 10% of chamber volume.

Fig. 3. Magnetic mill produced by Eltraf, Lubliniec.

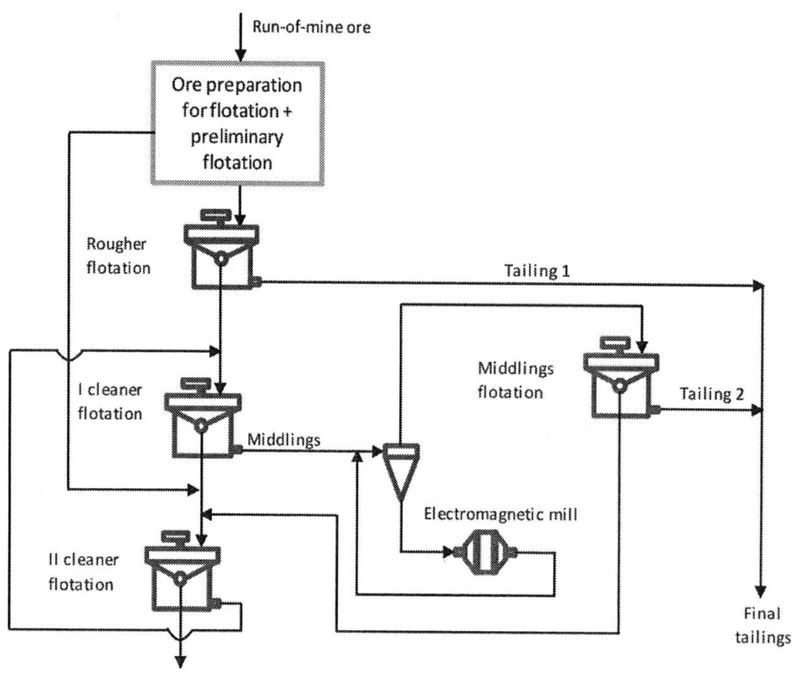

Fig. 2. Flowsheet of middlings processing.

 DOI: 10.1051/e3sconf/20171801003

3.1 Middlings regrinding

Carried out tests of fine grinding in a magnetic mill on a laboratory stand showed that the mill was characterized by high grinding intensity.

Grinding tests of middlings were performed:

• in a laboratory ball mill (BM) with an operation chamber volume of 2.6 dm^3 to determine the kinetics of the middlings grinding; durations of grinding were - 5, 10, 15, 20 min; mass of grinding media - about 5.9 kg

• in an electromagnetic mill (MM), the suspension of 1 dm^3 of about 1200 g. A grinding chamber volume of 0.76 dm^3. Suspension flow - 2, 4 and 7 dm^3/min; Weight of grinding media - 425 g.

Analyses of particle size distribution of the feed and grinding products are presented in Tables 3 and 4. The following equation was used to describe the course of the grinding process:

$$\gamma(d, t) = \gamma(d, 0)\exp(-k\,t^m)$$

t - grinding time
k - grinding speed
m - parameter of distribution.

An example of the changes in percentage of the retained products of the analyzed grain classes in the case of a magnetic and ball mill, depending on the duration of grinding is shown in Figures 4 and 5.

An evaluation of the grinding process can be made on the basis of the determined parameter $\boldsymbol{k}$ called the grinding speed factor and $\boldsymbol{m}$ which is the distribution factor. The factors $\boldsymbol{k}$ and $\boldsymbol{m}$ are given under the figures.

Grinding time of the middlings in MM is very short - several dozen times shorter than in the conventional ball mill (BM). It limits the phenomenon of overgrinding of the mineral grains.

This is confirmed by the calculated grinding speed factors $\boldsymbol{k}$ for fine grains (below 25 μm and below 35 μm) which are lower for grinding in MM than for grinding in BM.

Table 3. Results of grinding of middlings from Rudna Concentrator in a ball mill.

Product	Grain class [mm]	Ball mill, regrinding time									
		no regrinding		5 min		10 min		15 min		20 min	
		γ [%]	Σγ [%]	γ [%]	Σγ [%]	γ [%]	Σγ [%]	γ [%]	Σγ [%]	γ [%]	Σγ [%]
Sandstones	+0.10	35.51	35.51	7.05	7.05	2.56	2.56	0.14	0.14	-	-
	0.10-0.075	18.43	53.94	10.59	17.64	2.92	5.48	1.04	1.18	0.55	0.55
	0.075-0.045	17.80	71.74	27.20	44.84	15.98	21.46	10.02	11.20	6.12	6.67
	0.045-0.036	3.97	75.71	2.16	47.00	3.84	25.30	5.16	16.36	3.62	10.29
	0.036-0.025	5.64	81.35	7.97	54.97	10.53	35.83	10.88	27.24	15.26	25.55
	0.025-0	18.65	100.00	43.03	98.00	62.17	98.00	72.56	100.00	76.45	102.00
	Σ	100.00		98.00		98.00		100.00		102.00	
Carbonates	+0.10	22.93	22.93	3.25	3.25	0.49	0.49	0.10	0.10	-	-
	0.10-0.075	14.14	37.07	7.49	10.73	2.62	3.11	0.90	1.00	0.78	0.78
	0.075-0.045	20.09	57.16	24.14	34.88	17.03	20.14	10.56	11.57	7.02	7.80
	0.045-0.036	5.06	62.22	4.63	39.51	6.91	27.05	4.47	16.03	3.24	11.04
	0.036-0.025	7.97	70.19	6.12	45.63	9.86	36.91	9.15	25.18	11.43	22.47
	0.025-0	29.81	100.00	49.37	95.00	63.09	100.00	74.82	100.00	79.53	102.00
	Σ	100.00		95.00		100.00		100.00		102.00	

Table 4. Results of grinding of middlings from Rudna Concentrator in magnetic mill.

Product	Grain class [mm]	Electromagnetic mill, feed flow							
		no regrinding		7 dm^3/min		4 dm^3/min		2 dm^3/min	
		γ [%]	Σγ [%]	γ [%]	Σγ [%]	γ [%]	Σγ [%]	γ [%]	Σγ [%]
Sandstones	+0.10	35.51	35.51	13.83	13.83	8.22	8.22	4.19	4.19
	0.10-0.075	18.43	53.94	13.35	27.18	8.59	16.81	5.21	9.40
	0.075-0.045	17.80	71.74	21.42	48.60	18.60	35.41	14.32	23.72
	0.045-0.036	3.97	75.71	3.57	52.17	4.58	39.99	3.33	27.06
	0.036-0.025	5.64	81.35	6.98	59.15	8.21	48.20	10.37	37.43
	0.025-0	18.65	100.00	40.85	100.00	51.80	100.00	62.57	100.00
	Σ	100.00		100.00		100.00		100.00	
Carbonates	+0.10	22.93	22.93	10.03	10.03	6.31	6.31	3.07	3.07
	0.10-0.075	14.14	37.07	8.79	18.83	6.20	12.51	3.14	6.21
	0.075-0.045	20.09	57.16	19.90	38.73	15.41	27.93	8.88	15.09
	0.045-0.036	5.06	62.22	3.95	42.68	4.56	32.49	3.84	18.93
	0.036-0.025	7.97	70.19	8.93	51.61	10.37	42.86	9.66	28.59
	0.025-0	29.81	100.00	48.39	100.00	57.14	100.00	71.41	100.00
	Σ	100.00		100.00		100.00		100.00	

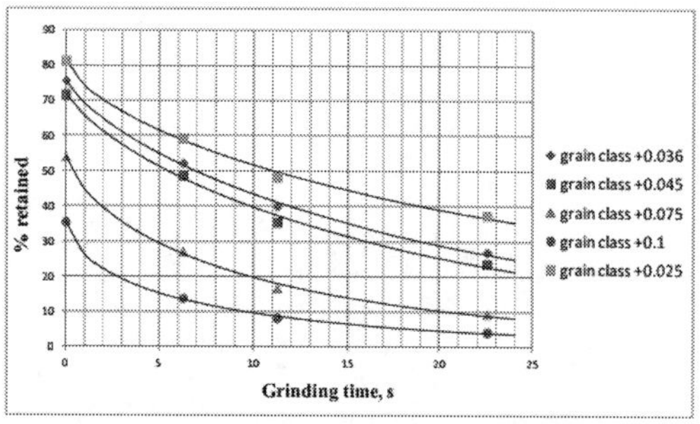

k – for grain class +0.025mm – 0.0924, m – 0.6924
k - for grain class +0.036 mm – 0.0895, m – 0.7914
k - for grain class +0.045mm – 0.0917, m – 0.8114
k - for grain class +0.075mm – 0.1875, m – 0.727
k - for grain class +0.1mm – 0.3002, m – 0.6367

Fig. 4. Grinding kinetics of ore of predominantly sandstone fraction from Rudna Concentrator in MM.

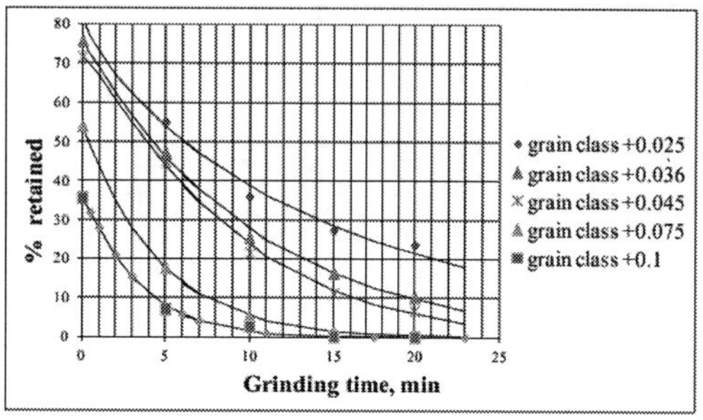

k – for grain class +0.025 mm – 0.106, m – 0.847
k – for grain class +0.036 mm – 0.0943, m – 1.0308
k – for grain class +0.045mm – 0.0735, m – 1.1813
k – for grain class +0.075mm – 0.2089, m – 1.0463
k – for grain class +0.1mm – 0.2451, m – 1.1223

Fig. 5. Grinding kinetics of ore of predominantly sandstone fraction from Rudna Concentrator in BM.

Mineralogical analyses of selected products after the grinding process in magnetic mill were performed using QEMSCAN® device. It is shown an exemplary mineralogical analysis result in which the share of copper in liberated sulphides was determined (degree of liberation 80-100%) after grinding time of about 6 and 11 seconds (Table 6). Large increase of liberated grains was observed in a short time as a result of grinding in MM. This mill can be a solution for future in the aspect of regrinding of middlings circulating in the enrichment circuits of non-ferrous metal ores.

Table 5. The share of copper in liberated Cu bearing minerals (QEMSCAN® analysis).

Mill	Grinding time	Cu [%]
MM	0 s	37.6
	6 s	55.4
	11 s	60.7

3.2 Influence of regrinding in magnetic field on flotation results

The effects of the middlings regrinding in MM on the results of flotation were investigated. The results of flotation of the reground in MM middlings were related to the results of flotation after grinding under standard conditions (in a ball mill).

As a result of grinding in BM (ball mill) and MM (magnetic mill) grain sizes d_{80} were as follows:
- grinding in BM

Middlings from Rudna Concentrator	Particle size of d_{80} product, μm					
	no regrinding	5 min	10 min	15 min	20 min	30 min
Sandstone fraction	144	71	47	34	32	30.4
Carbonate fraction	106	63	45	34	32	30.4

- grinding in MM

Middlings from Rudna Concentrator	Particle size of d_{80} product, μm			
	no regrinding	7 dm^3/min	4 dm^3/min	2 dm^3/min
Sandstone fraction	144	88	70	53
Carbonate fraction	106	73	60	36

Below the examples of flotation results are presented, obtained for middlings from the Rudna Concentrator, reground in BM and MM (Figures 6 and 7).

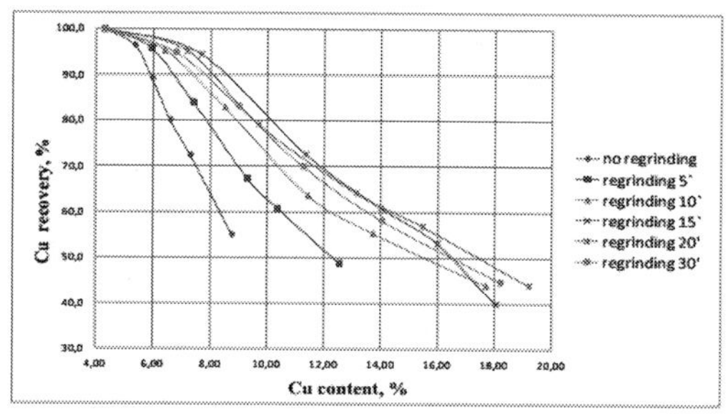

Fig. 6. Flotation results of the reground in BM carbonate middlings from Rudna Concentrator.

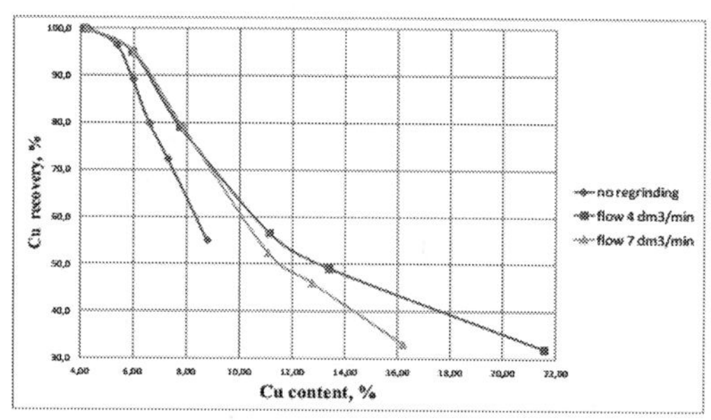

Fig. 7. Flotation results of the reground in MM carbonate middlings from Rudna Concentrator.

There was stated a positive effect of regrinding, both in the ball mill and the electromagnetic mill, on the results of the middlings flotation. There is a regrinding time limit and the associated degree of granulation of the tested middlings beyond which the flotation indexes are deteriorating.

DOI:10.1051/e3sconf/20171801003

On the basis of the flotation experiments optimum particle size (grain size d_{80}) was determined for the tested middlings to ensure a favourable copper recovery in the flotation process.

The flotation evaluation criterion was the quality of the concentrate (β) obtained for an equal recovery of 90%. It is presented for the middlings processed at the Rudna Concentrator as follows:

- carbonate circuit for particle size distribution determined by grain size d_{80} = 34 μm - β = 8%
- sandstone circuit for particle size distribution determined by grain size d_{80} =30 μm - β = 24%.

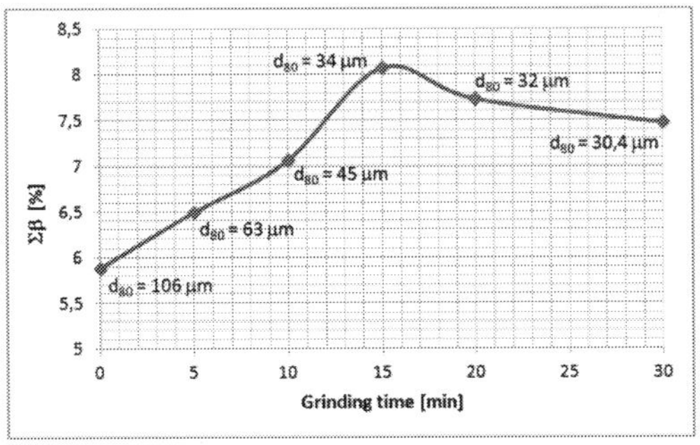

Fig. 8. Cu content in laboratory flotation concentrate as a function of grinding time and grain size d_{80}.

Figure 8 shows an exemplary copper recovery curve as a function of grinding time (d_{80}) for the carbonate middlings from the Rudna Concentrator.

The tests of regrinding and flotation of the middlings have confirmed the possibility and rightness of a separate middlings processing circuit including separate regrinding (e.g. in a magnetic mill) and the classification of I cleaner tailings in hydrocyclones followed by the flotation of these tailings (Figure 9).

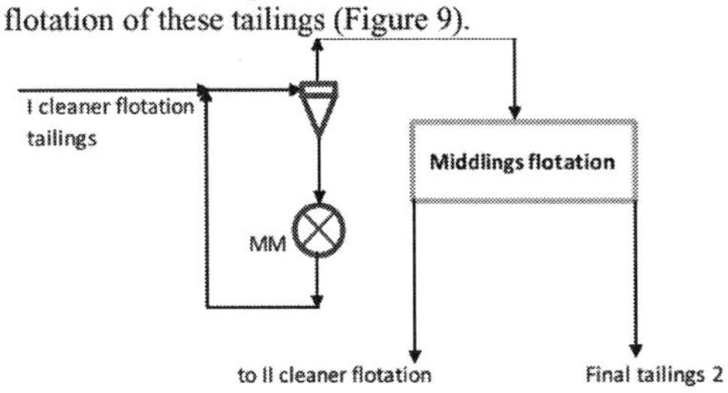

Fig. 9. The qualitative and quantitative diagram of the separate processing of the middlings at the Rudna Concentrator (carbonate circuit).

Another possible approach is flotation of only a hydrocyclone overflow carried out in the presence of depressant for organic carbon and regrinding of the hydrocyclone underflow returned to the main system (Figure 10).

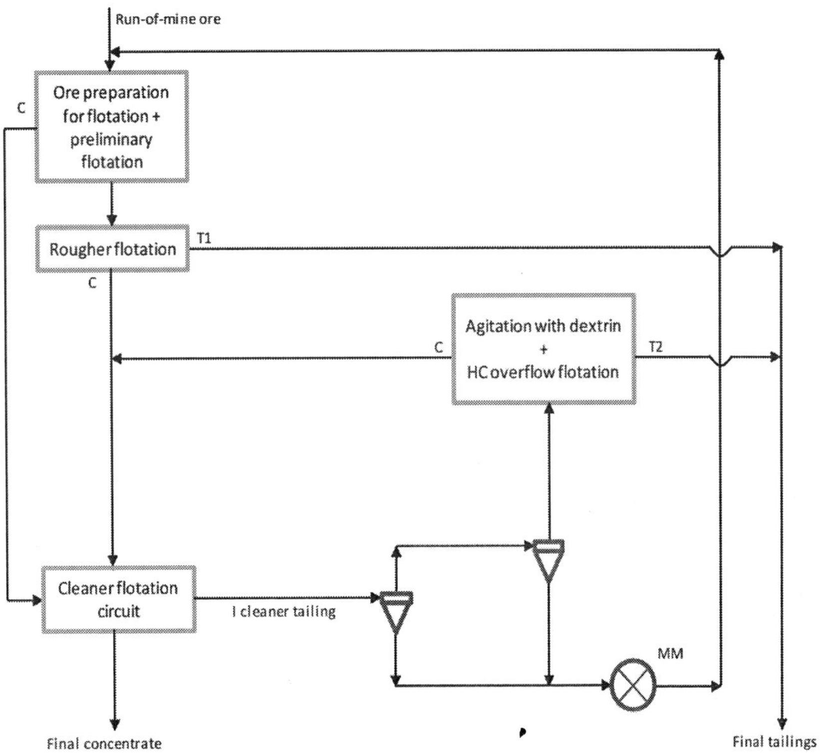

Fig. 10. Proposal of flowsheet for the production of concentrate with the reduced organic carbon content at the Rudna Concentrator (carbonate circuit).

4 Summary

Processing technology of middlings should be adapted to the mineral and petrographic properties of the processed ore. Therefore, in the case of Polish copper ores, which tend to produce large quantities of difficult to treat middlings, separate processing of these middlings should be preferred.

An important element of processing is to provide an optimal degree of granulation, which allows to obtain the most favourable flotation results. Comminution can be carried out using conventional grinding in tumbling mills, however, due to the required high degree of fineness, it would be advantageous to grind in mills designed for the fine grinding.

Among the well-known mills solutions for fine grinding worth noting is the construction of a magnetic mill that has existed on the Polish market for several years. The magnetic mill is characterized by high grinding intensity, so that the time of the grinding of the middlings is very short. This can have a positive effect on the granulation of the middlings, as the short grinding time limits the phenomenon of overgrinding of the grains.

Another important element of the middlings processing is flotation, which should be carried out under dynamic conditions, as this favours the fine grains flotation.

This approach will ensure better copper recovery from middlings and higher selectivity of the rougher flotation, which will contribute to the improvement of the index of copper recovery in the concentrate.

This article is based on the results of the project "Development of highly effective technology of Polish copper ore beneficiation" financed by KGHM PM SA and NCBiR.

References

1. T. Chmielewski, Physicochem. Probl. Miner. Process., **41,** 337-348 (2007)
2. St. Sobierajski et al, *Opracowanie koncepcji technologii wzbogacania rud z wykorzystaniem metod chemiczno – flotacyjnych*, Report 6316/W/06
3. E. Tabosa, J. Rubio, Minerals Engineering, **23**(15), 1198-1206 (2010)
4. B.A Wills, T. Napier-Munn, *Mineral Processing Technology*, (2006)
5. T. Lokiec, H. Lokiec, *Wzbudnik mlyna elektromagnetycznego,* UP 412389, (2016)
6. A. Luszczkiewicz, T. Chmielewski, Internat. J. Miner. Process., **88(1-2)**, 45-52 (2008)

Research on the Possibility of Sorting Application for Separation of Shale and/or Gangue from the Feed of Rudna Concentrator

Andrzej Grotowski[1], and *Kajetan* Witecki[1]

[1]KGHM Cuprum Ltd. Research and Development Center, 53-659 Wrocław, Poland

Abstract. Shale, which occurs in the copper ore deposits belonging to KGHM Polska Miedź S.A., is the reason for a number of difficulties, at the stage of not only processing but also smelting. Gangue, in turn, getting in a feed during mining is a useless load of a concentrator and also contributes to lowering concentrating indexes. Its content in a feed is being evaluated at 15-30%. The multiple attempts to solve those issues by the methods of conventional mineral processing or even selective mining failed. In the range of work, research on the lithological composition and Cu content in 300 individual particles (selected from Rudna feed) have been carried out. Using those results, the simulation of gangue separation with an application of sorting have been done. The positive results have been received: introduction of a sorting operation causes, theoretically, removing of approximately 20-30% sorting feed mass as final tailings with Cu losses not bigger than 5-10%. It means that the capacity of Rudna concentrator can be increased proportionally. To confirm those results, industrial sorting trials are necessary, when appropriate sorters will become available. Additionally, one should take also into account that the finest classes of feed (-12.5 mm) could not be concentrated in a sorter. In the range of work, the preliminary tests of the industrial sorter (PRO Secondary Color NIR) for separation of the shale concentrate from Rudna concentrator feed have been carried out. The shale concentrates were received both from 12.5-20 mm class and +20 mm class. The concentrates produced from the coarse classes, for both technological sides had shale content at the level of 48-49%, with recovery of 52.9-60%. In the case of the finer class, shale content in the concentrates for both technological sides amounts to 30.9-35%, at the slightly lower recoveries than for coarse classes. Cu and Corg behavior in the sorting process were checked also, however, the results turned out to be not very interesting. Because the results of shale concentrate production by sorting have a significant potential for improvement, the further researches in this direction have been recommended, however, making them start off from elaboration of a technology for shale concentrate processing and calculation of a total balance of a concentrating process involving flotation and separation.

1 Introduction

The presence of shale in Cu ore, which is processed in KGHM Polska Miedź S.A. concentrators, is the reason of a number of difficulties at the stage of not only processing [1] but also smelting [2-3]. Shale, and actually organic carbon contained in it, causes a higher reagent consumption in flotation process, lowering Cu content in a concentrate and lowering recovery [3-4]. The high Corg content (0.51-11.76%) in shale and – as a consequence – in concentrate, makes difficulties even at a smelting stage because of lowering capacity of a smelting furnace. Those unfavorable features of shale made that practically from the moment when the polish copper ore exploitation started in sixties of XX century, the research and tests on removal of shale from the ore or from the produced concentrates began and they are continued up till now. This issue has been explored using different methods only with selective mining, however, the most attention was paid to a flotation method. The research conducted both by research units (IMN, KGHM Cuprum), KGHM Polska Miedź S.A. concentrators, universities (Wrocław Technical University, AGH), and even foreign units, did not allow to elaborate a satisfactory technology. Analyzing properties of shale and flotation process it is unlikely that at this stage of technical development, the problem of shale (Corg) elimination could be solved by a flotation method. Today, the most satisfactory solution, albeit without defects, was found to be the roasting of the concentrate. In this context it is very reasonable to explore the possibility of using new enrichment techniques, such as sorting, for the elimination of slate.

Another issue connected to mining exploitation of deposits, including the polish Cu deposits, is contamination of feed by gangue, and related to this, its dilution. Dilution of feed is unavoidable and is a consequence of used mining methods and machines. For those reasons, it never could not be eliminated completely and the only reasonable solution is to keep it in acceptable limits [5]. According to the conservative evaluation,

* Corresponding author: agrotowski@cuprum.wroc

approximately 15-30% of the mining output in KGHM Polska Miedź S.A. constitutes the most probably gangue with Cu content lower than 0.05-0.10%. Its elimination from a concentrator feed would improve the throughput of the Cu concentrators and the current technological indexes [6-7].

A sorting process creates a chance to solve both above mentioned issues (shale and gangue presence in a feed). This is even more justified as the turbulent development of sorting techniques has led to the construction of a number of new sensors and sorting machines, resuming research in this field, and the introduction of many sorters into industrial practice [8].

The results of the research on the possibility of sorting application both for shale concentrate production and for gangue removal from the Cu ore processed in Rudna concentrator have been presented in this paper.

2 Sorting techniques

Sorting is separation of a material into two or more products with different required properties. Due to the sensor abilities to detect the properties of individual grains and technical possibilities of their effective separation, sorting is usually limited to particles with size from 10-15 mm to 250-300 mm [9-10]. Hence, sorting process can only be used for sufficiently coarse feed.

Although, manual sorting is used since the dawn of action, the first mechanical sorter was constructed only in XIX century [11]. This was made possible by the development of suitable sensors to enable efficient separation of particles. The first commercial sorter, which was based on the detection of optical properties, appeared only in 1931, in the food industry [12]. Since then, mechanical sorting has begun to supplant manual sorting, and in the late 1960s many branches of industry abandoned manual sorting [13-14].

However, even today in some situations manual sorting is irreplaceable, e.g. due to the specific characteristics of the raw materials (food industry, waste recycling) [15] economic conditions, small scale of production, special limitations, etc.

Regardless of sorting is done manually or mechanically, its effectiveness requires the following conditions [11-12, 16]: 1) proper preparation and feeding of a raw material for a separation process; 2) effective detection of a required parameter; 3) faultless and fast identification of particles with the desired properties and; 4) fast physical separation of a raw material for the required products.

In mining, sorting machines are used primarily in diamond mining, which since the 1960s has used optical sorters for this purpose. They are currently being displaced by X-ray luminescence (XRL) sensors and near-infrared (NIR) ones [8, 17-19]. However, the most promising solution, despite low throughput, seems to be the use of sensors that work on the principle of XRT [20]. The use of this technology allowed the separation of large diamonds with low luminescence, or closed in the gangue, and especially enabled the recovery of much larger diamonds than it was possible previously.

Sorting machines were introduced for the first time in the food industry which is still the source of new sorting technologies. More and more widespread use for food distribution in the food industry is the near infrared (NIR) technology [21-23]. Using this technology allows to capture the imperfection of the material or to obtain statistical correlations between a spectra obtained and a content of selected substances [23]. This solution is usually used in conjunction with optical sensors, allowing for easy and accurate separation of main components of homogeneous materials (e.g. copper ores). Currently, HSI (Hyperspectral Imaging) technology is at the stage of development and research. In general, HIS uses spectra of each pixel what makes it possible to create 3D models, showing the distribution of the components in individual particles [21, 24-25].

The pursuit of recycling (circular economy) requires the use of solutions that allow for the separation of waste products into the separate products with the required properties. Sorting machines have been successfully used by the Sense2Sort consortium, which combines Austin Al and Toratecnica to separate scrap aluminum from mixed metal scrap [26]. The proposed solution uses XRF and LIBD (Laser Induced Breakdown Spectroscopy) technology. The whole process runs automatically and delivers a product of up to 98% purity in ready-to-recycle form. Secondary aluminum recovery saves up to 95% energy, compared to the original production of this metal from bauxite [26]. Much more difficult is a separation of municipal wastes, which in order to be reused must be separated, first and foremost, into organic and inorganic components [27]. The authors used various methods for this purpose, resulting in the separation of leachates, plastics, metals, paper and organic fraction. The sorting methods used included both manual, magnetic, pneumatic and hydraulic separation. The necessity of using so many separation methods in a single technological chain was due to the complexity of the waste delivered. In the case of copper ores, the simplest way to eliminate the gangue would be to use a Cu content sensor, based on the X-ray fluorescence principle. However, according to the opinion of one of the major producers of sorters - Commodas Ultrasort [10], it is currently not possible to construct a sorting machine operating on this principle which could provide sufficiently efficient sorting and to obtain a low-Cu product. Therefore, the possibility of eradicating shale from Rudna ore was determined by examining the Cu content in selected grains and performing on this basis a sorting simulation [28] and the possibility of producing shale concentrate was verified by the use of an industrial sorter.

3 Materials and methods

The copper ore samples for gangue removal were collected in Rudna concentrator from belt conveyors transporting ore from the fine ore bins to the conveyors transporting ore to mills (I-st stage mills). The subject of the tests were both types of ore: the ore from A side (with increased content of sandstone) and the ore from B side (with elevated carbonates content). The collected samples, weighing about 250 kg, were individually averaged and reduced by the quartering method, finally receiving 4 samples: 2 samples weighting about 50-60 kg and 2 samples with approximately 5 kg each, representing A and B technological sides. Smaller samples, weighing about 5 kg each, were subjected to sieve analysis. The results are shown in Figure 1.

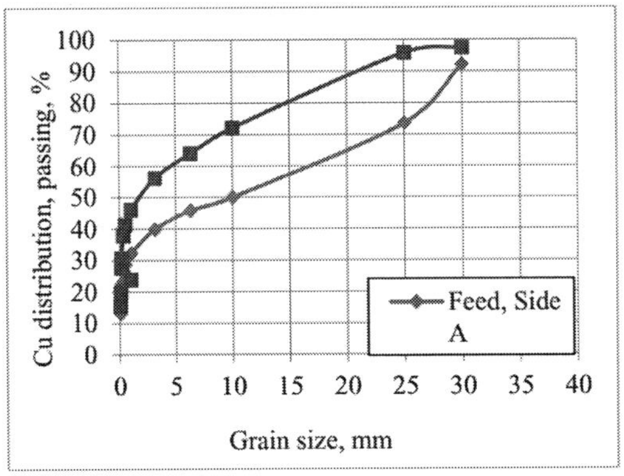

Fig. 1. The particle size distribution of the feeds for the individual technological sides, Rudna concentrator.

In the obtained particle classes, the Cu content was determined and Cu distribution was calculated on this basis. The results are displayed in Figure 2.

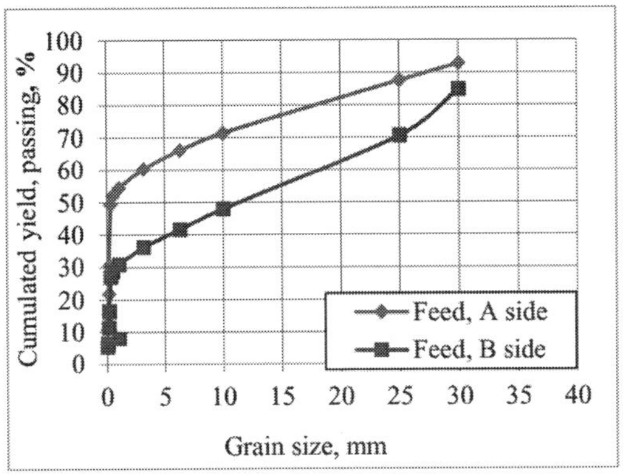

Fig. 2. Distribution of Cu in the particle classes of the feeds for the individual technological sides of Rudna concentrator.

Subsequently, the bigger samples were sieved through a 12.5 mm sieve to remove non-sortable material. Then each sample was manually sorted into shales, carbonates (limestone) and sandstones. In total, 300 particles representing all lithological ore components were extracted for A and B sides. Prior to the manual sorting, the particles were washed with water to clean the surface from dust in order to facilitate visual identification. After completion of manual sorting, the individual particles were crushed and sent to CBJ (Quality Research Centre) in Lubin to determine the Cu content by ICP-MS method.

The copper ore samples (from A and B sides) for shale separation in the industrial sorter were collected in Rudna concentrator too. The mass of each sample taken was approximately 1.5 Mg. Then the samples were sent to Commodas Ultrasort Test Center (now Outotec) in Wedel, Germany, where material was screened by 12.5 mm screen. The -12.5 mm class was saved for subsequent flotation test and +12.5 mm class was screened into 12.5-20 mm and +20 mm classes. Finished classes were rinsed with water to clean the surface in order to improve detection in the sorting process. PRO Secondary Color NIR sorter was used for all sorting tests. The sorter calibration was performed using the "rock-by-rock" results for 123 individual particles, selected from the ore samples and representing all lithological types. The lithological type of the individual particles was determined as well as Cu and other components content by the XRF method. Then, using such characterized particles, the possibility of using different sensors, based on such phenomena as NIR, color photometry, XRT and electromagnetism, to sort ore was investigated. The best results of detection for shale were obtained with NIR and photometric sensors and they have been selected for the sorting tests described further.

4 Results and discussion

4.1 Research on the elimination of gangue

Based on the analysis of Cu content in the individual particles of carbonates, sandstones and shales, which have been selected from the both sides of Rudna concentrator, the average Cu content in a given lithological type, the yield of a given lithological type, and the distribution of Cu in them were determined.

Table 1. Average Cu content in a given lithological type, lithological type yield and Cu distribution.

Lithological type	Cu cont., %		Yield, %		Cu distribution, %	
	Side		Side		Side	
	A	B	A	B	A	B
Shales	3.57	1.49	10.9	10.1	25.3	14.2
Sandstones	1.46	1.14	49.1	46.7	46.6	50.3
Carbonates	1.08	0.87	40.0	43.2	28.1	35.5

The shale content in the ore from both technological sides (A and B) fluctuates around the limit of about 10%, with the Cu content in it being significantly higher on side A. The largest share, on both sides, is observed for sandstones, but the share of carbonates is only slightly smaller. One can see minimal variation in the lithological composition between the individual sides.

Table 2 summarizes the basic statistics concerning the copper content in the individual grains of particular lithological types such as mean, minimum, maximum, standard deviation (SD) and relative standard deviation (RSD).

Shales on both technological sides are characterized by high variability, which can certainly be affected by the difficulty of identifying this lithological type, when compared to carbonates. However, the greatest variability was found for carbonates on B side. From the point of view of the possibilities for elimination of gangue, the most important feature of the presented results is the presence of particles with relatively low Cu content (0.04 to 0.11%).

Table 2. The basic statistics of Cu content in the particular lithological types of the feeds for the both technological sides, Rudna concentrator.

Statistics	Sandstone Side		Carbonate Side		Shales Side	
	A	B	A	B	A	B
Average	1.46	1.14	1.14	1.08	3.57	1.49
Min	0.04	0.06	0.06	0.08	0.10	0.04
Max	5.52	3.70	3.70	5.12	19.36	12.56
SD	1.19	0.80	0.80	2.02	4.70	2.44
RSD, %	81.6	69.9	69.9	186.5	131.6	164.41

The simulation of the sorting results was carried out using all the results of Cu content in individual particles, separately for A and B sides. For all cases, such limit concentrating indexes were set, that the total Cu losses in tailings were not higher as 5-10% (for the total Cu recovery in concentrate not lower than 90-95%). Under such assumptions, the total output of the gangue, read from the corresponding concentration curves for tailings, varies between 20-30%. Such mass of the ore can safely be discarded during a separation processes, without significantly impairing the flotation performance. The results of the simulated enrichment (sorting) curves for A and B sides are shown in Figures 3 and 4, respectively.

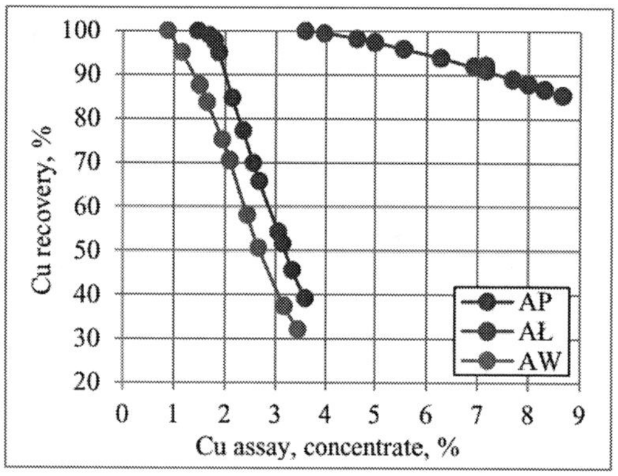

Fig. 3. Sorting simulation of Rudna concentrator feed – A side, class +12.5mm – Cu recovery for the individual lithological components vs Cu assay in concentrate (AP – sandstones, A side; AŁ - shale, A side; AW – carbonates, A side).

The presented results show that the sortability of carbonates (W) and sandstones (P) from both technological sides A and B is very similar, whereas the sortability of shale (Ł) - on both sides is quite far from the other components. However, if you judge it by the practical concentration index, shale turn out to be the best sortable of all three lithological components.

The obtained simulation results should be treated as the preliminary ones, because in the case of sorting in an actual sorter, the concentration indexes for a number of reasons (for example less efficient particle spreading, contamination of particle surface, imperfection of detection algorithm, and overall sorter efficiency compared to manual sorting) will be worse.

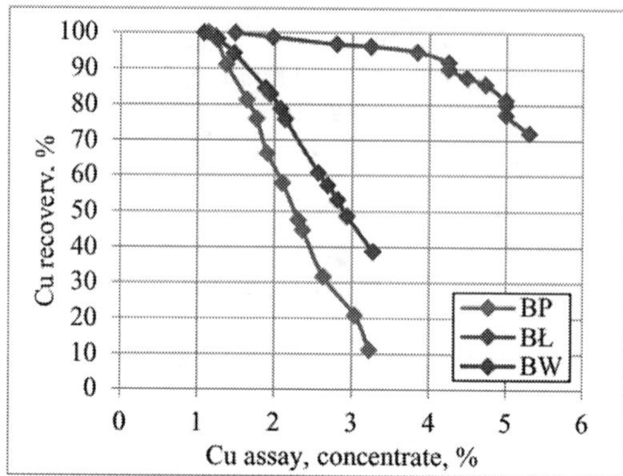

Fig. 4. Sorting simulation of Rudna concentrator feed – B side, class +12.5mm – Cu recovery for the individual lithological components vs Cu assay in concentrate (BP – sandstones, B side; BŁ - shale, B side, BW – carbonates, B side).

4.2 Research on the production of shale concentrate

Four products obtained from the Rudna concentrator, i.e. classes 12.5-20 mm from A and B sides A, classes + 20 mm also from A and B sides, were concentrated in the industrial sorter. The main purpose of the tests was to obtain a rich shale concentrate. The lithological composition of each product was determined as well as Cu and Corg content. The shale content in the sorting products was determined on the basis of macroscopic features, while the Cu and Corg content was determined in ICP-MS method by CBJ (Quality Researcg Centre) in Lubin. The sorting results are displayed in Table 3.

Generally, the richest concentrates, with a shale content close to 50%, were obtained for a grain class + 20mm. Also, they are characterized by the quite high concentrating ratios, amounting to, depending on the technological side, from 4.2 to 4.6. The yield of shale concentrate is not too high, but the general purpose of the tests was to obtain a possibly rich concentrate. For finer classes (12.5-20mm) the results are worse - in every respect - than for the coarser classes, what is in line with expectations.

The obtained concentrates did not show significantly increased copper content or concentrating ratio. The highest value of a concentration ratio, amounting to 2.1 was obtained under Test I, then for Test III and IV, it was 1.6 and 1.4, respectively. The least enriched in copper proved to be the concentrate from test II, where a concentration ratio amounted to 1.2 only. The highest value of a concentration ration, amounting to 2.1, was obtained under Test I, then for Test III and IV, it was 1.6 and 1.4, respectively. The least enriched in copper proved to be the concentrate from Test II, where a concentration ratio was 1.2.

The organic carbon content in the concentration products was also determined, to find how much shale concentration is correlated with this components. The separation results are shown in Table 4.

The results show that although the correlation between shale and Corg content exists, it is not sharp. In the concentrates obtained in tests I and II (coarse class, A and B sides), the shale content is close to 50% (See Tab. 3) but the Corg content in test I (coarse class, side A and B) amount to as low as 2.87% and 1.49%. On the phenomenon of low correlation of shale and Corg contents, especially in the industrial flotation circuits, attention has already been paid in the previous works [29-30].

Nevertheless, the analysis of the content of organic carbon in sorting products shows a markedly increased content of it in the shale concentrate. The highest concentration ratio of Corg, equal to 4.5, was achieved with Test I, in the successive tests the ratios amounted to 3.5 - Test II and III, and only 2.5 - Test IV.

Although the purpose of the tests in the industrial sorting machine was to determine the possibility of shale separating from copper ore, the Cu content was also determined in the products obtained. The results are listed in the same table. As expected, copper - both in terms of Cu content in concentrate, concentration ratio and yield - is concentrated significantly worse off than Corg. This is due to the presence forms of these two components in the Cu ore studied and the concentration method used.

Table 3. Sorting results of shale from Rudna Cu ore, A and B sides, particle class 12.5-20 mm and +20mm.

Test no	Side	Particle class, mm	Product	Yield		Shale content, %	Shale recovery, %	Concentr.ratio
				kg	%			
I	A	+20	Concentrate	16.1	11.57	49.0	52.9	4.6
			Tailing	123	88.43	5.7	47.1	-
			Feed	139.1	100.00	**10.7**	100.0	1.0
II	B	+20	Concentrate	31.5	14.35	48.7	60.9	4.2
			Tailing	188.0	85.65	5.2	39.1	-
			Feed	219.5	100.00	**11.5**	100.0	1.0
III	A	12.5-20	Concentrate	17.0	16.11	35.0	45.1	2.8
			Tailing	88.5	83.89	8.2	54.9	-
			Feed	105.5	100.00	**12.5**	100.0	1.0
IV	B	12.5-20	Concentrate	37.1	22.96	30.9	56.7	2.2
			Tailing	124.5	77.04	8.9	54.9	-
			Feed	161.6	100.00	**14.0**	111.6	1.0

Table 4. Sorting results of Corg and Cu from Rudna ore, A and B side, 12.5-20 mm and +20 mm.

Test no	Side	Grain class, mm	Produkt	Corg content, %	Corg recovery, %	Corg conc. ratio	Cu cont., %	Cu reco-very, %	Cu conc. ratio
I	A	+20	Concentrate	2.87	51.77	4.5	3.13	24.69	2.,1
			Tailing	0.35	48.23	-	1.25	75.31	-
			Feed	0.64	100.00	1.0	1.47	100.00	1.0
II	B	+20	Concentrate	1.49	50.00	3.5	1.82	1770	1.2
			Tailing	0.25	5000	-	1.42	82.30	-
			Feed	0.43	100.00	0.3	1.48	100.00	1.0
III	A	12.5-20	Concentrate	2.57	57.20	3.5	2.86	25.13	1.6
			Tailing	0,37	42.80	-	1.64	74.87	-
			Feed	0.73	100.00	1.0	1.84	100.00	1.0
IV	B	12.5-20	Concentrate	1.56	57.77	2.5	1.94	32.89	1.4
			Tailing	0.34	42.23	-	1.18	67.11	-
			Feed	0.62	100.00	1.0	1.35	100.00	1.0

5 Conclusions

1. The conducted tests and simulations of Cu ore sorting in order to extract the gangue (product with significantly reduced Cu content in comparison to Cu ore) have shown that this is theoretically possible. Using the results of the Cu content in individual ore particles, Cu losses in tailings were determined to be no more than 5-10%. It means that about 20-30% of ore can be removed as final tailing on the stage of sorting. As a result, the mass of feed to flotation will fall to 70-80% primary ore mass. It can be expected that the capacity of the flotation concentrator will increase by 20-30%, also. Determining the actual effect will require performing concentration tests in an actual sorting machine because its concentration efficiency will be lower than the simulation results. Also note that the smallest grades (-12.5 mm) cannot be enriched in the sorter. However, further tests will only be

justified when the possibilities for constructing sorting machines based on the Cu content criterion appear.

2. Preliminary tests using an industrial sorter to produce shale concentrate from the flotation feed have also shown that this is possible. It has been found that from the fed of Rudna concentrator, class +20 mm, it is possible to obtain the shale concentrate with a shale content of 48-49% and a concentrate yield of 52.9-60%. For the finer class (12.5-20mm) - for both technological lines - shale content in the concentrate amounts to 30.9-35%. Due to the existing potential for results improvement, further research in this direction could be justified, however, any possible work in this direction should be preceded by elaboration of a method of shale concentrate processing and execution of the overall balance of the ore concentrating process (sorting + flotation).

3. The sorting tests carried out have yielded very interesting results in terms of the applicability of this solution to the separation of Rudna concentrator feed into products with different lithological characteristics and different copper content. For the obtained sorting products it is planned to examine the content of the accompanying metals and determine their flotability. It is also necessary to check the flotability of the finest class (-12.5mm), to develop a complete balance of the enrichment process.

This paper is part of the collaborative research project Ecometals co-funded by German Federal Ministry of Education and Research (BMBF under the Grant ID 033RF001), French National Research Agency (ANR-13-RMNP-0006) and KGHM Polska Miedź S.A. (KGHM-BZ-U-0319-2014). The authors also would like to thank UVR-FIA GmBh, Freiberg, Germany for screening the ore to the big-scale tests and Tomra Sorting GmbH Wedel, Germany for the execution of the big-scale sorting tests.

References

1. A. Grotowski, K. Sztaba, A. Wieniewski, Cuprum Ore Min. Sci. Tech. Mag. **2(47)**, 105-124 (2008)
2. J. Czernecki, S. Sobierajski, Z. Śmieszek, *EMC `91,* (Elsevier Applied Science, London – New York, 1991)
3. B. Skorupska, A. Wieniewski, N. Kubacz, Min. Geol., **6(2)**, (2011)
4. Collective development, *Monografia KGHM Polska Miedź S.A.* (KGHM Polska Miedź S.A, Lubin, 2007).
5. E.A. Wright, *Erzmetall*, **36(1)**, 23-29 (1983)
6. T.R. Hughes, G. Cormack, , First International Future Mining Conference (Sydney, 1-8, 2008)
7. P. Whiteway, www.canadianminingreview.com/ 2011*/03/nickel-exploration.html* (2011)
6. S. Walker, Eng. Min. J. **215(4)**, 46 (2014)
7. B.A. Wills,: *Will's Mineral Processing Technology* (Butterworth-Heinemann, Queensland, 2006)
8. A. Grotowski, K. Witecki, *Personal data: Application of sorters in kimberlite ores enrichment* (Wedel, 2016).
9. J.D. Salter, N.P.G. Wyatt, *XVII Intern. Miner. Proc. Congr.*, (Dresden, **5**, 183-193, 1991a)
10. J.D. Salter, N.P.G. Wyatt, *Mineral Engineering*, **7(11)**,779-796. (1991b)
11. D.L. Carson, A.T. Moir, Proc. VII Comm. Mitt. Metall. Inst. Congr.. (Johannesburg. S.Afr. Inst. Min. Metall., 1961)
12. J.S. Freer, R.C. Bohme, Proc. Mintek 50, Intern. Conf. recent Advances in Miner. Sci. Technol., (Johannesburg, 36-37, 1984)
13 A. Procner, *Technologia gastronomiczna z towaroznawstwem* (Wydawnictwo Szkolne i Pedagogiczne, Warszawa, 1988)
14. H.R. Manouchehri, Proc. of Conference of Mineral Processing (Luleå, 1–17 (2003)
15. Anon., Min. Eng. J., **17** (1971)
16. M.G. Rylatt, G.M. Popplewell, Miner. Eng., **51**, 37-43 (1999)
17. T. Mahlangu, N. Meomise, M.M. Ramakokovhu, P.A. Olubambi, M.B. Shongwe, J. South. Afr. Inst. Min. Metall., **116**, 343-347 (2016)
18. L.M. van Niekerk, A Olivier, J. Armstrong, N.A. Sikwa, J. South. Afr. Inst. Min. Metall., **116**, 709-714 (2016)
19. R. Moscetti, W. Saeys, J.C. Keresztes, M. Goodarzi, M. Cecchini, M. Danilo, R. Massantini, Food Bioprocess Technol, **8**, 1593-1604 (2015)
20. X. Wu, B. Wu, J. Sun, M. Li, H. Du, Int. J. Food Prop., **19**, 1016-1028 (2016).
21. J.A. Cayuela, J.F. Garcia, J. Food Eng., **202**, 79-88 (2017)
22. M.F. Carlsohn, .J Real-Time Image Proc., **1**, 25-32 (2006)
23. G. Elmasry, M. Kamruzzaman, D.-W. Sun, P. Allen, Crit. Rev. Food. Sci. Nutr., **52(11)**, 999–1023 (2012)
24. E.-M. Gerosch, E. van Looy, Aluminium Int. Today, 25-27 (2016)
25. H. Wang, J. Xu, H. Yu, X. Liu, W. Yin, Y. Liu, Z. Liu b, T. Zhang, Renew. Susta. Energy Rev., **52**, 1881-1889 (2015)
26. A. Grotowski, A. Szubert, Cuprum Ore Min. Sci. Tech. Mag., **2 (75)**,123-136 (2016)
27. A. Grotowski, Konferencja „Perspektywy zastosowania technologii bioługowania do przerobu rud miedzi zawierających łupki – BIOPROCOP `06" (Lubin, 7-27, 2006)
28. P. Kijewski, R. Leszczyński, Zeszyty Naukowe Instytutu Gospodarki Surowcami Mineralnymi i Energi Polskiej Akademii Nauk, **79** (2010)

A multidimensional analysis and modelling of flotation process for selected Polish lithological copper ore types

Tomasz Niedoba[1], *Agnieszka* Surowiak[1],and *Paulina* Pięta[1,*]

[1]AGH - University of Science and Technology, Faculty of Mining and Geoengineering, Department of Environmental Engineering and Mineral Processing, Cracow, Poland

Abstract. The flotation of copper ore is a complex technological process that depends on many parameters. Therefore, it is necessary to take into account the complexity of this phenomenon by choosing a multidimensional data analysis. The paper presents the results of modelling and analysis of beneficiation process of sandstone copper ore. Considering the implementation of multidimensional statistical methods it was necessary to carry out a multi-level experiment, which included 4 parameters (size fraction, collector type and dosage, flotation time). The main aim of the paper was the preparation of flotation process models for the recovery and the content of the metal in products. A MANOVA was implemented to explore the relationship between dependent (β, ϑ, ε, η) and independent (d, t, c_d, c_t) variables. The design of models was based on linear and nonlinear regression. The results of the variation analysis indicated the high significance of all parameters for the process. The average degree of matching of linear models to experimental data was set at 49% and 33% for copper content in the concentrate and tailings and 47% for the recovery of copper minerals in the both. The results confirms the complexity and stochasticity of the Polish copper ore flotation.

1 Introduction

Flotation is a method of enrichment used to distribute grains, which differ in their surface properties. During the process the bubble-particle aggregates are formed and they are upraised to flotation froth. However, copper minerals do not exhibit natural floatability, which forces certain substances to modify their hydrophobicity.
The main factors that determine the final result in terms of high yield and content of the minerals in the concentrate can be divided into 3 groups: feed properties, flotation device characteristics and enrichment methods. An accurate analysis of these three aspects led to a set of more than100 different factors which simultaneously affect the final effects of the separation and they interact with one another. [Bazin, 2001; Aldrich, 2002; Drzymała, 2007; Brożek & Młynarczykowska, 2009].

What causes an additional difficulty of Polish copper ores beneficiation is their complex structure and the presence of three lithological types: dolomite, sandstone and shale, also varied internally. These lithological fractions appear simultaneously in most parts of the deposit, but in varying proportions. However, their exploitation does not currently allow to be selective. In addition, each type has different types and sizes of copper minerals. Other characteristics that differentiate the individual lithological fractions are the enrichment and copper content of the feed. Sandstone type exhibits the greatest floatability and the lowest content of copper minerals. The reverse situation is for the shale type, which is by far the hardest to enrich, but it is characterized by the highest average content of copper. On the other hand, dolomite copper ore has indirect properties in terms of the discussed features [Laskowski & Łuszczkiewicz, 1989; Kijewski & Jarosz, 2007; Kłapciński & Peryt, 2007; Nieć & Piestrzyński, 2007; Piestrzyński, 2007].

The complex structure of the phenomenon induces to choose multidimensional methods of data analysis that match the nature of the process and the properties of the feed. The use of universal, multi-parameter statistical methods allows a full analysis of empirical data, although their multiplicity may give rise to some interpretation difficulties. However, the implementation of mathematical statistics in the consideration of the flotation process and other issues related to the mineral processing is fully justified [Tumidajski, 2010; Tumidajski et al., 2012; Duchnowska et al., 2015].

2 Aims and experiment description

The aim of the study was analyzing and modelling the effects of the sandstone copper ore flotation in a Denver flotation machine in terms of yield and content of metals in concentrates and tailings in multi-parameter approach. The application of multidimensional methods of data analysis enabled a full examination of the complex flotation process and the determination of functional

* Corresponding author: ppieta@agh.edu.pl

dependency between selected separation indices and parameters.

The starting point for further consideration of the sandstone copper ore beneficiation was the results of laboratory tests that allowed the achievement of the objectives. Implementations of multivariate statistical methods and estimation of the process model were necessary to perform a multi-level experiment which assumed four analysed parameters (grain size, type of collector reagent, the dose of the reagent collecting the flotation time).

The material was taken from the technological line of the Copper Ore Preparation Plant Polkowice KGHM Polska Miedź SA. Sandstone copper ore was separated from the sample and subjected to further preparatory work. One of the aims of the analysis was to investigate the granulation influence on effects of copper ore flotation.Therefore, as a result of appropriate comminution and sizing processes, the following size fractions were separated: 125-200, 100-125, 71-100, 40-71, 20-40, 0-20 [μm], which constituted feed material for laboratory flotation tests in a Denver flotation machine. The average content of copper in individual size fractions were characterized in Tab. 1.

Experiments were carried out to maintain a constant concentration flotation slurry 200 g/dm3, an aerating and a rotor speed. Two types of flotation collector were used during the tests and they were added in two dosages:100 g/Mg and 150 g/Mg. They were an Hostaflot and a sodium ethyl xanthate aqueous solution. Nasfroth in dosage 50 [g/Mg] was used as a frother during the test. Before starting the single flotation test, sample was mixed in a flotation cell for 3 min, followed by addition of flotation reagents for the next 7 min. Samples of the froth concentrate were collected after 1, 2, 4, 6, 9, 12, 17, 22, 30 minutes. The flotation concentrates and tailings were dried and weighed and subjected to X-ray fluorescence to determine the copper content. According to the experimental plan, 24 major flotation tests were performed, which were repeated four times. The data set collected in this way made it possible to draw up a mathematical description of the process and to carry out the analysis. For each experiment, the following process evaluation indicators were determined: copper content in concentrate β (1), copper content in tailings (2), copper recovery in concentrate ε (3), copper recovery in tailings η (4), which are determined by the following formulas :

$$\beta_i = \frac{\sum_{i=1}^{n} \gamma_i \cdot \lambda_i}{\sum_{i=1}^{n} \gamma_i} \tag{1}$$

$$\vartheta_i = \frac{\sum_{i=1}^{n} \gamma_i \cdot \lambda_i}{1-\sum_{i=1}^{n} \gamma_i} \tag{2}$$

$$\varepsilon_i = \frac{\beta_i}{\alpha} \sum_{i=1}^{n} \gamma_i \tag{3}$$

$$\eta_i = \frac{\vartheta_i}{\alpha} \sum_{i=1}^{n} \gamma_i \tag{4}$$

where: α – a copper content in the feed,
$\gamma_i = \frac{m_i}{\sum_{i=1}^{n} m_i}$ – a yield of a product [%],
λ_i – content of copper in a product,
m_i – an amount of a product.

Table 1. The average copper content in size fractions.

Size fraction [μm]	Copper content [%]
0-20	1.50
20-40	1.51
40-71	1.10
71-100	0.42
100-125	0.30
125-200	0.21

Conducting tests in accordance with the requirements to the maintenance of representativeness of the data collection is time-consuming and requires considerable workload. The analyses, tests, or coefficients decreased in their simplicity compared to a 1 or 2-parameters experiments. For these reasons, multi-level experiments are rarely used in optimization, but they have undeniable benefits.

3 Calculations methods

The analysis and evaluation of the impact of the studied process parameters on the quality and composition of flotation products were based on the methods from the family of variance analysis. The multivariate analysis of variance (MANOVA) is a method that allows numerical analysis of at least two dependent variables while controlling any number of independent variables. Depending on the aim, it can be carried out as a one- and multivariate scheme. In this case, the multivariate vector of dependent variables ***Yi*** (5) is characterized by the following variables:$Y_1 = \beta_{Cu}$, $Y_2 = \vartheta_{Cu}$ $Y_3 = \varepsilon_{Cu}$, $Y_4 = \eta_{Cu}$. However, in the analysis, the vectors of mean values of the variables μ_i (6), where i=1, 2, ..., N for different systems of factors and their levels (6) were compared[Stanisz, 2007; Niedoba, 2013; Niedoba et al., 2016a, Niedoba et al., 2016b]. The relevance of the main effects among the investigated factors was considered for significance level equalled to 95%, which means that the model with a P-value lower than 0.05 could be considered.

$$Y_i = \begin{bmatrix} \beta_{iCu} \\ \vartheta_{iCu} \\ \varepsilon_{iCu} \\ \eta_{iCu} \end{bmatrix} \tag{5}$$

$$\mu_i = \begin{bmatrix} \mu_{\beta i} \\ \mu_{\vartheta i} \\ \mu_{\varepsilon i} \\ \mu_{\eta i} \end{bmatrix} \tag{6}$$

A linear (7) and linearized (8) model allowed to determine quantitative relationships between the measured parameters of the process ($x_1=d$, $x_2=t$, $x_3=c_t$, $x_4=c_d$,)and response variable (y_i). Equations (7) and (8) describe the general regression equations that form the basis for modelling.

$$Y_i = \beta_0 + \beta_1 \cdot x_1 + \beta_2 \cdot x_2 + \beta_3 \cdot x_3 + \beta_4 \cdot x_4 + \varepsilon_i \tag{7}$$

$$Y_i = \beta_0 + \beta_1 \cdot ln(x_1) + \beta_2 \cdot ln(x_2) + \beta_3 \cdot ln(x_3)$$
$$+\beta_4 \cdot ln(x_4) + \varepsilon_i \quad (8)$$

where: d – size fraction,
t – flotation time,
c_t – collector type,
c_d – collector dosage.
β_i – model parameter,
ε_i – standard estimation error.

In order to assess the accurateness of matching the regression functions to the experimental data, coefficients of determination R^2were used. It takes values from the range <0.1>.When the values of R2 are close to 1, the model images an appropriate explanation of the variability of experimental values to the predicted values [Jajuga, 1993; Güler & Akdemir, 2012, Masiya & Nheta, 2014; Nakhaei et al., 2015].

4 Results and discussion

The basis for the application of the MANOVA is preservation of the assumptions concerning the representativeness of the dataset and sampling, the normality of distributions of the dependent variables, and the homogeneity of the intergroup variance. However, by choosing the appropriate statistics, it is possible to refrain from adhering to these conditions in situations of minor deviations from the assumptions [Maxwell & Delaney, 2004]. Shapiro-Wilk normality and Brown-Forsyth homogeneity of variance tests were performed for the analyzed dataset, which in a few groups indicated a moderate deviation from normality. Relying on this finding Pillai's trace (9) was used in the MANOVA [Stanisz, 2007]:

$$V = \sum_{i=1}^{q} \frac{\lambda_i}{1+\lambda_i} \quad (9)$$

where: q – number of non-zero own values λ_i of matrix $\boldsymbol{I+BW^{-1}}$,
λ_i – own value of matrix $\boldsymbol{BW^{-1}}$,
$\boldsymbol{B}$ – matrix of intergroups of sums of squares and mixed products,
$\boldsymbol{W}$ – matrix of intersection sums of squares and mixed products.

Based on the results of the multivariate significance test (Tab. 2) it can be concluded that the combined effect of all tested independent variables (flotation time, dose and type of collector, feed granulation) on the effectiveness of the flotation process is significant. A one-dimensional analysis of variance (Tab. 3) confirmed these conclusionsin most cases. For single analyses only the lack of significant impact of the type of collector on the fianal effect was noted, as shown in Fig.2.

In order to analyze the effects of individual parameters on the flotation enrichment results, the mean standardized values of the separation rating indicators at different levels of independent variables are also shown in Fig. 1-4.

Table 2. The results of multivariate tests of significance using Pillai's trace.

Variable	V	F	p
Flotation time t	0.621	34.732	0.000
Collector type c_t	0.252	119.069	0.000
Collector dosage c_d	0.435	272.115	0.000
Size fraction d	0.793	76.472	0.000

Table 3. The results of one-dimensional tests of significance (Fisher-Snedecor test).

Variables	Copper recovery in tailings η		Copper content in tailings ϑ	
	F	p	F	p
Flotation time t	54.448	0.000	5.483	$1 \cdot 10^{-5}$
Collector type c_t	311.918	0.000	2.498	0.114
Collector dosage c_d	164.200	0.000	52.878	0.000
Size fraction d	100.039	0.000	108.903	0.000
Variables	**Copper recovery in concentrate ε**		**Copper content in concentrate β**	
	F	p	F	p
Flotation time t	54.451	0.000	54.451	0.000
Collector type c_t	311.934	0.000	311.934	0.000
Collector dosage c_d	164.203	0.000	164.203	0.000
Size fraction d	100.057	0.000	100.057	0.000

*p>0,05 no significant influence of the independent variable on the dependent variable

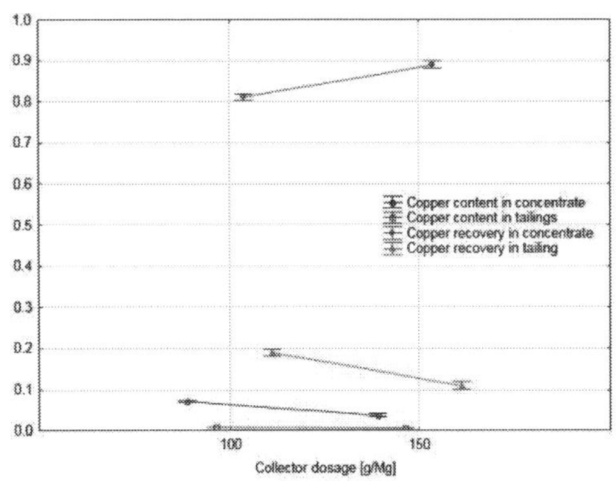

Fig. 1. Standardized average value of the dependent variable for the two dosages of the collector.

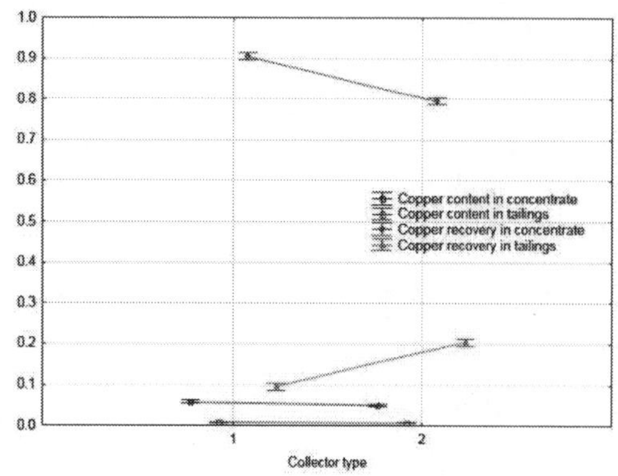

Fig. 2. Standardized average value of the dependent variable for the two types of the collector (1-sodium ethyl xanthate aqueous solution, 2-Hostaflot).

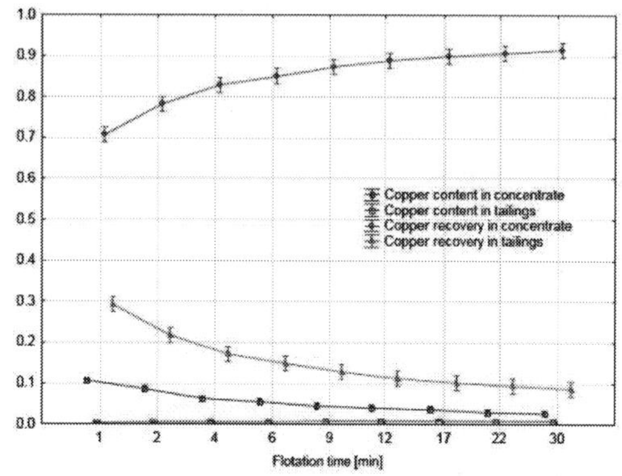

Fig. 3. Standardized average value of the dependent variable on time.

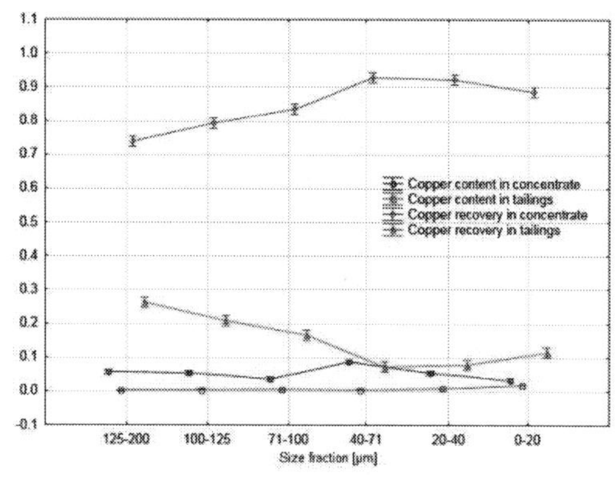

Fig. 4. Standardized average value of the dependent variable on size fractions.

It was found that the best effects of flotation were obtained for 40-71 and 20-40 [µm] size fractions. The use of xanthate mixtures clearly increased the metal recovery and contentin the froth product compared to the second collector. Higher copper recovery was also obtained with the introduction of an increased dosage (150 g/Mg) of the collector, which naturally degraded its quality.

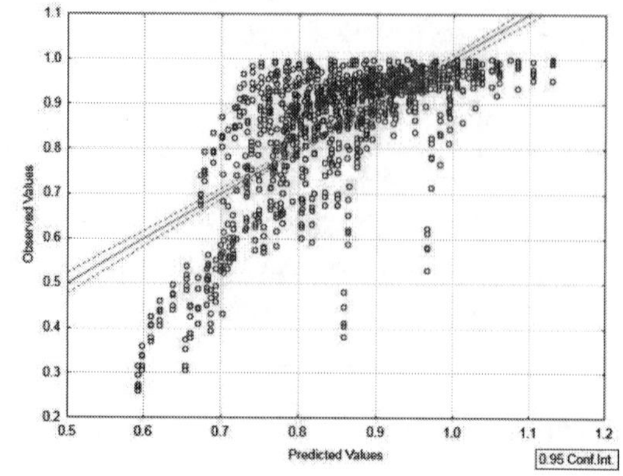

Fig. 5. Experimental and theoretical values for copper recovery in concentrate.

The construction of the model describing dependencies between the variables was based on linear (7) and nonlinear regression (8). Linear regression equations for copper recovery in the concentrate ε (10) and tailings η (11) included four independent variables that had a significant effect on the results of the flotation tests and the lack of mutual correlation. Fig. 5-6 show the regression function with the experimental points for the copper recovery in the concentrate. Their placement indicates the random nature of the process and therefore the difficulty in defining the model for ε and η. In both cases, dependent variables were explained by models only in 47%, hence confirming its poor match.

$$\varepsilon = 0{,}84446 + 0{,}0056 \cdot t - 0{,}10921 \cdot c_t + 0{,}00158 \cdot c_d - 1{,}2178 \cdot d \pm 0.112 \quad (10)$$

$$\eta = 0{,}15556 - 0{,}0056 \cdot t + 0{,}10921 \cdot c_t - 0{,}00158 \cdot c_d + 1{,}2178 \cdot d \pm 0.112 \quad (11)$$

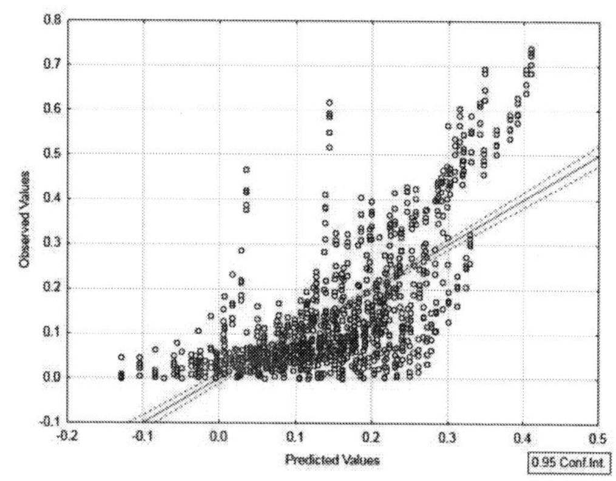

Fig. 6. Experimental and theoretical values for copper recovery in tailings.

For the other two parameters (β, ϑ) it was necessary to introduce linearized models, which, despite the shape of the empirical cloud, did not work well for copper recovery in both products. The linearized regression equation of the general form given by the formula (8) was obtained by making an elementary substitution of $d_i'=\ln(d_i)$, $t_i'=\ln(t_i)$, $c_{ti}'=\ln(c_{ti})$, $c_{di}'=\ln(c_{di})$. In this way, semilogarithmic functions were outlined, in which the curvilinear relation was reduced to a simpler linear form. According to the results of one-dimensional variance analysis, the copper content in concentrate model (12) included all the tested parameters, and the influence of collector type on separation results was omitted in the model of copper content in tailings (13). Based on the graphs of experimental and theoretical values (Fig. 7-8), significant differences can be noted for β and ϑ variables. The values of the determination coefficient R^2 confirms the conclusion. Evaluation of fit of all models is presented in Table 4.

$$\beta = 5.35713 + 0.21758 \cdot ln(d) - 0.42454 \cdot ln(t) -0.28347 \cdot ln(c_t) - 1.46706 \cdot ln(c_d) \pm 0.59207 \quad (12)$$

$$\vartheta = 0.02715 - 0.00526 \cdot ln(d) + 0.0013 \cdot ln(t) -0.00795 \cdot ln(c_d) \pm 0.00746 \quad (13)$$

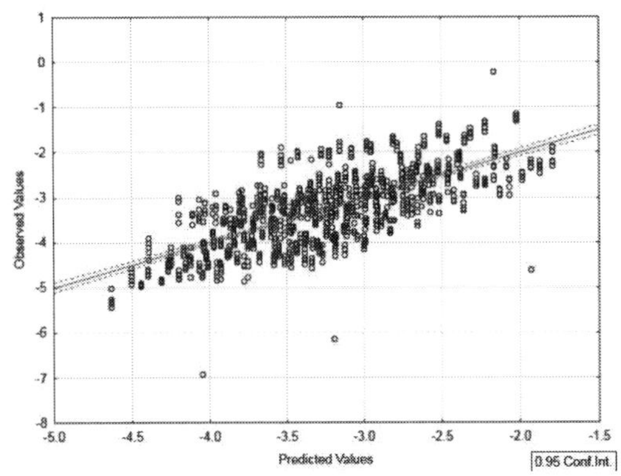

Fig. 7. Experimental and theoretical values for copper content in concentrate.

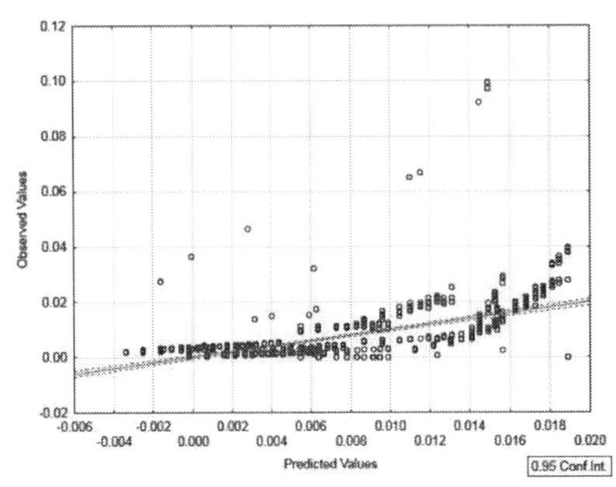

Fig. 8. Experimental and theoretical values for copper content in tailings.

Table 4. The coefficient of determination R^2 for models.

Dependent variables	Adjusted R^2
Copper content in concentrate	0.492
Copper recovery in concentrate	0.469
Copper content in tailings	0.336
Copper recovery in tailings	0.469

5 Conclusions

The paper presents the results of a multidimensional analysis and modelling of copper content and recovery in products of sandstone copper ore originating from Polkowice Copper Preparation Plant. The study concerned the influence of four controlled parameters (feed granulation, flotation time, dosage and type of collector) on the final effect of the process (recovery and content of copper in separation products).

Based on MANOVA results at a 95% confidence level, it was found that all investigated factors significantly affected the flotation of copper ore. Only the one-dimensional variance analysis performed for each dependent variable indicated some minor deviations that were included in the models.

Models of copper recovery in concentrate and tailings were generated by the linear regression method. The function of the relationship between the parameters involved all the variables, and the degree of model match to the experimental data in both cases was equal to 47%. In order to model the copper content in the concentrate and tailing, it was necessary to introduce nonlinear methods. The accomplishment of elementary substitutions $x_i'=\ln(x_i)$ enabled the preparation of semilogarithmic functions, in which the curvilinear relations were reduced to a simpler linear form. According to the results of the one-dimensional analysis of variance in the model of copper content in the concentrate, all investigated process factors were taken into account. On the other hand, the copper content in the tailing did not account for the impact of the collector type on the separation results. In all cases, the degree of explanation of the phenomenon by the models was unsatisfactory, not exceeding 50%. In addition, the distribution of empirical points indicated the random character of the copper ore flotation process, and therefore significant difficulties in determining the regression function. Flotation, as the main operation of enrichment of Polish copper ore, is characterized by its complex structure and randomization. Modelling such processes is extremely difficult, and in some cases even impossible.

The paper is the effect of the scientific project no. 15.11.100.021.

References

1. C. Aldrich, Process Metallurgy **12** (2002)
2. C. Bazin, M. Proulx, Int. J. Miner. Process. **61**, 1-12 (2001)
3. M. Brożek, A. Młynarczykowska, *Kinetics of flotation*. (IGSMiE PAN, Cracow, 2009) [in Polish]
4. J. Drzymała *Mineral Processing. Foundations of theory and practice of minerallurgy*. (Oficyna Wydawnicza Politechniki Wrocławskiej, Wroclaw, 2007) [in Polish]
5. M. Duchnowska, E. Kasińska-Pilut, A. Bakalarz, K. Konieczny, P. Kowalczuk, A. Łuszczkiewicz, CUPRUM, **2(75)**, 97-108 (2015) [in Polish]
6. T. GÜLER, U. AKDEMIR Trans. Nonferrous Met. Soc. China, **22**, 199-205 (2012).
7. K. Jajuga, *Multidimensional statistical analysis*. (PWN, Warsaw, 1993) [in Polish]
8. P. Kijewski, J. Jarosz, Monograph KGHM PM SA II.2.24 *Mineral properties*. (KGHM CUPRUM Sp. z o.o. CBR, Wroclaw, 244-245, 2007) [in Polish]
9. J. Kłapciński, T.M. Peryt, Monograph KGHM PM SA II.2.1*The geological structure of the Fore-Sudetic Mnocline* (KGHM CUPRUM Sp. z o.o. CBR, Wroclaw, 69-77 2007) [in Polish]
10. J. Laskowski, A. Łuszczkiewicz, *Mineral prcessing. Enrichment of mineral resources*. (Oficyna Wydawnicza Politechniki Wrocławskiej, Wroclaw, 1989) [in Polish]
11. T. Masiya, W. Nheta, *Flotation of nickel-copper sulphide ore: optimisation of process parameters using Taguchi method*. (Proceedings of the International Conference on Mining, Material and Metallurgical Engineering, Prague, Czech Republic, **113**, 1-11, 2014)
12. S.E. Maxwell, H.D. Delaney, *Designing experiments and analyzing data. A model comparison Perspective, 2-nd ed.* (Lawrence Erlbaum Associates, 2014)
13. F. Nakhaei, M. Irannajad, A. Sam, A. Jamalzadeh, Physicochemical. Problem. Mi., **52(1)**, 252-267 (2015)
14. M. Nieć, A. Piestrzyński, Monograph KGHM PM SA II. 2.19*Mineralisation*, (KGHM CUPRUM Sp. z o.o. CBR, Wroclaw, 167-197, 2007) [in Polish]
15. T. Niedoba, *Multidimensional characteristics of random variables in description of grained materials and their separation processes* (Mineral Resources Management, Cracow 2013)
16. T. Niedoba, A. Surowiak, P. Pięta, J. Pol. Miner. Eng. Soc., **17(1)**, 15-22 (2016a)
17. T. Niedoba, P. Pięta, A. Surowiak, D. Jamróz, *Multidimensional statistical and visualization methods in description of grained materials*. (MEC2016, E3S Web of Conferences, 8, **01036**, 2016b)
18. A. Piestrzyński Monograph KGHM PM SA II.2.19 *Mineralisation*. (KGHM CUPRUM Sp. z o.o. CBR, Wroclaw, 167-197, 2007) [in Polish]
19. A. Stanisz, *The accessible statistics course using STATISTICA PL on examples from medicine. Volume 2. Linear and nonlinear models* (StatSoft, Cracow, 2007) [in Polish]
20. T. Tumidajski, MRM, **23(6)**, 111-123 (2010).
21. T. Tumidajski, T. Niedoba, D. Saramak, AGH Journal of Mining and Geoengineering, **36 (4)**, 167-177 (2012)

Evaluation of flotation process course on the example of sulphide ores

Konrad Oleksik[1,*], *Daniel* Saramak[1], and *Anna* Młynarczykowska[1]

[1]AGH, University of Science and Technology, Faculty of Mining and Geoengineering, Department of Mineral Processing and Environmental Engineering

Abstract. The article concerns investigations over the influence of selected parameters of the flotation process course on the values of technological indices of beneficiation process assessment. Following parameters were under evaluation: the pulp density, pulp aeration (measured through the rotational speed of rotor) as well as dosage of flotation reagents. Four technological indices were registered: concentrate recovery, concentrate grade, yield of the concentrate and the tails grade. There were calculated correlation coefficients between selected parameters of flotation process course and the values of technological indices, along with determination of significance of these coefficient, at accepted probability level 0.95. In the next stage of investigations there were determined mathematical models with the concentrate recovery, concentrate grade, yield of concentrate and tails grade as the functions of pulp density, speed of the rotor and dosage of flotation reagent. On the basis of the investigations carried out, it was appeared that the dosage of flotation reagent was of the most significant influence on the values of technological indices of flotation process assessment. The least significant influence, in turn, was obtained for the pulp density.

1 Introduction

Flotation is the operation commonly applied in beneficiation of sulphide copper ores. The process is complex and its performance depends on a certain number of parameters, connected with both the feed material, and the process conditions. According to the literature, concerning the issue, following parameters are of significant importance on the obtained values of concentrate grade and recovery:

- pulp density,
- type and dosage of flotation reagents,
- pulp aeration,
- time and layout of flotation course,
- the grade of useful mineral liberation
- particle size of feed.

Pulp density influences the effectiveness of flotation. Together with increasing of the pulp density a linear increase in yields of concentrates can be observed, while the concentrate grades are decreasing [4].

The flotation reagents type and dosage has a strong effect on the course of flotation process. Proper analysis of type and dosages of collector and frother can improve the recovery of useful minerals [18]. Appropriate selection of the reagents regimen results in increasing the quality of flotation concentrates [9]. Results of numerous investigations show that increasing of collector dosage influences the higher concentrate recovery, but only to a certain level. Further increasing of dosage has very limited influence on improvement of flotation process effectiveness [6, 19, 21].

Together with increasing of aeration, the probability of collision of air bubble with particles of useful mineral increases too. Probability of adhesion of the air bubble and mineral particle depends on their volumes. Fine particles float more easily in the presence of fine air bubbles, while coarser particles require larger air bubbles [1-3, 11, 17]. While the rotor speed is constant, increasing of the air flow results with increasing the size of the bubbles Furthermore, aeration is influenced by the rotor speed. Increasing the rotor speed results in reduction of air bubbles size, while their number is increased [7, 8]. Furthermore, an increase in rotational speed of rotor results in increasing of concentrate recovery [4].

Results of various investigations [1, 2, 5, 11, 20] shows that depending on the type of ore under examination, the material particle size, for which the maximum values of recovery can be obtained, is within the size range 0,020-0,070 mm. Coarser particles might be too heavy to be able transferred into the froth product, while too fine particles, despite the high mineral liberation grade, might pass into waste, which result in decreasing the effectiveness of the flotation process course. [11, 15, 16]. It is worth mentioning that it is possible to obtain a high recovery of fine particles in flotation circuits, when they are designed particularly for the fines [10]. Furthermore, the mineral liberation grade is connected with the particle size distribution of feed and it affects the effectiveness of flotation process course. Liberation of useful mineral is strictly related to the course of comminution operations, the type of crushing and grinding devices applied and the layout of comminution

* Corresponding author: koleksik@agh.edu.pl

DOI:10.1051/e3sconf/20171801006

circuit [12, 13]. Properties of feed material also plays significant role in effectiveness of comminution process course [14].

The article presents selected problems connected with the course of flotation process for various values of the selected parameters, that influence its effectiveness. The effectiveness was measured through the obtained values of concentrate recovery, concentrate grade, as well as tails grade.

2 Scope of investigations

The aim of the paper was to test and determine optimal parameters of flotation process course, for sulphide copper ore. Several parameters were selected for testing:

- pulp density,
- rotor speed,
- dosage of flotation collector.

As it was mentioned the above, only the selected parameters were under investigation, and they were accepted for tests on the basis of literature studies. It is worth to underline that the material particle size was not tested in this investigative programme. It was assumed, and accepted, that maximum particle size for the material in flotation tests is below 0,071 mm. The feed material was a copper ore with following lithological composition: carbonates 70%, sandstones 20%, shales 10%, and average feed copper grade (α) was 1,5%. The ore was crushed and grinded for flotation test in a two-stage laboratory-scale comminution circuit:

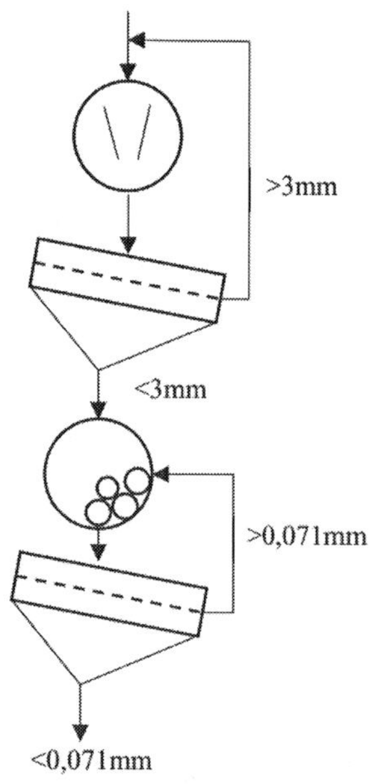

Fig. 1. Scheme of comminution circuit.

The final product from comminution circuits was then divided into 12 representative samples for flotation process.

Flotation was performed in a laboratory pneumatic-mechanical machine, Denver type, with 1dm^3 volume, as the rough flotation with a single fractioned flotation. Froth products were collected after 1, 3, 5, 10 and 15 minutes. Scheme of flotation results was presented in Figure 2.

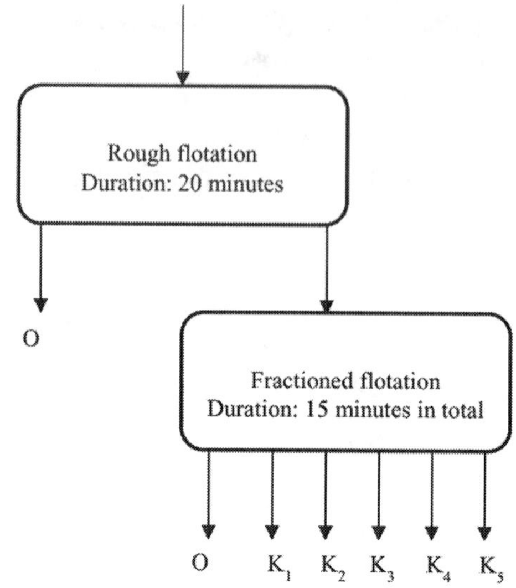

Fig. 2. Scheme of time and layout of flotation course.

The ethyl xanthate sodium was used as a collector in the process, and Nasfroth as a frother reagent. Twelve flotation tests were performed in total, and Table 1. summarizes the values of variable process parameters under investigation. The remaining conditions of the process were kept constant and the dosage of frother reagent was 30 g/Mg for each flotation test.

Table 1. Variable values of parameters for each flotation test.

Number of flotation test	Pulp density [g/dm^3]	Rotor speed [rpm]	Dosage of collector [g/Mg]
1	1100	1850	85
2	1100	1850	100
3	1100	1850	115
4	1200	2450	85
5	1200	2450	100
6	1200	2450	115
7	1100	2450	85
8	1100	2450	100
9	1100	2450	115
10	1200	1850	85
11	1200	1850	100
12	1200	1850	115

3 Analysis of results

For each flotation test there were determined the yields of concentrates (γ), concentrate grades (β), flotation recoveries (ε_k), as well as tail grades (ϑ). The XRF analyzer was used for determination of copper contents in flotation concentrates and tails. Complete results are presented in Table 2.

DOI:10.1051/e3sconf/20171801006

Table 2. Values of selected technological indicators for copper ore flotation.

Number of flotation test	ε_k [%]	β [%]	ϑ [%]	γ[%]
1	81.27	10.99	0.396	17.44
2	86.30	9.29	0.314	24.15
3	86.56	7.07	0.345	31.7
4	80.69	10.89	0.415	19.35
5	84.41	8.94	0.396	27.57
6	87.15	7.91	0.317	29.17
7	81.67	10.72	0.418	21.33
8	84.84	8.22	0.357	27.63
9	84.14	7.98	0.376	28.11
10	83.46	9.07	0.374	22.60
11	87.00	7.74	0.327	29.36
12	88.95	5.56	0.317	39.99

In the next stage of investigation there were determined quantitative relationships among parameters of process course and the obtained results. Suitable models were calculated with using the multiple regression methods.

Prior to determination of the models, there were calculated the values of correlation coefficients between the selected parameters of flotation process course and values of technological indicators for flotation process run (Table 3.). Statistically significant coefficients at accepted probability level $\alpha = 0,05$, were underlined. The results presented in Table 3 show that the collector dosage is of the strongest influence on values of all technological indicators, that is: the yields of concentrates (γ), concentrates grades (β), flotation recoveries (ε_k), and tail grades (ϑ). Speed of rotor show less strong correlation, while the pulp density has the lowest influence on values of technological indicators.

Table 3. Coefficients of correlation between all variables.

	ε_k [%]	β [%]	ϑ [%]	γ[%]
Pulp density [g/dm^3]	0.230	-0.221	-0.135	0.252
Rotor speed [rpm]	-0.356	0.263	0.463	-0.172
Dosage of collector [g/Mg]	0.808	-0.856	-0.683	0.843

An influence of dosage of collector and the rotational speed of rotor at the pulp density of 1100 g/dm^3, were presented in Figs. 3 – 6. Analysing the results it can be noticed that increase in the dosage of collector results in increase of flotation recovery (ε_k) and concentrate yield (γ) but causes decrease of concentrate (β) and tail (ϑ) grades

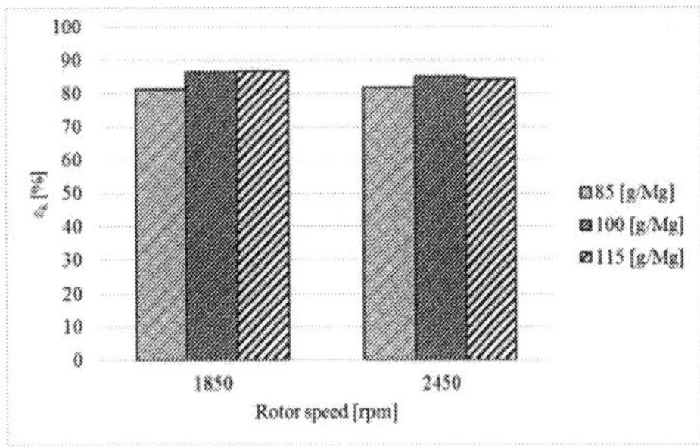

Fig. 3. Flotation recoveries (ε_k) in relationship with various values of rotor speed and dosage of flotation collector for pulp density 1100g/dm^3.

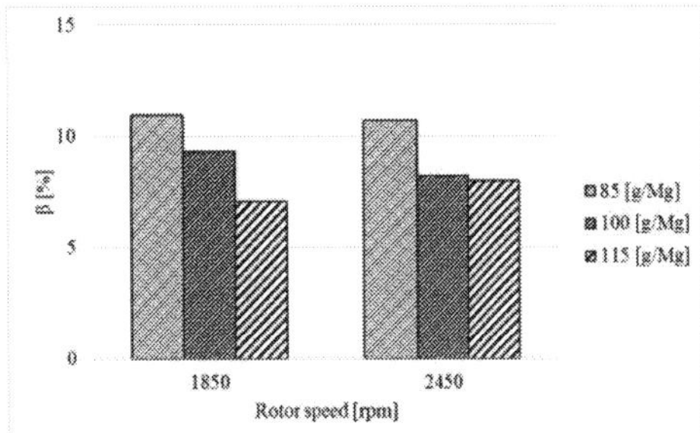

Fig. 4. Concentrates grades (β) in relationship with various values of rotor speed and dosage of flotation collector for pulp density 1100g/dm^3.

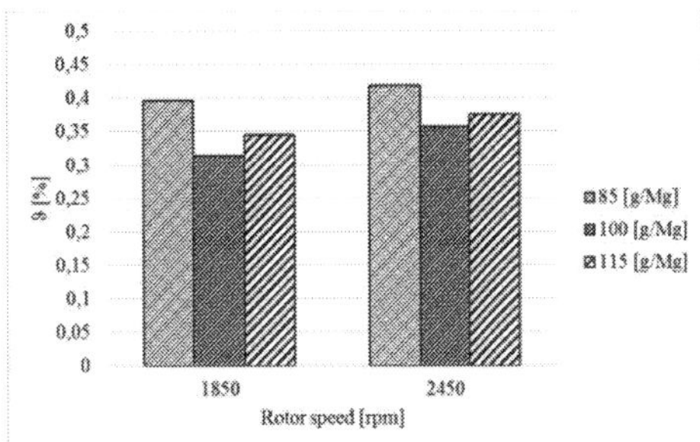

Fig. 5. Tail grades (ϑ) in relationship with various values of rotor speed and dosage of flotation collector for pulp density 1100g/dm^3.

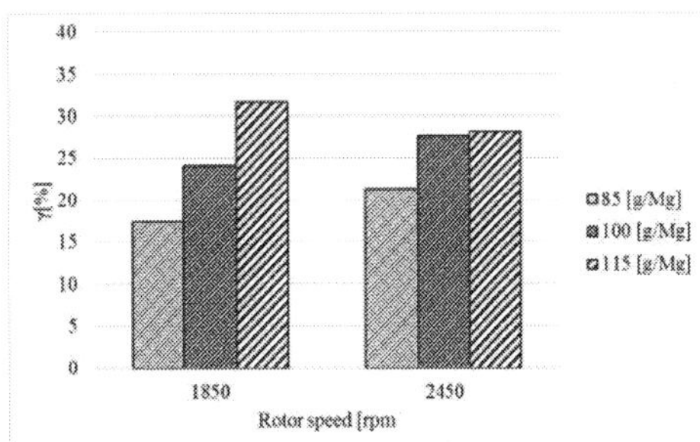

Fig. 6. Yields of concentrates (γ), in relationship with various values of rotor speed and dosage of flotation collector for pulp density 1100g/dm^3.

DOI:10.1051/e3sconf/20171801006

The models, linking the values of parameters of flotation process course with the values of selected technological indicators, were determined with using the multiple regression method. The pulp density, rotor speed and dosage of collector were independent variables in the model. Separate models were calculated for the concentrate recovery, concentrate grade, as well as tails grade. Values statistically significant were underlined.

$$\varepsilon_k = 61{,}45 + 0{,}0115a - \underline{0{,}0030b} + \underline{0{,}1643c} \quad (1)$$

$$\beta = 24{,}68 - 0{,}0069a + 0{,}0014b - \underline{0{,}1096c} \quad (2)$$

$$\vartheta = 0{,}56 - 0{,}0001a + \underline{0{,}00006b} - \underline{0{,}0021c} \quad (3)$$

$$\gamma = -40{,}35 + 0{,}0295a - \underline{0{,}0034b} + 0{,}4021c \quad (4)$$

where:

- a – Pulp density [g/dm^3]
- b – Rotor speed [rpm]
- c – Dosage of collector [g/Mg]

There were also determined values of R^2 that characterize the fitting of model:

- $R^2 = 83{,}3\%$ for equation (1)
- $R^2 = 85{,}1\%$ for equation (2)
- $R^2 = 69{,}9\%$ for equation (3)
- $R^2 = 80{,}3\%$ for equation (4)

Both in model one (ε_k) and two (β) there were achieved the fitting on satisfactory level, while for model number three (ϑ) the fitting level was lower.

4. Summary

The results of flotation tests and statistical analyzes confirm that the tested factors influence the effectiveness of the flotation process. The most significant effect shows the dose of collecting reagent. Its growth causes an increase in flotation recovery (ε_k) and yield of concentrate (γ) and decrease in concentrates (β) and tail grades (θ). However, it should be pointed out that, according to the literature [6, 19, 21, 22], this effect is only visible to a certain boundary level of the collector, which was not achieved in these tests.

Concentrates with the highest content of the useful ingredient (β) were obtained at a collection reagent level of 85 g/Mg, while the highest recovery (ε_k) and yield (γ) were obtained for the highest dose of collector. The obtained results are consistent with the results presented in literature within the scope.

This work was done as a part of the University of Science and Technology scientific grant No. 15.11.100.079

References

1. M. Brozek, A. Mlynarczykowska, Physicochem. Probl. Miner. Process. **44**, 23 (2010)
2. M. Brozek, A. Mlynarczykowska, Arch. Min. Sci., **Vol. 57**, No 3, p. 729-740 (2012)
3. J.A. Kitchener, Dev Miner. Process., **12**, 1-39 (1992)
4. J. Konieczny, W. Pawlos, K.Ksiezniak, M. Krzeminska, E. Kasinska-Pilut, P.Piwowar, CUPRUM, **2(75)**, 87-95 (2015)
5. B.P. Kowalczuk, O. Sahbaz, J. Drzymala, . Minerals Engineering **24**, 766-771 (2011)
6. A. J. Lynch, N. Johnson, E. V. Manlapig, C. G. Thorne, Dev. Miner. Process, **3** (1981)
7. A. Mlynarczykowska, A. Nyrek, K. Oleksik, IM, **1(35)**, 181-188 (2015)
8. A. Mlynarczykowska, A. Nyrek, K. Oleksik, IM, **18(1)**, 257–262 (2017)
9. M. K. Mpongo, E. Siame, AJST, **7(1)** (2006)
10. J.D. Pease, D.C. Curry, M.F. Young, Minerals Engineering, **19**, 831–840 (2006)
11. A. Potulska, doctoral thesis (Politechnika Wroclawska, Wydzial Geoinzynierii, Gornictwa i Geologii, Wroclaw, 2008)
12. D. Saramak, Mineral Resources Management , **28(4)**, 87-99, (2012)
13. D. Saramak, Physicochem. Probl. Miner. Process. **48(1)**, 219-226 (2012)
14. D. Saramak, R.A. Kleiv, Minerals Engineering, **43-44**, 105-111, (2013)
15. D. Saramak, A. Mlynarczykowska, A. Krawczykowska, Archives of Metallurgy and Materials., **59(3)**, 951-955, (2014)
16. D. Saramak, A. Krawczykowska, A. Mlynarczykowska, Arch. Min. Sci., **59(3)**, 731–740 (2014)
17. J. A. Solari, R. J. Gochin, , Dev Miner. Process., **12**, 361–399 (1992)
18. R. Sousa, A. Futuro, C. Setas Pires, M. Machado Leite, Physicochem. Probl. Minerl. Process. **53(2)**, 758-769 (2017)
19. G. C. Thorne, E. V. Manlapig, J. S. Hall, A. J. Lynch, Flotation A. M Gaudin memorial, **2**,725–752 (1976)
20. R. Tomanec, M. Blagojev, IM, **18(1)**, 165-172 (2017)
21. W.J. Trahar, Int. J. Miner. Process., **8(2)**, 289-32 (1981)
22. X. Luo, B. Feng, C. Wong, J. Miao, B. Ma, H. Zhou, JMRT, **5(2)**, 131 – 135 (2016)

 DOI:10.1051/e3sconf/20171801007

The effect of process water salinity on flotation of copper ore from Lubin mining region (SW Poland)

Alicja Bakalarz[1], *Magdalena* Duchnowska[1], and *Andrzej* Luszczkiewicz[1,*]

[1]Wroclaw University of Science and Technology, Wybrzeze Wyspianskiego 27, 50-370 Wroclaw, Poland

Abstract. The process water used for the flotation of sedimentary copper ore in ore concentration plants in KGHM Polska Miedz S.A. were characterized. The process water used in the flotation circuits is heavily saline. It contains between 25 and 45 g/dm^3 of soluble components, and the main constituent, in about 75%, is NaCl. Process water used for flotation consists of reclaimed water from the tailing dam and mine water. The effect of process water salinity on the processes of copper flotation from the Lubin mine area was described. The results of laboratory flotation experiments conducted in tap water and in water of different salinity levels were compared. The effect of the salinity of water within specified concentration limits was generally found to be beneficial for upgrading of the examined ore.

1 Introduction

1.1 The effect of process water salinity on the flotation processes

The physicochemical properties of the process water, i.e. its ionic composition and the amount of dissolved substances, are one of the most important factors influencing the course of flotation processes. It is known from industrial practice that the concentration of soluble salts in flotation slurries affects the course of flotation processes. Because of the widespread freshwater shortage in industrial areas, the use of saline water in the flotation process is a necessity. Since the 1930s saline sea water has been used for flotation [1]. Laskowski [2] described the beneficial effect of salts soluble in flotation solutions on the course and effectiveness of flotation. This was, however, mainly related to naturally very hydrophobic minerals and hard coal. In such raw materials there is a 2-3 times increase in flotation rate compared to non-saline water. Pugh et al [3] groups electrolytes that affect the efficiency of graphite flotation into three groups. Group A electrolytes of small, strongly charged and hydrated cations (divalent and trivalent), such as $MgCl_2$, $CaCl_2$, Na_2SO_4, $MgSO_4$, $LiCl_3$ have the most favourable impact. In the presence of these cations, the flotation occurs at a concentration of 0.02 M and reaches the maximum at a concentration of approximately 0.06 to 1.0 M. Group B, comprising NaCl, LiCl, KCl, CsCl, NH_4Cl, shows the most favourable, however, slightly weaker effects on flotation at concentrations from 0.05 to 0.1 M. The last, C group of electrolytes ($NaClO_4$, $HClO_4$, HCl, H_2SO_4, $LiClO_4$, CH_3COONa) has a very poor effect on the flotation recovery and effectiveness substantially regardless of the concentration measured to approximately 3.0 M. The effect of the mentioned mono- and divalent metal salts in sulphide minerals flotation differs considerably. Many scholars have unambiguously stated that in the case of chalcopyrite monovalent metal salts, e.g. NaCl and KCl, have a beneficial effect on flotation upgradeability. The presence of calcium and magnesium salts, on the other hand, causes respectively a weak and strong effect of the depression of flotation of this mineral [4,5]. Many authors point to the beneficial effect of sulphides flotation in saline water against low mineralization waters, e.g. tap water [6].

Flotation of metal ore in saline water has been the subject of many studies conducted in recent years due to the increasing shortage of freshwater or lack of it in major mining areas, especially of non-ferrous metal ore. The described problems most often concern the use of seawater and, in general, closed circuit water including water recycling from thickening and filtration of the concentrate and, first of all, from the system of processing plant – settling pond of flotation tailing [7]. The quoted authors recommend ongoing monitoring of the physicochemical state of circulating water, e.g. ionic composition, Eh-pH potential, suspended colloidal particles, and the adapting of the technological regime to changing water properties.

The effect of process water salinity on the processes of flotation of ores mined in Lubin Mining Region has been observed since the start of the concentrating plants operation. From the study of Zmudzinski and Lekki [8], analyzed by Karpisz with upgrading curves [9], it results that the artificial addition of NaCl to tap water up to 100 g/dm^3 has a positive influence on the flotation of carbonate and shale ore, while the addition above 6 g/dm^3 adversely affected the flotation of sandstone ore. In the studies of Kowalska [10,11], the influence of dilution of

* Corresponding author: andrzej.luszczkiewicz@pwr.edu.pl

DOI:10.1051/e3sconf/20171801007

process water of Polkowice Concentrator with low-saline water collected in those days from the Odra River was found to be unfavorable.

The problem of the effect of process water salinity on the flotation process was also recently investigated by Luszczkiewicz and co-workers [12]. The studies have shown that dilution of process water does not alter the results or leads to improvement of flotation efficiency. By contrast, increasing of salt concentration of process water by evaporation results in a significant deterioration of the flotation efficiency. In turn, the purification of process water with the use of activated carbon most often also leads to a deterioration in the selectivity of flotation. The deliberate addition of NaCl to process water in a very different way influenced the results of flotation. This generally did not change the selectivity of Cu and C_{org} flotation but slightly decreased the flotation selectivity of Ag and Pb.

1.2 Water used in technological processes of ore upgrading in KGHM Polska Miedz S.A.

The process water used in ore upgrading in KGHM Polska Miedz S.A. consists of reclaimed water from the tailing dam which is fed with mine water. The latter significantly differs in the amount of the soluble components in the particular mining areas associated with processing plants. Water circulating in concentrating plants is additionally supplied with soluble components washed out of the treated ores. As a result, water in the flotation circuits of the KGHM Polska Miedz S.A. processing plants is characterized by high content of inorganic salts. According to Luszczkiewicz and others [12], the concentration of soluble components in recycled water from the Zelazny Most tailings dam, which feeds predominantly flotation systems, varies within 23–33 g/dm^3. The recycled water from the "Zelazny Most" pond is mixed with mine water in the processing plants, carrying different amounts of soluble salts, therefore, in different areas of the Concentrator, flotation process water varies and, depending on the region, contains between 20 and 40 g/dm^3 of soluble components (Table 1).

As noted in the study of Luszczkiewicz et al. [13], data from Table 1 show that the ionic product of calcium and sulphate ions is $0.77 \cdot 10^{-3}$ up to $1.2 \cdot 10^{-3}$ and it is significantly higher than the solubility equilibrium product of various forms of calcium sulphate (below $3.1 \cdot 10^{-5}$). This means that the KGHM process water is supersaturated with calcium sulphate (gypsum and anhydrite) and precipitation of these substances is expected. This supersaturation is based on the presence of other salts that shift chemical equilibrium and the so-called activity coefficients and thus solubility. The source of sulphate ions, besides gypsum or anhydrite, is also magnesium sulphate. The $MgSO_4$ stoichiometry (SO_4^{2-} = 96 g/mol, M_{Mg+2} = 24.3 g/mol, M_{MgSO4} = 120.3 g/mol) shows that if in process water of the Concentrator is at least 250 mg/dm^{3}mg^2 this is accompanied by about 1000 mg/dm^3 sulphate ions. In turn, the discussed water contains approximately 3000 mg/dm^3 SO_4^{2-} ions, i.e. 2000 mg/dm^3 comes from $CaSO_4$ (theoretically approx. 1700 mg/dm^3).

Table 1. Results of determination of the major chemical components of process water of the Lubin, Polkowice and Rudna Concentrators [13].

Symbol		unit	Lubin	Polkowice	Rudna
Fe		mg/dm^3	0.22	0.66	0.61
Cu		mg/dm^3	0.202	4.80	2.02
Zn		mg/dm^3	0.168	0.320	0.672
Pb		mg/dm^3	<0.005	0.478	1.50
As		mg/dm^3	<0.005	0.038	<0,005
Mg		mg/dm^3	265	253	121.50
		mol/dm^3	0.011	0.010	0.005
Ca		mg/ dm^3	1325	1614	994
		mol/dm^3	0.0330	0.0403	0.0248
Na		mg/dm^3	6 162	7 827	12 171
		mol/dm^3	0.268	0.340	0.529
SO_4^{2-}		mg/dm^3	2 810	2 829	2 994
		mol/dm^3	0.0293	0.0295	0.0312
Cl^-		mg/dm^3	11 150	13 438	19 728
		mol/dm^3	0.315	0.379	0.556
Soluble components		mg/dm^3	24 180	29 820	43 450
Total suspended matter		mg/dm^3	9.0	18.0	56.0
Dry residue		mg/dm^3	24 440	29 690	41 330
Dry residue	CaO	%	4.44	4.71	3.58
	MgO	%	1.19	1.28	1.04
	NaCl	%	74.82	86.04	62.43
Na+Cl in dissolved components		%	71.6	71.32	73.4

Groundwater (mine water), depending on the area of the deposit and the depth of the mining, varies considerably. Kleczkowski et al. [14] report that the increase in mineralization (dry residue after evaporation) is observed with depth. Mineralization in the mining areas of the deposit in years 2003–2006 in the southern areas ranged from 0.79 to 4.25 g/dm^3 on average 2.12 g/dm^3, while in the northern areas from 22.55 to 328.80 g/dm^3. The average mineralization of water in this area in the region of the deepest mining works is approx. 161 g/dm^3 including concentration of chlorides over 100 g/dm^3. In the NE direction with depth, the mineralization increases even to more than 330 g/dm^3. The increase in mineralization is related to the increase in chlorides, since the concentration of sulphate already at a depth of about 600 m is stabilized at about 2 g/dm^3. The water in the northern area of the deposit is saline of type Cl-Na-Ca, in which the presence of iodides and bromides and significant amounts of iron are found. In the southern areas there is saline of type SO_4-Cl-Ca and SO_4-Cl-Na [14].

Process water used in flotation in different regions of the concentrating plant differs in composition. This is due to the fact of mixing of recycled water from the tailing dam with mine drainage water, which in the particular mining areas varies in both amount and concentration. As shown in Table 1, the salinity considered as the amount of substances dissolved in water varies between 24 and 43 g/dm^3 depending on the Concentrator's connected with the mining area. The dominant components of this water are Na^+ and Cl^- ions, therefore, the main component is sodium chloride, whose share in total water salinity in all three plants is similar, at 71–73%. The calcium and sulphate ions present in the process water are mostly derived from the dissolution of the anhydrite found in the ore. The sulphate concentration is maintained at a constant level of 3 g/dm^3. The source of sulphate ions is also magnesium sulphate.

The analysis of salt content in process water in all three regions of the Concentrator observed over many years shows that chlorine concentration in process water has been steadily increasing [12].

The increase in salinity of flotation water has been observed since the beginning of the development of the Lubin-Glogow Copper Basin (LGOM) deposits. The salinity is systematically increasing from around 6 g/dm^3 of soluble substances in reclaimed water from the settling pond "Gilów" [15] to the present amount of approx. 35–40 g/dm^3. Currently, in the reclaimed water of the "Zelazny Most" tailings dam, compared to data from the late 1960s, the content of soluble components and chloride ions increased approximately fivefold/sixfold and the amount of sulfate ions increased more than one and a half times with a tendency toward equilibrium at approx. 3000 mg/dm^3.

2 Experimental

2.1. Research materials and methodology

The research material was a sample of ore from the Lubin Mine area. The ore sample represented crushed raw feed from current production. It was collected from the Lubin Concentrator's carbonaceous ore stream, directed to the 1st stage of grinding. The sample was crushed in a laboratory jaw crusher to a grain size <1 mm and subjected to grinding in a laboratory to a grain size P_{90}<0.071 mm (Table 2). The ground ore was the material for flotation experiments. Chemical analyzes showed the following content of balanced components: 0.97% Cu, 50 g/Mg Ag, 0.99% C_{org}.

The composition of process water used in the flotation experiments, taken from the plant technological circuit during the collection of test ore, is summarized in Table 3. Table 3 also contains the composition of tap water used in experiments.

The experiments were performed in a laboratory Mechanobr type flotation machine with 1 dm^3 cell volume. Standard flotation reagents used in the Lubin Concentrator were used in the experiments: the collector was sodium ethyl and isobutyl xanthate, mixed with Hostaflot LET at 80:20, dosed at 74 g/Mg of the ore; a mixture of Corflot with Nasfroth in ratio 40:60 in the amount of 16 g/Mg was used as frother.

Table 2. Particle size analysis of feed for flotation experiments.

Grain class, mm	Yield, γ,%	100-$\Sigma\gamma$, %	Cu		Ag	
			λ, %	ε, %	λ, g/Mg	ε, %
> 0.100	2.5	100.0	0.56	1.4	60	3.1
0.071 ÷ 0.100	7.4	97.5	0.62	4.8	36	5.5
0.040 ÷ 0.071	25.0	90.1	0.93	24.2	44	22.7
0.020 ÷ 0.040	21.7	65.0	1.16	26.2	40	17.9
<0.020	43.3	43.3	0.96	43.3	57	50.8
Calculated feed	100.00		0.96	100.0	49	100.0
Assayed feed			0.97		50	

λ – component content in the product, %
ε – component distribution in the product, %

Table 3. Results of chemical analysis of process and tap water used in flotation experiments.

Component	Content mg/dm^3	
	Process water	Tap water
Total iron	<0.05	<0.02
Copper	0.028	not found
Zinc	0.096	not found
Lead	0.013	not found
Arsenic	<0.005	not found
Magnesium	394	13.3
Calcium	1687	93
Sodium	8866	22.3
Sulphates	2962	105
Nitrates	35.9	17.7
Nitric nitrogen	8.12	not found
Chlorides	15295	49.8
$CaCO_3$	5839	288
Soluble components	32070	350
Dry residue	32320	

The experiments were performed according to the flowsheet shown in Fig. 1. Four series of tests with different types of water were made:

- with natural process water taken in the Lubin Concentrator where the ore material came from (Table 3). This water contained approx. 32 g/dm^3 of dissolved components,
- with process water diluted with tap water in the proportion of 30% of process water and 70% of tap water. This water therefore contained approx. 10 g/dm^3 of dissolved components,
- with process water, in which pure sodium chloride was additionally dissolved. The amount added was equal to 100% of the amount of this salt present in the natural state in process water. The content of NaCl in the natural state of process water was calculated stoichiometrically assuming that the total Cl^- ion is bounded in the form of NaCl. 18.4 g/dm^3 of crystalline NaCl was added to natural process water, resulting in its total content of 43.9 g/dm^3 of soluble components, including 37.8 g/dm^3 of dissolved NaCl,
- with tap water.

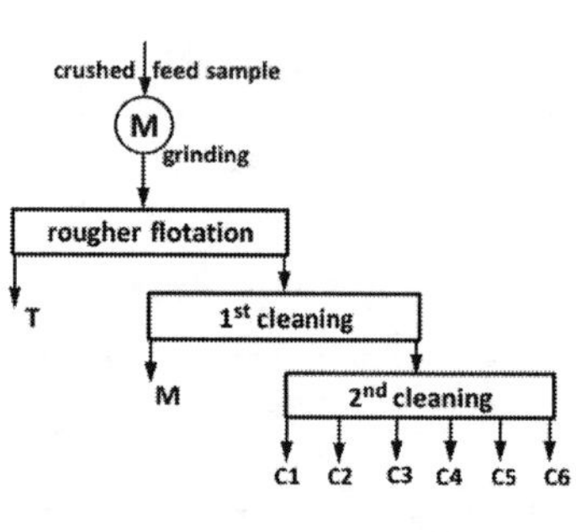

Fig. 1. General flowsheet of flotation experiments.

2.2. Discussion of results

Fig. 2-4 show the results of the flotation experiments of the examined ore in the form of recovery-grade and recovery-recovery upgrading curves obtained in various types of water, balanced for copper, silver and organic carbon.

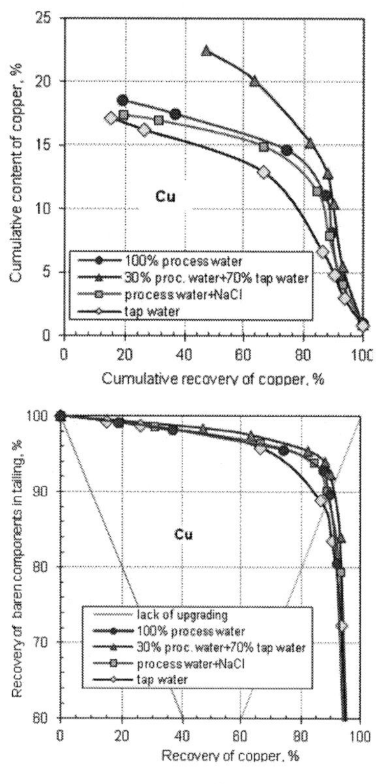

Fig. 2. Results of ore flotation according to Cu balance in different types of water.

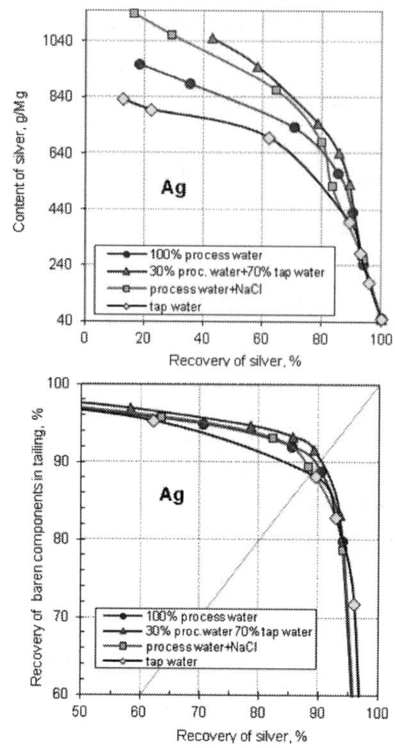

Fig. 3. Results of flotation experiments balanced by silver

Analysis of upgrading curves showed that water salinity has a positive effect on the floatability of all three balanced components. The worst results of the process is observed in the case of tap water. With the increase of salts dissolved in the flotation water (30% of the saline process water with 70% of tap water) used in the tests, the results of upgrading were most favorable in the series of tests described. Flotation in the naturally saline process water was significantly worse than flotation in diluted process water, but significantly better than in tap water. Increasing the salinity of process water by dissolving NaCl in it, slightly affected the results of flotation according to the Cu balance (Fig. 2). In contrast, this addition had a better effect on silver upgrading (Fig. 3). By comparing the floatability of copper and silver in all four types of water (Fig. 5), one notices a slightly worse silver upgradeability. With the increase in salinity (NaCl addition) these differences become smaller.

Upgrading of organic carbon as an unfavorable component in copper concentrates (Fig. 4) shows a downward trend with increasing water salinity in flotation. The strong increase in salinity, however, causes a significant increase in the content of this component in the obtained concentrates. Comparison of the behavior of organic carbon in the analyzed water types (Fig. 5) clearly indicates that this component is the worst upgraded in tap water, and any increase in salinity has a positive effect on its flotation.

DOI:10.1051/e3sconf/20171801007

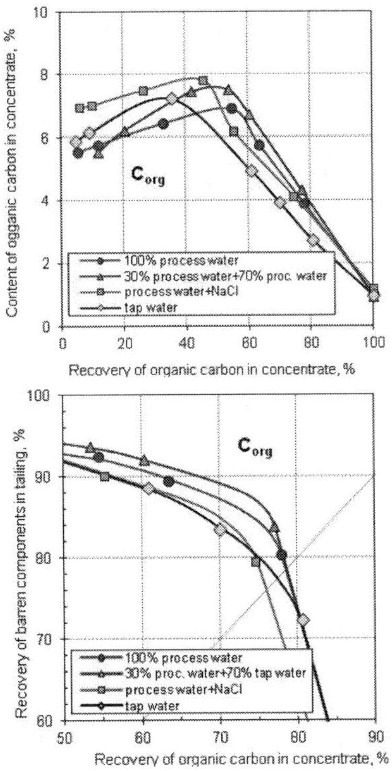

Fig. 4. Results of flotation experiments balanced by organic carbon.

The results of copper ore flotation experiments in flotation water of different salinity levels have been confirmed by numerous literature data on the effect of sea water on the efficiency of upgrading of these ores. The copper ore from the Lubin Concentrator used in experiments is of the polymineral character. The metal carriers in this ore are mainly bornite, chalcopyrite and to a lesser amount chalcocite. Mineralogical analyzes of flotation products indicated a beneficial effect of salinity on the upgrading of chalcopyrite and bornite in the concentrate, whereas chalcocite clearly exhibited poorer floatability. It can be assumed that the mechanism of the beneficial effect of salinity on the flotation of sulphide minerals is in accordance with the literature [5,6]. In case of presence of clay minerals in the flotation suspension Zhao and Peng [16] explain the deterioration of water flotation in tap water by the electrochemical impact of mineral surface resulting in easier mulching in tap water on sulphide minerals, especially on chalcocite.

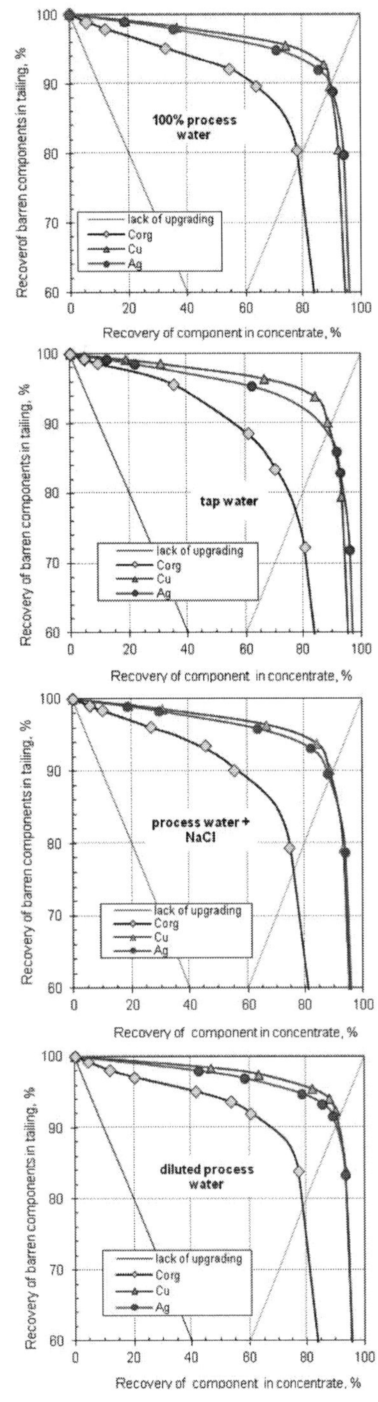

Fig. 5. Comparison of the results of flotation experiments in the analysed types of water.

3 Conclusion

The balances of flotation experiments of copper ore from the Lubin mining region, analyzed in terms of copper, silver and organic carbon indicate the beneficial effect of the salinity of the water used on the upgradeability of the balanced components. Dilution of process water with tap water in a ratio of 70 to 30% resulted in the most favorable flotation conditions for all balanced components: copper, silver and organic carbon. Flotation in process water indicates more favorable flotation effects of all balanced components compared to flotation in tap water.

When comparing the flotability of copper and silver in all four analysed types of water, a slightly weaker silver upgradeability is noted. With the increase in salinity (NaCl added at 100% of its original amount in process water), these differences become smaller. On the other hand, the disadvantage of organic carbon is that it is poorly upgraded in tap water, and any increase in salinity has a positive effect on its flotation.

The article was co-financed by funds of statutory research no. 0402/0161/16 carried out at the Faculty of Geoengineering, Mining and Geology of Wroclaw University of Science and Technology.

References

1. S. Castro, I. Venegas, A. Landero, J.S. Laskowski, *Frothing in seawater flotation systems*. In Proceedings of XXV International Mineral Processing Congress, Brisbane, Australia. The Australasian Institute of Mining and Metallurgy, 4039-4047 (2010)
2. J. Laskowski, *Chemia fizyczna w procesach mechanicznej przeróbki kopalin* (1969)
3. R.J. Pugh, P. Weissenborn, O. Paulson, 1997. Int. J. Miner. Process. **51(1-4)**, 125-138 (1997)
4. T. Hirajima, WSE Suyantara, O. Ichikawa, A.M. Elmahdy, H. Miki, K. Sasaki, 2016. Miner. Eng. **96**, 83-93 (2016)
5. F. Rao, I. Lázaro, L.A. Ibarra, Miner. Process. Extr. Metall. Rev. **126(3)**, 139-145 (2017)
6. F. Ikumapayi, K.H. Rao, 2015. Miner. Process. Extr. Metall. Rev. **36(1)**, 45-64 (2015)
7. G. Levay, R.S.C. Smart, W.M. Sinner, J. South. Afr. Inst. Min. Metall. **101(2)**, 69-75 (2001)
8. K. Zmudziński, J. Lekki, *Badania nad wpływem składu wód kopalnianych w legnicko- głogowskim okręgu miedziowym na flotację rud miedzi*. Report No. 1216/67, Department of Ore Processing, Institute of Non-Ferrous Metals, Gliwice (1967)
9. A. Karpisz, *Wpływ parametrów fizykochemicznych wody technologicznej na flotowalność rud miedzi*. Diploma thesis, Faculty of Geoengineering, Mining and Geology, Wroclaw University of Sience and Technology, Wroclaw (2010)
10. M. Kowalska, *Badanie wpływu wody z Odry na wyniki flotacji rudy polkowickiej*. ZD Cuprum Report in Lubin No. 3/TP/76 (1976)
11. M. Kowalska, *Wpływ jakości wody technologicznej na wskaźniki wzbogacania w ZG ZWR Rudna*. ZD Cuprum report in Lubin (1978)
12. A. Luszczkiewicz, J. Drzymala, T. Henc, Ż. Konopacka, M. Duchnowska, *Określenie wpływu związków chemicznych zawartych w wodach przemysłowych na proces wzbogacania w O/ZWR*. Study Report: Report No. S-42/2011, Wroclaw University of Science and Technology, Institute of Mining (2011)
13. A. Luszczkiewicz, A. Konieczny, E. Kasinska-Pilut, J. Drzymala, *Charakterystyka wód stosowanych do flotacji w O/ZWR KGHM Polska Miedź S.A*. Materiały III Polskiego Kongresu Górniczego, Mineralurgia i wykorzystanie surowców mineralnych, Drzymała J., Kowalczuk P.B. (eds), 14-16 September 2015, Wrocław, 28-34 (2015)
14. A. Kleczkowski, S. Downorowicz, W. Zimny, R. Becker, *KGHM Polska Miedź S.A. Monograph*, multiple authors, eds A. Piestrzyński, A. Banaszak, M. Zalewska-Kusmierczyk, KGHM Cuprum sp. z o.o. CBR, Wrocław, 133–138 (2007)
15. L. Dobrowolski, B. Nowakowska, *Ustalenie przyczyn obniżania się wskaźnika zużycia odczynnika pianotwórczego w procesie wzbogacania rudy miedzi w ZPM w Lubinie*. Report No TGE / TW / 69, part. 2, KGHM Experimental Plant in Lubin (1969)
16. S. Zhao, Y. Peng, Miner. Eng. **66**, 152-156 (2014)

Possibilities for reusing the waste from the process of Zn-Pb ore beneficiation

Krystian Cichy[1,*], *Kornel* Tobiczyk[1], *Arkadiusz* Mańka[2], *Grażyna* Pajor[2], and *Cezary* Reguła[2]

[1]Institute of Non-Ferrous Metals, Sowińskiego 5, 44-100 Gliwice, Poland
[2]Zakłady Górniczo-Hutnicze „Bolesław" S.A, Kolejowa 37, 32-332 Bukowno, Poland

Abstract. This paper discusses the areas of storage, resources, and granulometric and chemical characteristics of old Zn-Pb tailings stored in heaps in the city of Bytom area. It presents the results of laboratory tests for development of the technological flowsheet for transformation of the material into Zn-Pb sulfide concentrates and the results of trials in an experimental system of the beneficiation flowsheet which was developed. In the further part of the paper, the results of the research work on preparation of the tailings with reduced metal content for further use are presented.

1 Introduction

In years 2015 – 2017 IMN carried out research within the project called "Waste-free technology for processing of the Bytom area heaps with the recovery of concentrates for Zn-Pb production". The project concerned the development of a waste-free technology for processing of the material from old heaps of tailings generated in XX century during operation of beneficiation plant for zinc-lead ores of "Orzeł Biały" mine and smelter. The idea of the project was to develop a technological flowsheet for production of selective or collective sulfide concentrates of zinc and lead out of the tailings and the testing of the remaining material (secondary tailings) for possible use in mining technologies, production of aggregates and fillers, as well as engineering works.

The existing wastes of Zn-Pb ore beneficiation processes in this area could be a valuable raw material for the production of zinc and lead concentrates for ZGH "Bolesław" or HC "Miasteczko Śl" zinc smelters. Due to the large distance to both plants, the optimum method of use would be processing them in a specially designed beneficiation plant located on site.

2 Determination of the volumes of resources and metal content in the examined waste

The main goal of the project was first to identify the currently available waste resources from the examined area and then to develop a flowsheet of waste processing for the intended production of selective or collective concentrates of Zn and Pb. The examination covered four waste heaps: OG/81, OG/ 88, OG/86 and OG/85.

The test material was composed of samples from individual heaps extracted by the drilling. Based on them the detailed inventory of the examined sites was performed. The volume of heaps and the average metal content in them were presented in Table 1.

The presented data indicate that, from an economical point of view, the most interested is heap OG/81, with its resources being approx. 3.173 million tons, which represents 84.8% of the examined resources of waste in the area. The second largest heap is OG/88 with a total waste amount of approx. 330.5 k tons, which is 8.84% of the total resources.

Because of low content of lead in the sulfide form in the range of 0.24% Pb(S), the potential for production of selective Pb concentrates is very limited. Therefore, examination of the flotation process has been performed with the focus on production of collective Zn-Pb concentrates.

3 Determination of the optimum conditions for Zn-Pb collective flotation

The basic parameters of the flotation process covered by the laboratory tests were: feed grain size distribution, amount of added reagents and concentration of solid particles in the feed material.

Preparation of the materials for the flotation trials started with crushing to the grain size of < 2 mm; next, the samples were milled in a wet state in a laboratory rod mill to the grain size of < 0.2 mm, < 0.15 mm and < 0.1 mm. Flotation tests have been performed in a laboratory flotation machine (Mechanobr type) with a cell capacity of 1L. Other flotation parameters were as follows:

- Solid particles concentration in the feed – 200 g in 1 L, 350 g in 1 L, 500 g in 1 L
- Doses of $CuSO_4$ – 350 g/t; Hostaflot X 23 – 20 g/t, 25 g/t, 30 g/t; Amyl xanthate – 25 g/t, 30

* Corresponding author: krystianc@imn.gliwice.pl

g/t, 35 g/t; Ethyl xanthate – 15 g/t, 20 g/t, 25 g/t; Corflot – 15 g/t

- Flotation time – 15 minutes (3 min + 5 min + 7 min)

During the process of flotation, three froth products were taken: the first one after 3 minutes of flotation, the second one after 8 minutes of flotation and the third one after 15 minutes of flotation. The froth products and waste product (tailings left in the flotation cell) were dried and then weighed to determine their yield. Next, the samples were analyzed to determine the content of Zn, Pb and Fe.

Based on the obtained mass and metal content in individual products, flotation process mass balances were made for zinc, lead and iron. Based on the balance data, a beneficiation capability curve for the analyzed metal as the recovery versus its content in the froth product was presented. Pictures 1 –3 present zinc beneficiation curves depending on the feed material grain distribution, solid particles concentration in the feed and the amount of collectors added to the flotation, for the example, of the largest heap OG/81.

With the tailings from heap OG/85, completely negative flotation results were obtained in the test conditions, as the froth products brought practically no zinc beneficiation. Due to the lack of real possibilities of enrichment of the material with the flotation method, it was excluded from further tests.

In terms of size distribution, the flotation process is optimal when the waste is milled down to < 0.15 mm. Finer milling of the feed material does not lead to a significant increase of the zinc recovery, while reduction of zinc content in the froth product is observed, which results mainly from mechanical transfer of the finest grains. The second parameter which was tested was the concentration of solid particles in the flotation feed. The optimum value is approx. 350 g in 1 L. Although flotation at the lowest concentration allows to obtain froth products with higher zinc content, it reduces the recovery. The last parameter examined was the amount of collectors added to the flotation. As shown by the presented beneficiation curves (fig. 3.3), the optimum amount of added collectors in this case is approx. 75 g/t (X 23 - 25 g/t + KXamyl- 30 g/t and KXetyl + 20 g/t). Increasing the dose to 90 g/t does not lead to a significant increase in the zinc recovery, while a reduction in the amount of zinc in the froth product is observed. The results also show that in the process of the flotation, an important factor can be the amount of the finest particles in the feed, which can be a major obstacle for obtaining commercial collective Zn-Pb concentrates.

Table 1. Volumes of resources and metal content in the examined heaps of tailings.

	Heap OG/81	**Heap OG/86**	**Heap OG/86**	**Heap OG/85**	**Total**
Area, ha	11.685	1.038	0.977	0.923	**14.623**
Cubic volume, m^3	1 670 000	171 270	47 288	76 000	**1 964 558**
Bulk density, t/m^3	1.90	1.93	1.94	1.94	**1.91**
Resources, t	3 173 000	330 551	88 901	147 440	**3 798 431**
Zn, %	4.01	5.16	5.18	7.30	**4.27**
Zn(O), %	2.06	2.69	2.70	4.22	**2.22**
Zn(S), %	1.95	2.47	2.48	3.08	**2.05**
Pb, %	0.81	1.25	0.93	1.21	**0.87**
Pb(O), %	0.56	0.93	0.53	0.69	**0.60**
Pb(S), %	0.24	0.31	0.39	0.52	**0.27**
Fe, %	9.38	11.50	7.93	18.81	**9.90**

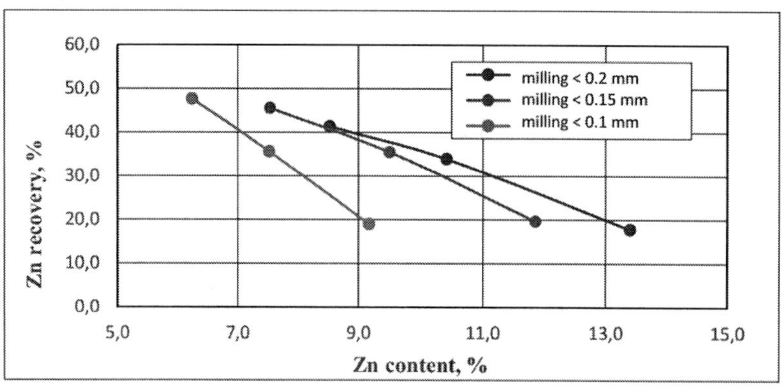

Fig. 1. Heap OG/81. Zinc recovery depending on its content in the froth product at different grain distributions of the feed for flotation.

 DOI:10.1051/e3sconf/20171801008

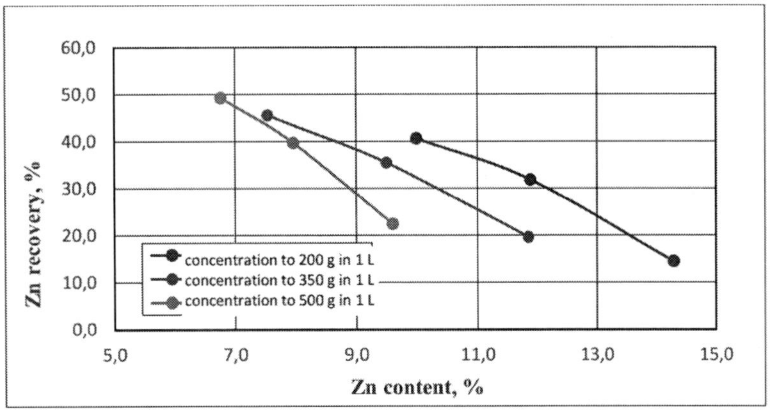

Fig. 2. Heap OG/81. Zinc recovery depending on its content in the froth product at various solid particles concentration in the feed for flotation.

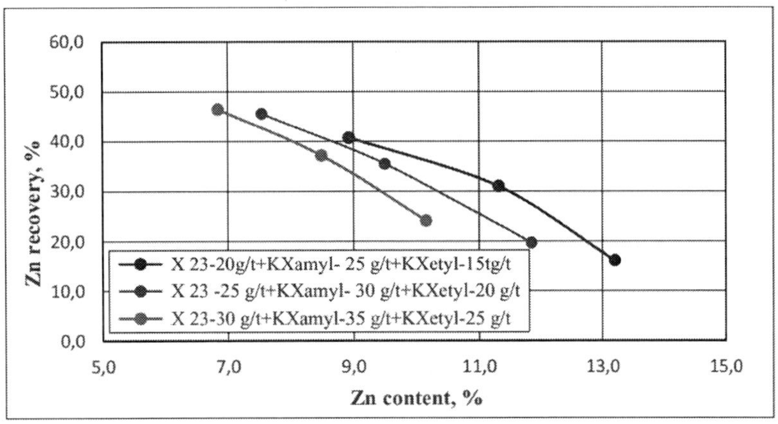

Fig. 3. Heap OG/81. Zinc recovery depending on its content in the froth product at different doses of flotation collectors.

4. Flotation testing in a flowsheet with cleaning flotation

The objective of this stage of testing was to determine the conditions which would enable the production of commercial Zn-Pb concentrates containing Zn+Pb > 50%. The final product requires a series of cleaning flotation trials according to an adequate beneficiation flowsheet.

In the first case, the simplest beneficiation flowsheet, in which waste flotation was split into two stages covering the rougher and scavenger flotation of Zn-Pb is tested. The froth product from the rougher flotation showing the highest content of zinc is fed to the two-stage cleaning flotation, where the final product will be obtained. The froth product from the scavenger flotation showing a relatively low content of zinc is returned to the start of the rougher flotation. This flotation is also fed with the waste product from the first cleaning flotation, while the waste product of the second cleaning flotation returns to the preceding operation, i.e. the first stage of cleaning.

Mass balances of the flotation trials in this flowsheet are presented in Table 2.

The presented balance data show that in this flowsheet the collective concentrates with the required content of Zn+Pb > 50% cannot be obtained. Their amount is only 36.44% to 42.6%. This flowsheet shows a high contribution of technological reverts which are fed to the start of the rougher flotation, being 20% to approx. 32.5% of the fresh feed.

5 Testing in a flowsheet with a separate flotation of the mud fraction

Obtaining commercial concentrates of zinc and lead containing > 50% Zn+Pb would require the introduction of more cleaning stages. In practice, such a solution is disadvantageous, as it leads to a significant reduction in the recovery of metals in the concentrate. It is believed that the reason for poor beneficiation in this flowsheet can be the high content of the finest grains in the flotation feed.

In order to check if these grains have such a high influence on the beneficiation parameters, flotation trials were performed in the most extended beneficiation flowsheet covering the classification of the flotation feed into mud fraction and sand fraction, followed by separate flotation cycles.

In this case, the samples of waste milled down to <0.15 mm were separated in a laboratory hydrocyclone, with a diameter of 30 mm, into mud and sand fractions, where mass outputs of the overflow fractions in the range of 28 – 33% were obtained. The mud fraction was processed in flotation with the reduced concentration of solids at 140 – 160 g in 1 L, a higher dose of collectors and the addition of sodium silicate as the depressor of the finest grains of the gangue. For the sand fraction, flotation conditions were not changed.

Results of tests with a separate flotation run for the mud fraction are presented on fig. 4 as zinc beneficiation curves. For comparison, the figure also includes zinc beneficiation curves for the feed with a solid particles concentration of 350 g in 1 L which were obtained in conditions without a separate mud fraction flotation cycle. The comparison of beneficiation curves show that in all cases there was a significant improvement of zinc flotation results with a separate mud fraction flotation cycle. The significant improvement of Zn content in the froth product and the increase of Zn recovery were also observed. So, in order to obtain qualified collective concentrates containing > 50% Zn-Pb from the tested wastes it will be necessary to expand the technological flowsheet for beneficiation with the mud fraction flotation system.

Table 2. Mass balances of the beneficiation of waste from the Bytom area in a flowsheet with cleaning flotation.

Product	Yield	Content, %			Recovery, %		
	%	Zn	Pb	Fe	Zn	Pb	Fe
Heap OG/81							
Feed	100.00	3.96	0.97	9.62	100.00	100.00	100.00
Conc. Zn-Pb	3.64	35.12	1.32	18.40	32.27	4.95	6.96
Waste	96.36	2.78	0.96	9.29	67.73	95.05	93.04
Heap OG/86							
Feed	100.00	5.18	0.94	8.05	100.00	100.00	100.00
Conc. Zn-Pb	3.86	41.28	1.31	13.25	30.73	5.37	6.35
Waste	96.14	3.73	0.93	7.84	69.27	94.63	93.65
Heap OG/88							
Feed	100.00	5.21	1.18	11.65	100.00	100.00	100.00
Conc. Zn-Pb	4.35	37.64	1.54	15.20	31.44	5.68	5.68
Waste	95.65	3.73	1.16	11.49	68.56	94.32	94.32

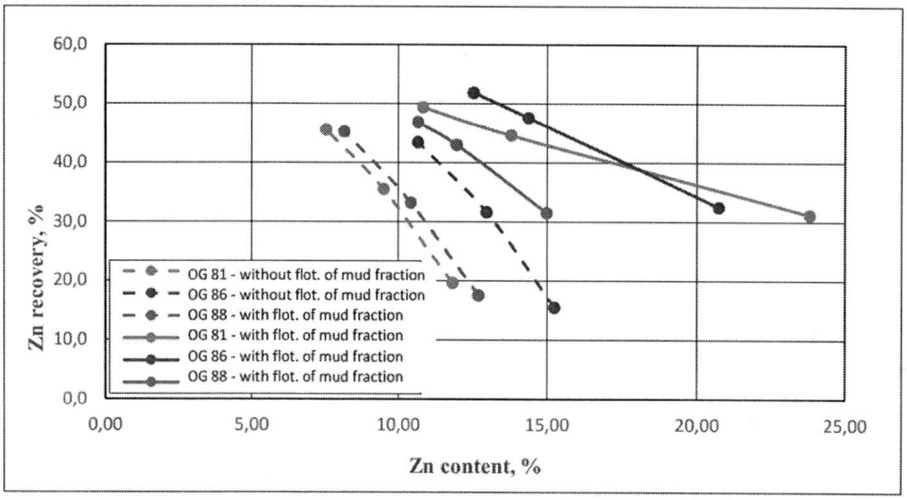

Fig. 4. Relation of zinc recovery to Zn content in the froth product in flotation with and without separate mud fraction flotation cycles.

6 Verification of the developed beneficiation flowsheet in an experimental system

The beneficiation flowsheet developed on the laboratory scale was verified in an experimental system in continuous operation in conditions representing industrial production. During the trials, it was confirmed that in order to reduce the amount of returned material to the start of the rougher flotation of the mud fraction, it would be beneficial to implement an additional operation, i.e. separating flotation performed at a higher pH of the suspension, to which depleted froth products from the stages of the scavenger flotation of the mud and sand fractions, containing large amounts of marcasite will be fed. In this case, the mud fraction rougher flotation will be fed only with the returned froth product from the

 DOI:10.1051/e3sconf/20171801008

separating flotation enriched with zinc, while the cell product containing mainly marcasite will be fed directly to the final waste stream. This operation is very important in the beneficiation of materials containing high amounts of sulfide iron minerals, as it reduces the volume of their revert to the start of flotation.

Mass balances of the flotations performed are presented in Table 3. Compared to the results obtained in the previous flowsheet, the introduction of the separating flotation enabled higher Zn content in the concentrate and a slight increase of Zn recovery in the concentrate. This is particularly seen in waste from heap OG/88, which contains the highest amounts of sulfide iron minerals. In this case, the content of zinc in the concentrate increased from 48.6% to 50.8% Zn and its recovery increased from 31.5% to 32.16%.

Table 3. Mass balances of the beneficiation tests in the experimental system.

Product	Yield	Content, %			Recovery, %		
	%	Zn	Pb	Fe	Zn	Pb	Fe
Heap OG/81							
Feed	100.00	3.93	0.93	8.43	100.00	100.00	100.00
Concentrate	**2.71**	**51.60**	**1.23**	**10.30**	**35.64**	**3.60**	**3.32**
Waste	97.29	2.60	0.92	8.38	64.36	96.40	96.68
Heap OG/86							
Feed	100.00	5.19	0.94	8.09	100.00	100.00	100.00
Concentrate	**3.25**	**53.26**	**1.83**	**7.62**	**33.38**	**6.33**	**3.06**
Waste	96.75	3.57	0.91	8.11	66.62	93.67	96.94
Heap OG/88							
Product	**Yield**	**Content, %**			**Output, %**		
	%	**Zn**	**Pb**	**Fe**	**Zn**	**Pb**	**Fe**
Feed	100.00	5.21	1.19	11.71	100.00	100.00	100.00
Concentrate	**3.30**	**50.82**	**1.64**	**9.35**	**32.16**	**4.54**	**2.63**
Waste	96.70	3.65	1.17	11.79	67.84	95.46	97.37

Table 4. Mass balances of the waste thickening after flotation of tailings from heaps with the use of a double classification system in hydrocyclones.

Product	Yield %	Qs, kg/min	V l/min	Weight 1 l g	Density of solids g/l
Flotation waste	100.00	15.00	62.50	1160.0	240.0
HC discharge 2nd stage of discharge classification	77.76	11.66	7.68	2012.4	1518.6
Discharge of the lamella clarifier	21.92	3.29	6.26	1350.0	525.0
Overflow of the lamella clarifier	**0.32**	**0.05**	**48.56**	**1000.7**	**1.0**
HC discharge 2nd stage of discharge classification + discharge of the lamella clarifier	**99.68**	**14.95**	**13.94**	**1714.9**	**1072.3**

7 Tests of the process of thickening and dewatering of flotation tailings

Prior to its further use, the waste product ('secondary tailings') requires first thickening and, if necessary, dewatering, if it is used for ground leveling and similar purposes. The easiest way to prepare the thickened waste would be to feed it to the Dorr settler, where the discharged product would make a material ready for further applications or for dewatering, while the clarified water obtained as the overflow product would be returned to the flotation circuit.

The waste which was tested shows a very low sedimentation rate, which results from the significant presence of the finest grains <15□m. The average sedimentation rate is v = $1.884 \cdot 10^{-5}$m/s, which results in 6.8 cm thick layer of clarified water in 1 hour. With the assumed quantity of processed waste being 42.3 t/h, the required surface area of clarification would be 2540.0 m^2. For this surface area, the minimum diameter of Dorr settler is D = 56.9 m. From a technical point of view, such a high diameter is unacceptable. To solve this problem, trials with thickened waste in an extended flowsheet with the use of hydrocyclones were performed in the experimental system.

The first stage of classification occurs in the hydrocyclone, the overflow product of which is fed to the second stage of classification in the hydrocyclone. The overflow product from this hydrocyclone with solids at approx. 60 g/l is thickened in a lamella clarifier, where clarified circulating water and thickened sludge, which is the second component of the thickened waste, are produced. The discharge product from the first stage of classification is subject to secondary thickening in the hydrocyclone. The discharge product from this stage is a thickened final product and together with the sludge of the

DOI:10.1051/e3sconf/20171801008

lamella clarifier it represents the thickened material intended for further use.

The mass balance of the process of waste thickening in this flowsheet is presented in Table 4.

In the last stage, dewatering trials with the thickened waste were performed.This results from the need to prepare them in a form that allows safe road transportation if they are to be used as materials for ground leveling. The examination covered the testing of thickened waste from the experimental system with a solids content of approx. 1000 g/dm^3.

The tests of suspension filtration were performed in a laboratory pressure type filter press with plate dimensions 0.63 x 0.63 m and 20 mm thick filtering chamber. Results of the completed trials were as follows:

1. The sludge shows a lump structure with a moisture content of 17 – 21% H_2O,
2. Specific density of the cake in a wet state is 2.2 kg/dm^3.

This data allows us to define the volume of filter chambers for the target plant with the capacity of 41- 42 t/h and initial thickening of approx. 1000 g/dm^3 with the following parameters:

1. Filtration time – 20 minutes
2. Number of cycles per hour – 3

The calculated volume of filtering chambers for the obtained moisture levels is 5 500 dm^3 to 6 500 dm^3.

8 Conclusions

1. Based on the geological drillings, a detailed inventory of the examined heaps of tailings generated in Zn-Pb ore beneficiation in the Bytom area was made. The total waste resources are 3 798 431 tons and the average metal content is: Zn – 4.27%, Zn(O) - 2.22%, Zn(S) - 2.05 5%, Pb –0.87%, Pb(O) – 0.87%, Pb(S) – 0.27%, Fe – 9.9%.
2. Among the three examined beneficiation flowsheets, only the one with a separate flotation for the mud fraction and the segregating flotation brings collective concentrates with the required content of Zn+Pb. For the waste from heap OG/81 containing 3.93% Zn, the obtained concentrate contained 51.6% Zn and 1.23% Pb with a zinc recovery of 35.61%, for the waste from heap OG/88 containing 5.2% Zn, the obtained concentrate contained 50.82% Zn and 1.64% Pb with a zinc recovery of 32.2%, for the waste from heap OG/86 containing 5.2% Zn, the obtained concentrate contained 53.26% Zn and 1.83% Pb with a zinc recovery of 33,4%.
3. Implementation into industrial practice of the waste-free technology of processing of Zn-Pb wastes from the Bytom area, i.e. application of the waste for production of sulfide concentrates of zinc and lead and then using the generated secondary tailings with depleted zinc in engineering works, would require development of an efficient method of thickening and dewatering to ensure safe transportation to the destination.
4. The completed tests show that due to the unfavorable size distribution, i.e. presence of approx. 66% of the waste in the finest grain category < 0.036 mm, the process of waste thickening in natural conditions is ineffective with its low sedimentation rates. This problem was solved by application of a system with two stages of classification in hydrocyclones, plus a lamella clarifier for thickening the overflow product from the hydrocyclone of the 2nd stage. This system delivers thickened waste material and clarified circulating water being the overflow product of the lamella clarifier. After treatment of the feed with solids thickened to 240 g in 1L a final product with solids thickened to 1070 g in 1L and circulating water with solids content < 1 g in 1 L are produced.
5. The performed test of filtrations of the thickened waste product confirmed the possibility of product dewatering to an average moisture of approx. 19% H2O. For the target system of 41- 42 t/h capacity and initial suspension density of 1000 g/dm^3, the calculated volume of filtering chambers is 5500 dm^3 to 6 500 dm^3.

The study was funded by the National Center of Research and Development and National Fund for Environmental Protection and Water Management within the scope of GEKON programme.

E3S Web of Conferences **18**, 01009 (2017) DOI:10.1051/e3sconf/20171801009
MEC2017

Evaluation of frothers based on properties of the solids/liquid/gas system

Jan Drzymala[1,*]

[1]Wroclaw University of Science and Technology, Faculty of Geoengineering, Mining and Geology, Wybrzeze Wyspianskiego 27, 50-370 Wroclaw, Poland

Abstract. A methodology of selection of frothers for flotation process is presented in the paper. It is based on performing a series of fractionating flotation tests of a considered material at different doses of a selected frother and calculation of the mass balance of the process. Next, three steps are performed. The first one is determination of the frother dose and kinetic constant or specific rate, using the flotation limits plot relating maximum recovery and process rate, both at a selected maximum recovery of the considered material component. The second step is finding selectivity of separation by using the upgrading Fuerstenau plot, while the third element is frothers comparison plot in the form of selectivity versus process kinetics, at a selected maximum recovery, and also versus frother dose, for choosing the most suitable frother.

1 Introduction

Frothers play a vital role in flotation of ores, wastes and byproducts. Their role is to form froth and modify the properties of films between solid particles and bubbles as well as between bubbles [1, 2, 3, 4, 5, 6, 7]. Action of a frother depends on the type of material subjected to flotation. Literature data indicate that for highly hydrophobic solids frothers can sometimes be harmful while for medium hydrophobic is very effective and for weakly hydrophobic and hydrophilic is usually neutral [8]. In flotation other reagents are used, especially collectors. Their main role is to regulate hydrophobicity of the processed material, though some interactions between frother and collector take place [9, 10, 11].

There are many classification of frothers which are used for evaluation flotation frothers. According to Bhattacharya and Dey [12] presently no accepted criteria for selection of frothers exist. Common sense suggests that the essential parameters which should be used for selection of frothers for industrial application are frother dose, flotation rate and process selectivity. These three parameters determine the cost of the beneficiation process. Therefore, the goal of this work is to propose a methodology which can be used for determination of the three essential process parameters, that is selectivity, power and kinesis provided by a frother in the solids/water/gas system during flotation process. Three plots will be used for this purpose, that is flotation limits graph, Fuerstenau upgrading curve and comparison relation.

2 Mass balance of a hypothetical flotation process

The base of all mineral processing considerations is the mass balance. The mass balance of a process can be calculated utilizing the grades of the feed (α), flotation concentrates (β) and tailings (ν). The formulas for calculation of concentrate yield γ is:

$$\gamma = 100\% \frac{\alpha-\nu}{\beta-\nu}, \tag{1}$$

for recovery of a considered component in concentrates is

$$\varepsilon = 100\% \frac{\alpha-\nu}{\beta-\nu} \cdot \frac{\beta}{\alpha}, \tag{2}$$

while for recovery of remaining components in the tailing

$$\varepsilon_r = (100-\gamma)\frac{\nu}{(100-\alpha)} \tag{3}$$

The mass balance of a hypothetical flotation process in which frother 1 was used is given in Table 1.

* Corresponding author: jan.drzymala@pwr.edu.pl

DOI:10.1051/e3sconf/20171801009

Table 1. Mass balance of a hypothetical flotation process.

Product	Flotation time, min	Yield γ, %	Grade β, %	Recovery ε, %	Remaining components recovery in tailing, ε_r, %
1	1.1	3.46	30	52	98
1+2	2.3	5.76	25	72	95
1+2+3	4.1	15.45	11	85	85
1+2+3+4	14.0	39.20	5	98	62
Feed		100	2 (α)	100	0

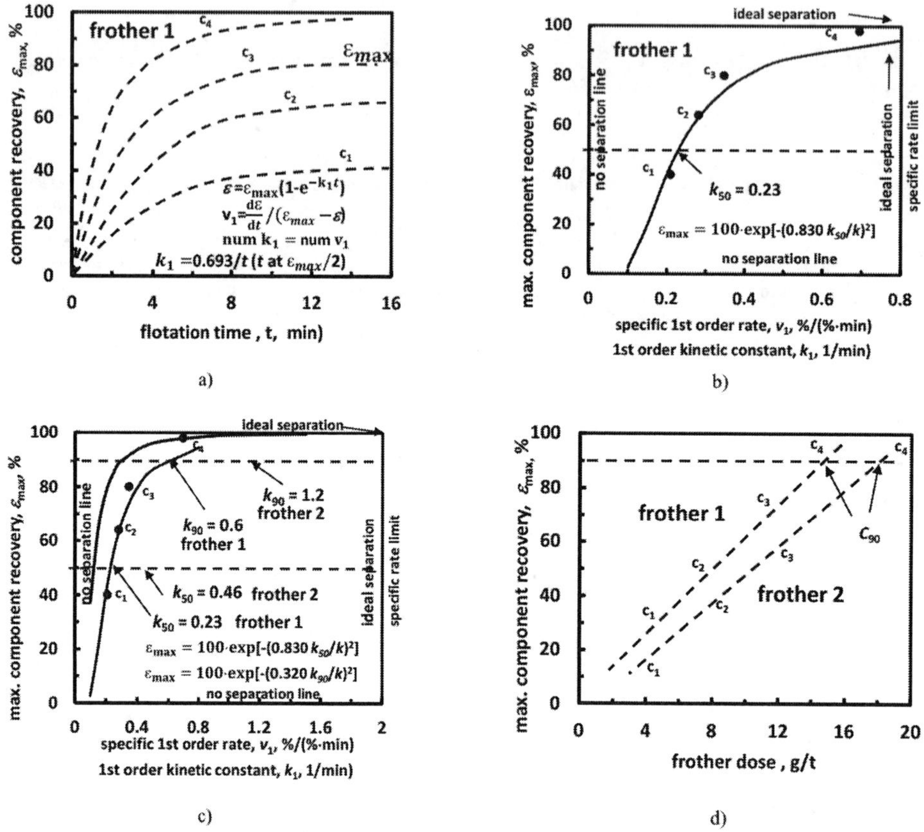

Fig. 1. Evaluation of kinetics and dose of flotation frothers: a) kinetic curves, b) flotation limits curve, c) determination of kinesis that is k at ε_{max}=90%, d) determination of frother dose $C_{\varepsilon max}$ at ε_{max}=90%, that is C_{90}.

3 Determination of kinesis of flotation and frother dose

Frother must provide appropriate rate of the flotation process at a reasonable dose during the process. To determine both the frother dose and process rate, a series of flotation experiment at different doses of frother ought to be performed. Next, a separation balance should be performed as shown in Table 1. The mass balance is good base for plotting the recovery versus flotation time, which is presented in Fig. 1a. The experimental data are further processed by approximation of the experimental data with a mathematical equation having one fitting parameter as shown in Fig.1a. Very useful are the first order kinetic equations, both in differential and integral forms. The equations for the specific rate (v) and rate constant (k) are given in Fig. 1a. Index 1 for k indicates that these parameters follow the 1st order kinetics. Both v_1 and k_1 are numerically identical, though the specific rate is a more informative parameter because it has a physical meaning, that is indicates the speed of the process in the form of amount of solids of the considered component in concentrate per amount of solids in the flotation cell per unit time. Other kinetic equations can also be used [13, 14].

The kinetic plot (Fig. 1a) provides two parameters, that is maximum recovery (ε_{max}) and a constant (v or k) along with information at what frother dose (C) they were obtained. Next, the limits flotation curve [14] can be plotted as shown in Fig. 1b. The limits flotation curve indicates the maximum recovery ε_{max} that can be obtained

 DOI:10.1051/e3sconf/20171801009

at a specific flotation rate (v or k) and frother dose C. The advantage of the limits flotation curve is that one curve characterizes a flotation frother with just one curve. The course of the line can be approximated with an equation which is given in Fig. 1b. Since the goal is to compare different frothers, it is useful to draw the limits flotation curves for all of the considered frothers (Fig. 1c).

The dose of frothers ($C_{\varepsilon max}$) depends on the required maximum recovery, which is equivalent to a certain specific rate of the process. It is useful to set up the level of the maximum recovery. In Figure 1c two levels (50% and 90%) are indicated. The value of k (or v), at a certain level of ε_{max}, is called the kinesis of the process $k_{\varepsilon max}$ [15]. If necessary, an addition plot can be created to determine precisely the dose for each frother at certain ε_{max} (Fig. 1d). Thus, the limits flotation curve provide the dose and kinetics (kinesis) of the process.

4 Determination of selectivity of flotation and frother dose

The selectivity of flotation and also the frother dose can be determined with the help of the upgrading plot, that is graphs relating quality and quantity of the separation process products. Very useful for this purpose is the Fuerstenau upgrading plot, relating recovery of a flotation feed component in flotation products [16, 17]. There are four versions of the Fuerstenau upgrading curve [5]. The most useful modification is the one relating recovery of the considered component in concentrate ε versus recovery of the remaining components in the tailing ε_r. The procedure of determination of the frother selectivity and dose, starts with the same mass balance of the process experimental data (Table 1). Next, the Fuerstenau plot is drawn (Fig. 2a). The experimental data has to be approximated with appropriate mathematical equations [18]. The most frequently used is the equation with one fitting parameter a. It has three forms:

$$a = \frac{\varepsilon \cdot \varepsilon_r}{\varepsilon + \varepsilon_r - 100}, \quad (4)$$

$$a \cdot A = 100\%, \quad (5)$$

$$F^2 - 2aF + 100a = 0. \quad (6)$$

In this paper selectivity parameter F is used, as shown in Fig. 2a.

The Fuerstenau curve provides two parameters [19]: power (P), and selectivity of the frother (F). The dose of frother depends on the criterion imposed on the flotation system. Since the frother dose ($C_{\varepsilon max}$) has been already established by means of the limits cure for ε_{max}=90%, this parameter can be marked, as shown, in Fig. 2a. Another option is to read-off the frother dose from Fig. 2a at the cross point of the upgrading curve with the ascending diagonal. This parameter is called frother power (P).

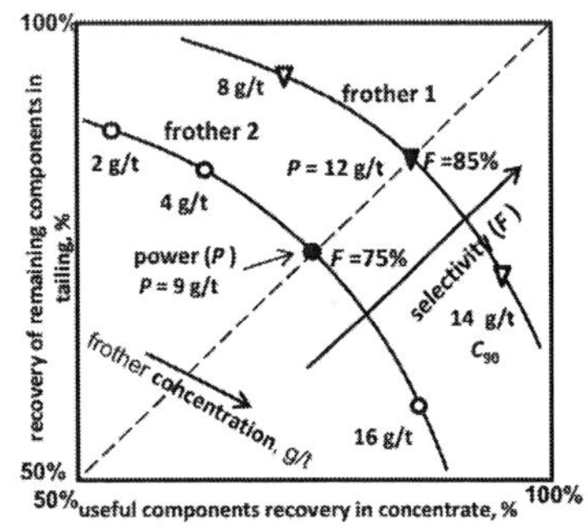

a)

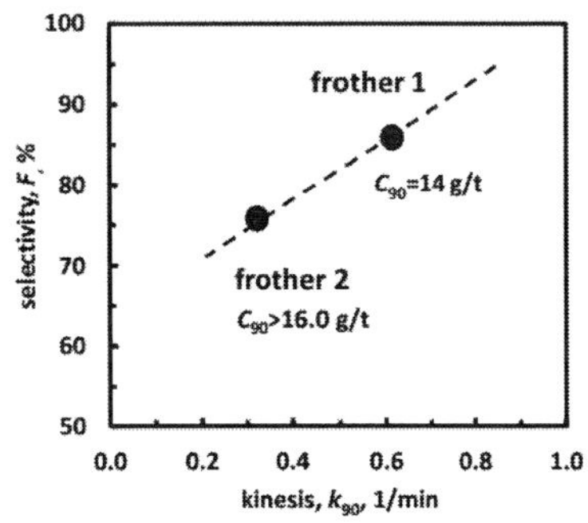

b)

Fig. 2a. Procedure of determination of selectivity F (%) and power P (g/t) of flotation frothers using the Fuerstenau upgrading curve [19]. If used, the concentration of collector is constant.

Fig. 2b. A proposed graph for comparison of three parameters (kinesis k_{90}, selectivity F, and dose C_{90}) of each frother.

5 Comparison of frothers

Once the three most important flotation process results parameters, that is kinesis, dose (power) and selectivity of different frothers are determined, they can be compered for decision making purpose, leading to proper frother selection. Figure 2b represents a proposition of such a graphical comparison method. According to Fig. 2b, hypothetical frother 1 is faster (greater k_{90}), has a smaller dose (C_{90}) and better selectivity (greater F) than frother 2. Thus this frother should be selected for the planned flotation process. In more complex cases, additional criteria may be needed for the final frother selection.

6 Conclusions

A selection of a frother for flotation process is possible by using three plots, that is limits plot used for determination of the frother dose and kinesis, the upgrading plot for selectivity determination and finally a comparison plot for a final decision.

The work was financed by the Polish Statutory Research Grant PWR/W6 for the year 2017/8.

References

1. J. Czarnecki, K. Malysa, A. Pomianowski, J. Colloid Interface Sci. **86(2)**, 570–572, (1982)
2. D. Mobius, R. Miller, (Eds.), *Studies in interface science, Novel methods to study interfacial layers*, **11**, 1-521, (Elsevier Science, 2001)
3. Y.S. Cho, J.S. Laskowski, Int. J. Miner. Process. **64(2-3),** 69-80, (2002).
4. M.C. Fuerstenau, G.J. Jameson, R.H. Yoon, *Froth flotation: A century of innovation.* (SME Inc., Littleton, 2007)
5. J. Drzymala, *Mineral Processing. Foundations of theory and practice of minerallurgy*, 1st English Edition, (Oficyna Wydawnicza Politechniki Wroclawskiej, Wroclaw, 2007)
6. D. Kosior, J. Zawala, M. Krasowska, K. Malysa, Phys. Chem. Chem. Phys. **15,** 2586-2595, (2013)
7. B.A. Wills, J.A. Finch, *Wills' Mineral Processing Technology, 8th Ed., An introduction to the practical aspects of ore treatment and mineral recovery* (Elsevier Ltd, Amsterdam, 2016)
8. J. Drzymala, P.B. Kowalczuk, *A system of flotation frothers classifications,* (in preparation, 2017)
9. J. Leja, J. H. Schulman, Trans. AIME **199**, 221, (1954)
10. J. Lekki, J. Laskowski, Trans. IMM, Sec. C., **80**, 174-180, (1971)
11. D.J. Bradshaw, P.J. Harris, C.T. O'Connor, SAIMM, **98(4)** 189-194, (1989)
12. S. Bhattacharya, S. Dey, Miner. Process. Extr. Metall. Rev., **29**, 275-298, (2008)
13. X. Bu, G. Xie, Y. Peng, L. Ge, C. Ni, Physicochem. Probl. Miner. Process. **53(1),** 342–365, (2017)
14. J. Drzymala, T. Ratajczak, P.B. Kowalczuk, Physicochem. Probl. Miner. Process. **53(2),** 983–995, (2017)
15. J. Drzymała, M. Stodulski, *Łupek miedzionośny III*, Kowalczuk P.B., Drzymała J. (eds), (WGGG PWr, Wrocław, 167-171, 2017)
16. D.W. Fuerstenau, et al., *Coal surface control for advanced fine coal flotation,* (Final Report, University of California, Berkeley, Final Report DOE/PC/88878-T13, DE92 015625 for U.S. Dept. of Energy. Prepared by Univ. California, Columbia Univ., Univ. of Utah, and Praxis Engineers Inc., 1988-1992)
17. J. Drzymala, Physicochem. Probl. Miner. Process., **40,** 19–29, (2006).
18. J. Drzymala, H.A.M. Ahmed, Int. J. Miner. Process. **76(1-2),** 55–65, (2005)
19. P.B. Kowalczuk, J. Drzymala, Physicochem. Probl. Miner. Process. **53(1)**, 515–523, (2017)

Dynamic aspects of design and maintenance of the rotating machinery applied in the mining industry

Tomasz Szolc[1,*], *Robert* Konowrocki[1], *Andrzej* Pochanke[2] and *Maciej* Michajłow[1]

[1]Institute of Fundamental Technological Research of the Polish Academy of Sciences, ul. Pawińskiego 5B, 02-106 Warsaw, Poland
[2]Department of Electrical Engineering of the Warsaw University of Technology, Plac Politechniki 1, 00-661 Warsaw, Poland

Abstract. In the paper a dynamic behaviour of the selected typical group of rotating machines applied in the mining industry is investigated. These are the beater mills and crushers as well as blowers and compressors, all driven by the asynchronous motors. In particular, there is considered an influence of various types of machine working tool loadings on the system lateral steady-state dynamic responses as well as a mutual torsional electromechanical interaction between the driving motor and the driven machine in transient operational conditions. The theoretical calculations have been performed by means of the advanced structural mechanical models. The conclusions drawn from computational results can be very useful during design phase of these devices as well as helpful for their users during regular maintenance.

1 Introduction

An operation of many machines and devices applied in the mining industry is characterized by rotational motion of their fundamental working tools. To this broad group belong several types of beater mills and beater crushers used in main production processes as well as blowers, rotor pumps, compressors and others regarded as auxiliary units. Majority of them are driven by electric motors by means of flexible couplings and by spur- or planetary gears. The still increasing work efficiency requirements to these machines, commonly observed nowadays, result in more and more dynamic character of their nominal operation. Namely, all these machines transmit now greater and greater power, they work with bigger and bigger rotational speeds, with greater and greater flow velocities and pressures as well as with raising impact force values caused by the comminution processes. These operational performances cause high steady-state or transient dynamic over-loadings which usually excite several types of mechanical vibrations. These vibrations often lead to more or less detrimental exploitation consequences in the form of dangerous material fatigue defects and too fast tribological wear of the responsible structural elements as well as to harmful noise generation, transmission of damaging oscillations to the surrounding environment and to many others. Thus, these dynamic aspects and particularly the vibratory effects should be more and more seriously taken into consideration during design and maintenance phases of such machines, in addition to the common traditional engineering routines applied till present. This postulated modernized approach ought to reduce to thorough and complete dynamic analyses associated with experimental verifications in the frameworks of design and prototyping phases as well as to advanced on-line monitoring and fault detections during a regular maintenance. It is worth emphasizing that for this reason there should be applied modern mechatronic methods based on more and more advanced control and diagnostic algorithms using several sensors, controllers, actuators, transducers and other electronic devices.

As it follows from practical engineering observations and scientific literature studies, e.g. in [1-3], such modern approaches to design and maintenance are mainly practiced in the case of typical flow rotating machines, i.e. for compressors, turbines, pumps, fans and blowers. But very little attention for this purpose seems to be focused to beater mills and crushers which can be also regarded as classical rotor-machines, because of their highly dynamic character of operation. According to the above, in this paper some important aspects of dynamic analyses necessary to achieve a correct design and possibly trouble-free maintenance of the high-speed beater mills, crushers and other rotor-machines are going to be considered.

2 Objects of investigations

The dynamic investigations will be focused on the typical high-speed beater mills, beater and hammer crushers as well as to the medium-speed blowers commonly applied as auxiliary devices for the ball-, rod- and roller-bowl mills. All these units are usually characterized by mutually similar structures. Namely, their working-tool parts consist of heavy beater wheels or impellers attached to the drive shaft in the form of an

* Corresponding author: tszolc@ippt.pan.pl

overhung rotor creating in this way a rotor-shaft. This rotor-shaft is commonly supported by rolling bearings and driven by an electric motor by means of a flexible coupling only, but without any gear stage. The schematic view of such object is presented in Fig. 1. It is worth noting the characteristic feature of these devices, namely

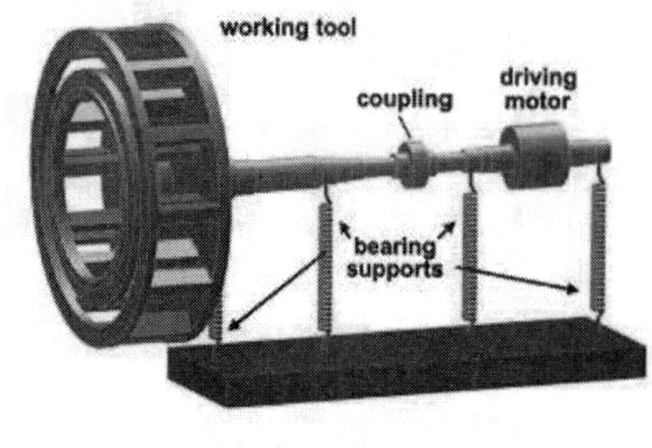

Fig. 1. Scheme of the rotor-machine.

a huge mass of the beater mill or impeller in comparison with the total mass of the entire unit as well as a much greater polar mass moment of inertia of the whole driven part than the polar mass moment of inertia of the driving electric motor rotor. These machines work with maximal rotational speeds reaching 3000 rpm or even more. Due to the mentioned above features a character of their operation is highly dynamic, which results in excitation of bending, torsional and axial steady-state or transient vibrations, where the two former ones seem to be predominant. As it follows from fundamentals of applied mechanics and vibration theory, because of the abovementioned lack of gear stages in the considered devices, their bending and torsional vibrations can be investigated separately, since eventual possible weak couplings between them are negligible.

Fig. 2. The 3D FEM model of the rotor-machine.

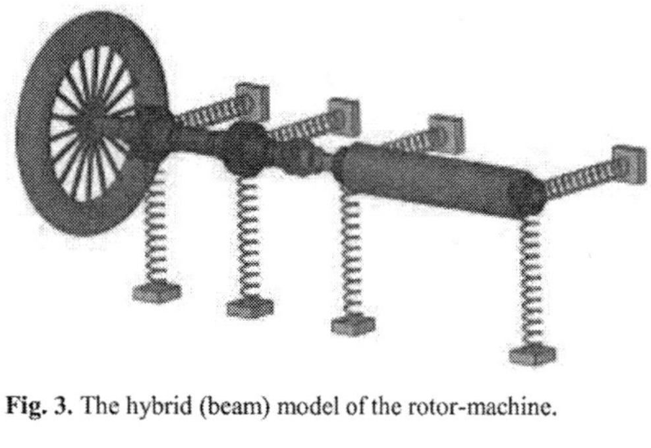

Fig. 3. The hybrid (beam) model of the rotor-machine.

For the aim of possibly reliable vibration analyses two structural mechanical models of the discussed objects have been assumed. These are the 3-D finite element model presented in Fig. 2 as well as the beam continuous finite element (hybrid) model shown in Fig. 3. Both of them thoroughly take into consideration system geometrical shapes, material parameters and bearing support visco-elastic properties. An expected mutual similarity of results of eigenvibration analyses and simulations of forced vibrations obtained by means of these models is a criterion of their computational reliability for practical engineering purposes.

3 Investigation of bending vibrations

The kinematic and elastic properties of the flexible coupling which connects the machine with the driving motor enable us to regard the driven shaft and the electric motor rotor as dynamically separated from the viewpoint of their lateral motions. Thus, for the aim of bending vibration analysis investigations can be focused on the driven machine part only, while the lateral dynamics problem of the electric motor can be assumed as more or less completely solved by a producer.

Here, the lateral motion, associated with bending vibrations of the rotor-shaft-bearing system represented by the two abovementioned structural models shown in Fig. 2 and 3, can be generally described by the following set of linear ordinary differential equations:

$$\mathbf{M}\cdot\ddot{\mathbf{r}}(t)+\mathbf{C}(\Omega)\cdot\dot{\mathbf{r}}(t)+\mathbf{K}(\Omega)\cdot\mathbf{r}(t)=\mathbf{F}(t,\Omega^2), \qquad (1)$$

where:

$$\mathbf{C}(\Omega)=\mathbf{C}_0+\Omega\cdot\mathbf{C}_g(\Omega) \text{ and } \mathbf{K}(\Omega)=\mathbf{K}_0+\Omega\cdot\mathbf{K}_d(\Omega).$$

The symbols $\mathbf{M}$ and $\mathbf{K}_0$ denote respectively the mass and stiffness matrices, $\mathbf{C}_0$ is the damping matrix containing the damping coefficients of the bearing supports and $\mathbf{C}_g(\Omega)$ denotes the skew-symmetrical matrix of gyroscopic effects. Anti-symmetrical effects due to the standard body material damping model of the rotating shaft are expressed by the skew-symmetrical matrix $\mathbf{K}_d(\Omega)$. The symbol $\mathbf{F}(t,\Omega^2)$ denotes the external excitation vector due to the dynamic loadings induced by the working process, unbalances and gravitational forces. In the case of 3D FEM model vector $\mathbf{r}(t)$ contains generalized coordinates. However, motion of the hybrid model is described by the modal coordinates contained in vector $\mathbf{r}(t)$ as well. The number of equations (1) corresponds to the total number of degrees of freedom of the FEM model or to the number of bending eigenmodes taken into consideration in the frequency range of interest, if the hybrid model is used.

It is worth reminding that bending natural frequencies, eigenmodes and magnitudes of forced lateral vibrations of rotor-shaft systems are strongly dependent on visco-elastic properties of the bearing supports. Since the considered rotating object is assumed to be suspended on relatively hard rolling element bearings, a flexibility of the bearing housings and of their foundation must be thoroughly taken into consideration. For this purpose the finite element method has been applied for a three-dimensional modelling of these structures in order to determine their equivalent vertical and horizontal static and dynamic stiffness characteristics. The 3D-FEM model with 88561 degrees of freedom of the steel foundation frame of the entire unit is presented in Fig. 4. Here, the proportional 4% damping of the hysteresis type has been assumed for further computations.

Fig. 4. The 3D FEM model of the machine foundation frame.

Since the 3D FEM model shown in Fig. 2 very precisely describes geometrical shapes of the beater wheel or blower impeller, but on the other hand is characterized by a relatively great number of degrees of freedom, i.e. exceeding 33000, it has been applied as a reliable tool for eigenvibration analyses. For this purpose matrices $\mathbf{C}_0$, $\mathbf{K}_d(\Omega)$ and vector $\mathbf{F}(t,\Omega^2)$ could be omitted in Eqs. (1) for computations. Here, natural frequency values and mode shapes determined by means of this model were used as a benchmark for calculation of selected structural parameters of the hybrid model which is simpler, but much more effective for simulations of forced vibrations.

In order to investigate the qualitative dynamic properties of the considered rotating object, the eigenvalue problem had to be solved first by means of the two mechanical models described above. In Fig. 5 there are plotted the fundamental natural frequency characteristics expressed as rotational speed functions in the form of Campbell diagram with marked critical speeds corresponding to forward whirls of the successive eigenmodes. For a better clarity, on the left-hand side of this diagram also the respective bending eigenfunctions are depicted. As it follows from this figure, within the entire rotational speed range of the considered machine, i.e. 0-3000 rpm corresponding to 0-50 rev/s, three eigenmodes occur. Due to the overhung structure of the rotor-shaft, where the beater-wheel/impeller mass is equal to ca. 80% of the entire rotating mass, its first two eigenmodes are significantly sensitive to the gyroscopic effects with the natural frequency values within ca. 20-65 Hz for the first mode and within 100-145 Hz for the second one. However, the backward and forward precession branches of the third eigenform with the natural frequencies 415 and 419 Hz are almost independent of the shaft rotational speed, as shown in Fig. 5. It is worth noting here that the mutual differences

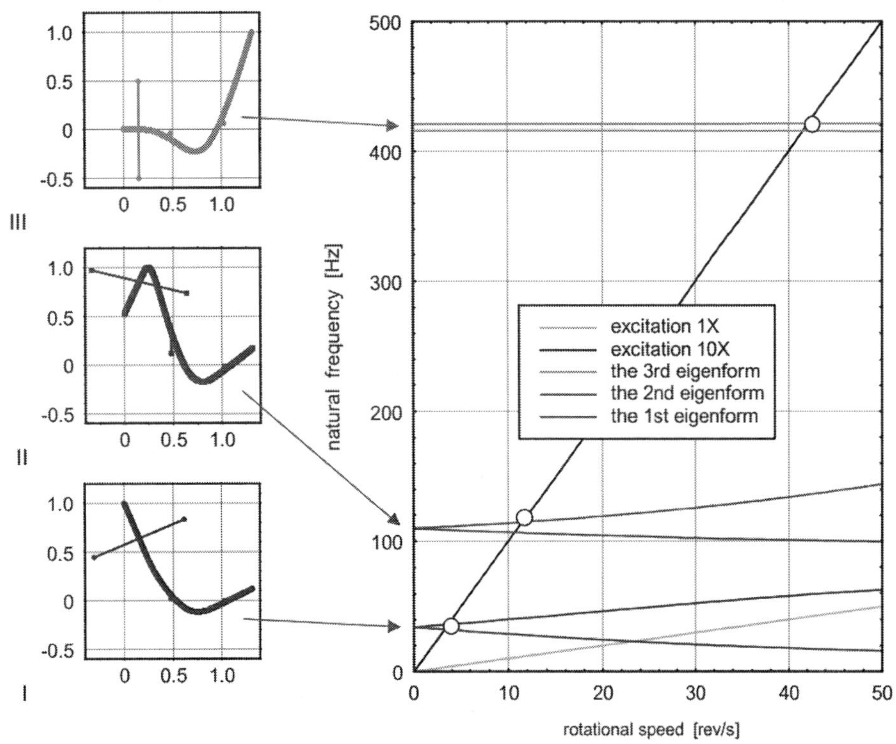

Fig. 5. The Campbell diagram for bending vibrations of the overhung rotor-shaft

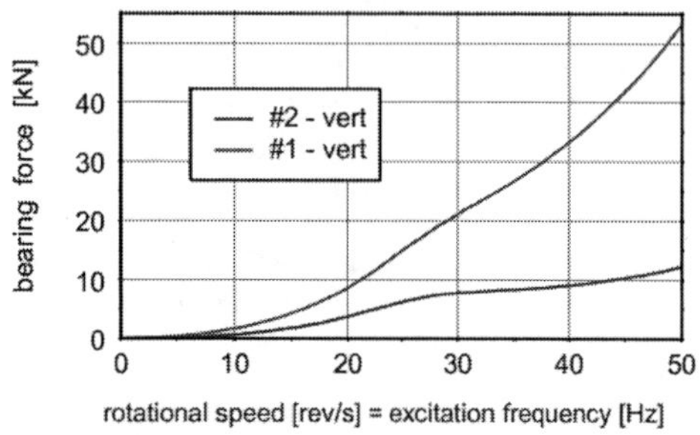

Fig. 6. Bearing force amplitude characteristics due to beater wheel unbalances.

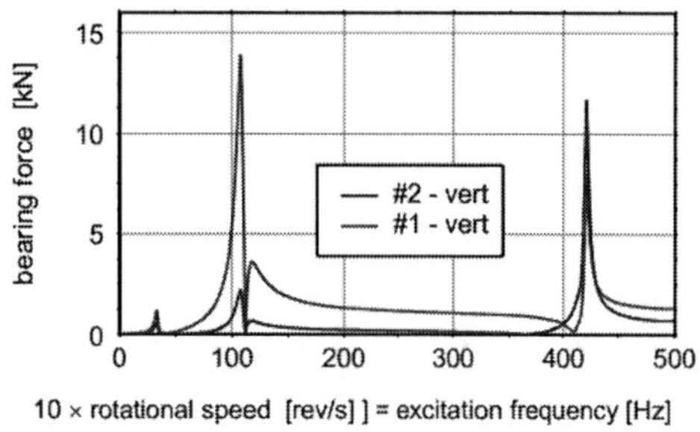

Fig. 7. Bearing force amplitude characteristics due to beater wheel operational dynamic loadings.

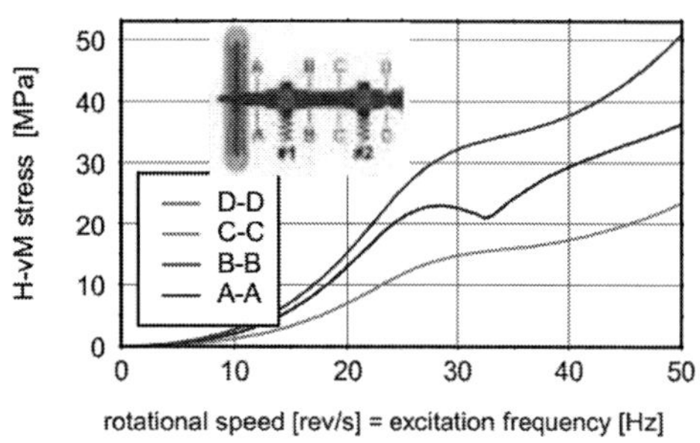

Fig. 8. Huber–von Mises stress amplitude characteristics due to beater wheel unbalances.

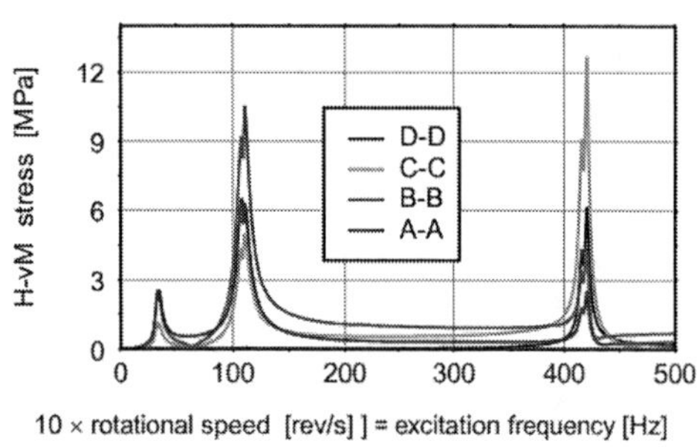

Fig. 9. Huber–von Mises stress amplitude characteristics due to beater wheel operational dynamic loadings.

between the corresponding natural frequency values determined using the 3D-FEM model and by means of the hybrid one did not exceed 2.9%. Moreover, form the obtained Campbell diagram it follows that the investigated machine operates as an under-critical rotor shaft within its entire rotational speed range under synchronous excitation of frequency 1X. However, if an excitation of frequency 10X, i.e. ten times rotational speed value, is assumed, e.g. due to an operation of the beater wheel with 10 beaters, then three critical speeds are observed, which can be a source of possible resonances, see Fig. 5.

According to the above, two types of external excitation for the investigated rotating machine will be assumed for the forced bending vibration analysis. These are: the synchronous harmonic excitation 1X in the frequency range 0-50 Hz and the harmonic excitation 10X with the frequency ten times greater than that following from the rotational speed in the entire range 0-500 Hz. The former can be caused not only by the beater mill residual unbalances, but also by a non-proportionate mass change of the beaters due to their fast wear or random adhesion of the comminuted raw material to the working surfaces. However, the latter type of excitation can be associated with the comminution process resulting in repeatable mutual impacts of the raw material particles to the beaters within each one full beater wheel rotation.

For the aim of possibly reliable assessment of a dynamic load affecting the investigated rotor-shaft system an equivalent unbalance of the beater-wheel/impeller is applied, which overestimates commonly admissible values in order to take into consideration additional effects caused by the abovementioned fast wear or random raw-material adhesion. Here, the magnitude of this equivalent unbalance has been assumed ~18.5 times greater than that following form an application of the ISO 1940-1, [4], and ISO 11342, [5], norms. In Figs. 6-9 there are plotted the amplitude-frequency characteristics in the form of Bode diagrams obtained for steady-state dynamic lateral responses of the rotor-shaft-bearing system. Fig. 6 and 7 present plots of bearing reaction force amplitudes, where in Fig. 6 the response due to the synchronous excitation 1X by the assumed equivalent unbalance is illustrated and in Fig. 7 there is shown the analogous response induced by the harmonic excitation of frequency 10X caused e.g. by the comminution dynamic effects. Respectively, for the same excitations 1X and 10X in Figs. 8 and 9 the Bode diagrams of the Huber – von Mises stress in the most responsible rotor-shaft cross-sections are presented. The plots demonstrated in Fig. 6 and 8 indicate clearly under-critical character of the dynamic responses due to the synchronous excitation 1X within the admissible rotational speed range. This fact is qualitatively confirmed by the Campbell diagram shown in Fig. 5. Here, the bearing force- and stress amplitudes significantly raise with the rotational speed, but they do not reach a resonance peak with the first system eigenform. However, for the harmonic excitation 10X in Fig. 7 and 9 there are observed three resonance peaks

* Corresponding author: tszolc@ippt.pan.pl

corresponding to the successive three system eigenforms presented in Fig. 5. Nevertheless, it is worth to emphasize that the all extreme force and stress values of these peaks are respectively many times smaller than the maximal force and stress amplitudes presented in Figs. 6, 8 and obtained for the under-critical external excitation. This fact can be substantiated by a much greater sensitivity of the considered rotor-shaft system to relatively low-frequency harmonic excitations with amplitudes proportional to squares of the actual rotational speed values in comparison with the excitations characterized by correspondingly the same amplitudes and appropriately ten times higher excitation frequencies. In addition, it is to remember that structural damping of such mechanical systems attenuates more intensively extreme dynamic response magnitudes with greater fluctuation frequencies.

4 Investigation of torsional vibrations

Contrary to the bending vibration analysis performed above, for investigations of torsional oscillation problem for the studied object the entire system, i.e. including the electric motor rotor, must be considered. Namely here, the qualitative and quantitative elastic properties of the flexible coupling play very important role for an electro-mechanical interaction between the driving motor and the driven machine.

Here, torsional motion of the rotor-shaft-motor system represented by the two abovementioned structural models shown in Fig. 2 and 3 can be generally described by the following set of non-linear ordinary differential equations:

$$\mathbf{M}\ddot{\mathbf{r}}(t)+\mathbf{C}(\Delta\theta(t))\dot{\mathbf{r}}(t)+\mathbf{K}(\Delta\theta(t))\cdot\mathbf{r}(t)=\mathbf{T}(t,\Omega(t),\theta(t)), \quad (2)$$

where:

$$\mathbf{C}(\Delta\theta(t))=\mathbf{C}_0+\mathbf{C}_c(\Delta\theta(t)),\quad \mathbf{K}(\Delta\theta(t))=\mathbf{K}_0+\mathbf{K}_c(\Delta\theta(t)).$$

The symbols $\mathbf{M}$, $\mathbf{K}_0$ denote respectively the mass and stiffness matrices, $\mathbf{C}_0$ is the damping matrix and $\mathbf{C}_c(\Delta\theta(t))$ denotes the matrix containing the non-linear damping coefficients of the flexible coupling, where $\Delta\theta(t)$ is the current coupling twist angle. The non-linear stiffness coefficients of this flexible coupling are contained matrix $\mathbf{K}_c(\Delta\theta(t))$. The symbol $\mathbf{T}(t,\Omega(t),\theta(t))$ denotes the vector of external excitation due to the electromagnetic torque generated by the driving motor as well as due to retarding torques imposed to the driven machine working tool and expressed both as functions of the current rotational speed value $\Omega(t)$ and the system dynamic response $\theta(t)$. Similarly as before, in the case of 3D FEM model vector $\mathbf{r}(t)$ contains generalized coordinates and motion of the hybrid model is described by the modal coordinates contained in vector $\mathbf{r}(t)$ as well. Similarly as before, the number of equations (2) also corresponds to the total number of degrees of freedom of the FEM model or to the number of torsional eigenmodes taken into consideration in the frequency range of interest, when the hybrid model is used.

As already mentioned above, actual torsional stiffness values of the flexible coupling very essentially influence natural frequencies of the considered system. The flexible coupling applied here is non-linear with the following progressive characteristic determined by its producer and depicted in Fig. 10. In the investigated case, the fundamental first torsional natural frequency value significantly depends on the working point

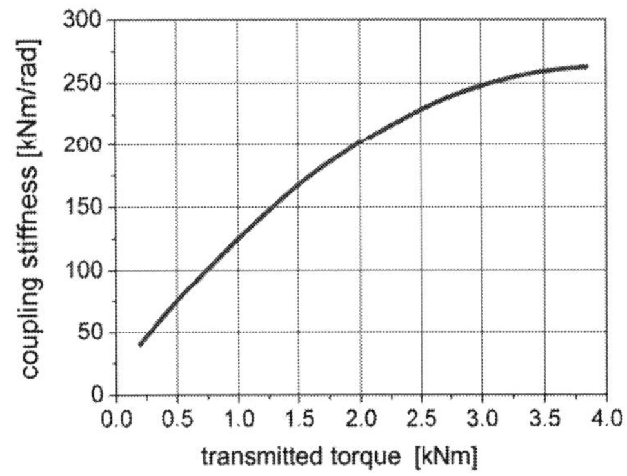

Fig. 10. Torsional stiffness characteristic of the flexible coupling.

determined by the current torque transmitted by this coupling, according to the characteristic presented in Fig. 10. Namely, for zero initial twist of the coupling the first natural frequency calculated by means of the 3D-FEM model and using the hybrid model is equal to ca. 15 Hz, where the mutual difference did not exceed 0.5% But for the maximal admissible torsional loading the appropriately increased coupling stiffness results in the first natural frequency exceeding 50 Hz. As it will turn out later, this property shall have a significant influence on the system transient torsional vibrations. Nevertheless, within the entire range of twisting torques transmitted by this coupling, i.e. for all the respective first natural frequency values, the corresponding eigenmode function has the following shape presented in Fig. 11. This eigenfunction is characterized by the whole driven machine part (on the left hand side) almost equal to zero and by the motor part (on the right hand side) equal to ~ −1. This means that dynamic excitations imposed only to the motor rotor are able to induce torsional vibrations of the considered system. From the results of eigenproblem analysis performed by these two mechanical models it follows that the next eigenmodes occur above 500 Hz and they are very hardly excitable by dynamic retarding torques loading the beater-wheel or impeller. These features follow from the relatively large torsional stiffness of the overhung rotor-shaft structure as well as because in such machines the polar mass moment of inertia of the driven part is much greater than that of the driving motor rotor. In the studied case this ratio exceeds the value of 41.

For the need of simulation of the rotor-machine torsional vibrations as well as in order to investigate electromechanical coupling effects in the considered system the properly advanced circuit model of the

 DOI:10.1051/e3sconf/20171801010

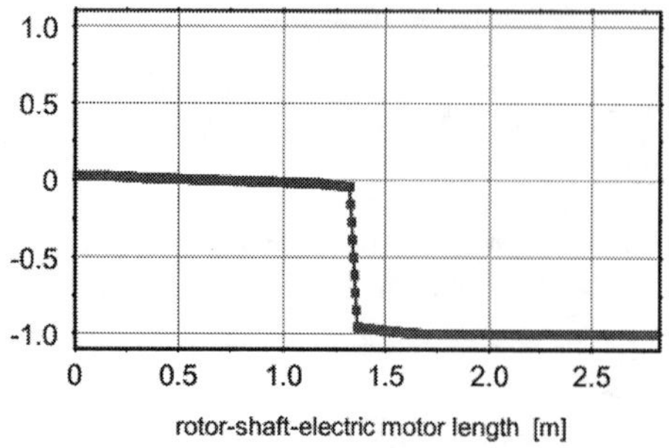

Fig. 11. The first torsional eigenmode function of the rotor-shaft system.

electric motor seems to be sufficiently accurate. In the case of the symmetrical three-phase asynchronous motor electric current oscillations in its windings are described by the six circuit voltage equations. Then, according e.g. to [6], by the use of Clark's transformation they are reduced into four voltage equations in the so called 'αβ' reference system. This approach enables us more effective numerical computations without any loss of accuracy, together with a possibility of taking into consideration a variable resistance of the rotor windings as well as a convenient control of the motor operation. Then, the electromagnetic torque generated by such a motor can be expressed by the following formula:

$$T_{el} = \frac{3}{2} pM \left[\cos(p\vartheta(t)) \cdot \left(i^{s}_{\beta} \cdot i^{r}_{\alpha} - i^{s}_{\alpha} \cdot i^{r}_{\beta} \right) - \right. \tag{3}$$

$$\left. - \sin(p\vartheta(t)) \cdot \left(i^{s}_{\alpha} \cdot i^{r}_{\alpha} - i^{s}_{\beta} \cdot i^{r}_{\beta} \right) \right],$$

where M denotes the relative rotor-to-stator coil inductance, p is the number of pairs of the motor magnetic poles, $\vartheta(t)$ is the current rotor angular displacement including the average and vibratory component and i_α^s, i_β^s are the electric currents in the stator windings reduced to the electric field equivalent axes α and β and i_α^r, i_β^r are the electric currents in the rotor windings reduced to the electric field equivalent axes α and β, [6,7]. The static characteristics of the considered asynchronous motor, determined accoriding to [7,8], are shown in Fig. 12. Here, the blue line

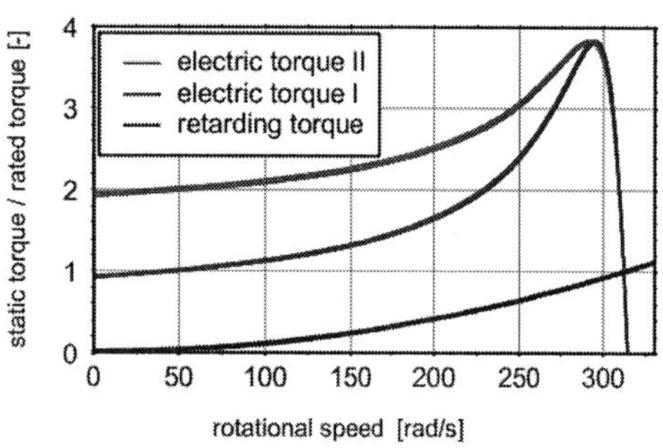

Fig. 12. Static characteristic of the asynchronous motor and of the retarding torque.

corresponds to Variant I characterized by the smaller staring torque. By means of the red line there is plotted the characteristic for Variant II with the greater staring torque caused by an increased rotor resistance.

As it followed from results of introductory simulations, because of the abovementioned relatively huge mass moment of inertia of the machine working tool in the form of a beater wheel or blower/compressor impeller, and because of the correspondingly small magnitude of the retarding torque variable components, in steady-state operating conditions possible torsional dynamic excitations turn out to be negligible. Then, the entire object works under almost static twist. According to the above, the retarding torques generated by the working tool can be regarded as of the aerodynamic type and modelled by means of the parabolic functions of the current rotational speed values. However, in transient operating conditions, e.g. during start-ups, the considered rotating machine drive system can be affected by severe torsional vibrations excited by the electromagnetic torque generated by the asynchronous motor described by Eq. (3). It is particularly the case, when in order to decrease the machine price any inverters are applied. Then, each start-up has to be carried out using an usually troublesome dischargeable clutch. Otherwise, such run-ups may lead to too high overloadings and damage of the flexible coupling which connects the driven machine with the motor.

In order to illustrate such a situation, the introductory numerical simulations have been performed for classical start-ups of the considered machine, i.e. without any electronic control of the asynchronous motor. In the first computational example the stiffness characteristic of the flexible coupling has been linearized around the working point corresponding to the nominal steady-state operating conditions with the average rotational speed 2980 rpm. Then, the first torsional natural frequency of the machine drive system associated with the eigenmode function shown in Fig. 11 is equal to 34.3 Hz. In Fig. 13 there are plotted time-histories of the system dynamic response due to the start-up from a standstill to the nominal operation. In this figure by the black line the time-history of the current rotational speed is marked. The red line corresponds to the electromagnetic motor torque and using the blue line there is illustrated a parabolic rise of the working tool retarding torque. The

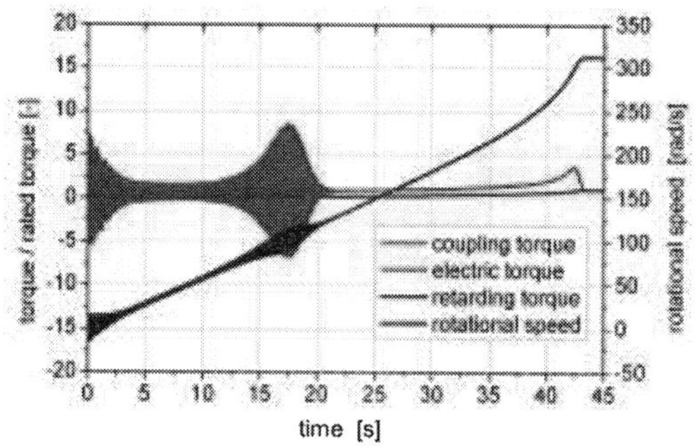

Fig. 13. Torsional transient dynamic response due to the start-up with the linearized flexible coupling characteristic.

DOI:10.1051/e3sconf/20171801010

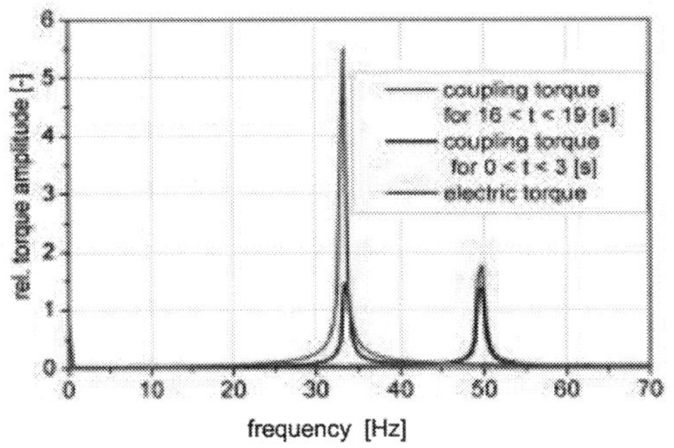

Fig. 14. Amplitude spectrum of the dynamic response during start-up for the system with linearized coupling.

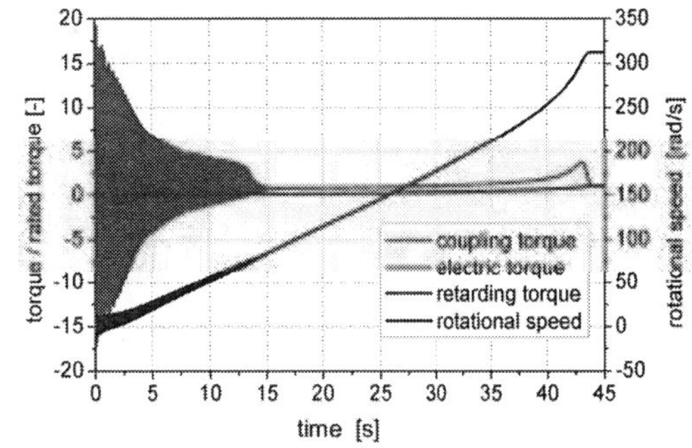

Fig. 15. Torsional transient dynamic response due to the start-up with the nonlinear flexible coupling characteristic.

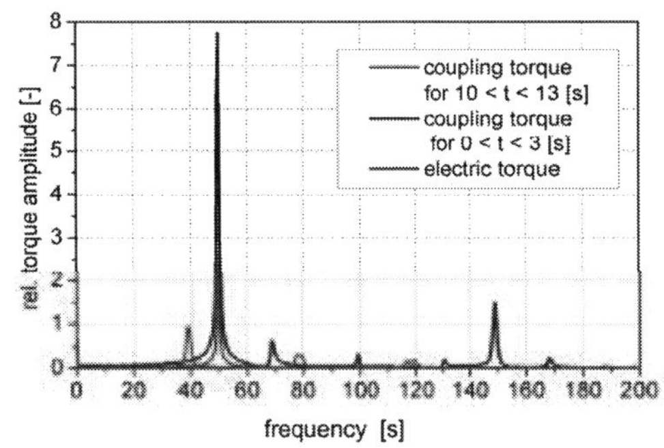

Fig. 16. Amplitude spectrum of the dynamic response during start-up for the system with nonlinear coupling..

time-history of the dynamic torque transmitted by the flexible coupling has been marked by the green line. From the plots depicted in Fig. 13 it follows that the considered machine was runned-up during ca. 43 s. At the beginning of this process the time-history of the driving motor torque is characterized by the typical, severe fluctuating component of the supply frequency 50 Hz quickly decaying with time and causing a relatively weak amplification of the dynamic torque transmitted by the flexible coupling and other shaft segments. From the results of the FFT analysis performed for the first three seconds of the run-up and depicted in Fig. 14 it follows that the system dynamic response consists of two fundamental components. The first one of the supply frequency 50 Hz results from the direct external excitation produced by the driving motor. However, the second one of the first system natural frequency 34.3 Hz is a typical transient state component induced by the driving torque generated by the asynchronous motor. Then, this component slowly decays with time, but after the 15th second of the start-up it becomes significantly amplified, when the rotating system passes through the rotational speed value ca. 16 rev/s corresponding approximately to one half of the 1st natural frequency. Since in the considered electromechanical system the rotational speed is affected by the fluctuating component of this frequency, the asynchronous motor electro-magnetic torque becomes also additionally affected by this rotational speed fluctuating component. Then, in the case of ca. 34 Hz the severe resonance is observed, which leads to amplitudes of the flexible coupling dynamic torque exceeding 8 times the nominal torque value, see Fig. 13 and 14. Although these resonant vibrations also quite quickly decay with time, but they are severe enough to damage gradually the drive system elements particularly sensitive to dynamic over-loadings during successively repeated machine start-ups.

In the second computational example the stiffness characteristic of the flexible coupling has been assumed nonlinear according to the plot presented in Fig. 10. Then, the analogous run-up simulation has been carried out and its computational results in time domain are depicted in Fig. 15 in the identical way as in the previously considered case. As it follows from the obtained plots, here the system dynamic response has a completely different qualitative and quantitative character in comparison with that illustrated in Fig. 13. Here, during start-up, together with the rise of the transmitted mean driving torque, the flexible coupling stiffness gradually increases and then the considered drive system changes its first natural frequency from the mentioned above ca. 15 Hz till more than 50 Hz at the maximal admissible transmitted torque value. Thus, the drive system consequently changes its dynamic properties, which prevents a possibility of resonance excitation. But at the beginning of the start-up process the dynamic amplification caused by the motor electromagnetic torque is much stronger than in the previous example. Here, the amplitudes of the dynamic torque transmitted by the flexible coupling are more than 15 times greater than the nominal torque value. Hence, despite of a lack of resonance effect, the run-up of the rotor machine equipped by the flexible coupling with such non-linear, progressive characteristics can result in a relatively fast damage of this drive system element expected to transmit dynamic torques exceeding the nominal value only 6 times. From the results of FFT analysis demonstrated in Fig. 16 it follows that during the first 3 seconds of the run-up the excitation by the asynchronous motor with frequency 50 Hz causes a severe dynamic amplification of the dynamic torque transmitted by the flexible coupling with the fundamental component of the same frequency 50 Hz as well as with smaller super-harmonics of frequencies ca. 70, 100, 130, 150 and 170 Hz. However, during the final phase of transient vibrations, i.e. within the range of 10-

DOI:10.1051/e3sconf/20171801010

13 s of the start-up, the time-history of the coupling dynamic torque consists of the much smaller fundamental component of frequency ~40 Hz together with the remarkable super-harmonics of frequencies ca. 80 and 120 Hz. According to the above, in order to avoid these detrimental transient oscillations the asynchronous motor of the considered rotor machine must be equipped with a proper inverter enabling us secure and fluent start-ups by means of driving electromagnetic torques possibly free of forcing fluctuation components.

An operation of majority of asynchronous motor inverters usually reduces to a gradual increase of the supply voltage in order to achieve the required rotational speed at the given loading of the runned-up machine. In the idealized case, when the voltage slowly rises from zero to the nominal value, for the smaller starting torque Variant I the start-up of the considered object lasts almost two times longer than in the previous examples, i.e. without an inverter. But then, the entire run-up process is completely free of torsional oscillations, which follows from the respective plots in Fig. 17 demonstrating the system dynamic response in an identical way as in Figs. 13 and 15. However, in order to overcome some natural static retarding torques in the drive system, when the supply voltage generated by the inverter starts from e.g. ca. 10% of the nominal value, some slight initial torsional vibrations are excited by the motor. As shown in Fig. 18, they gradually decay with time and when the system with the linear flexible coupling slowly passes the resonance frequency 34.3 Hz,

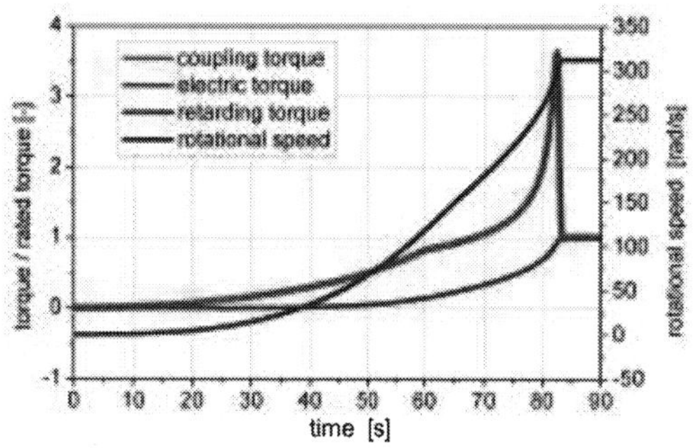

Fig. 17. Torsional transient dynamic response due to the start-up with the linearized flexible coupling characteristic for Variant I and zero initial supply voltage.

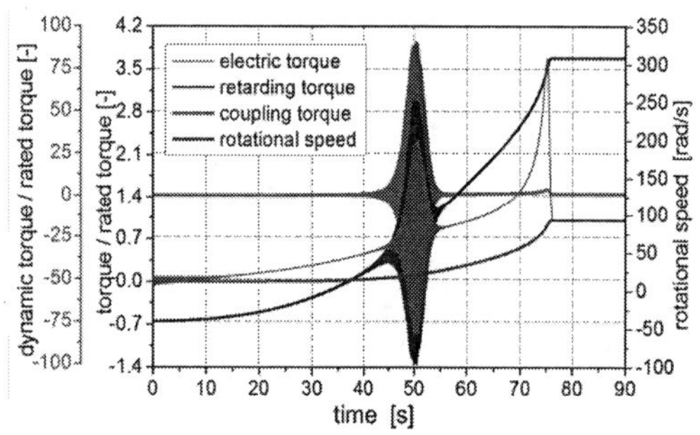

Fig. 18. Torsional transient dynamic response due to the start-up with the linearized flexible coupling characteristic for Variant I and non-zero initial supply voltage.

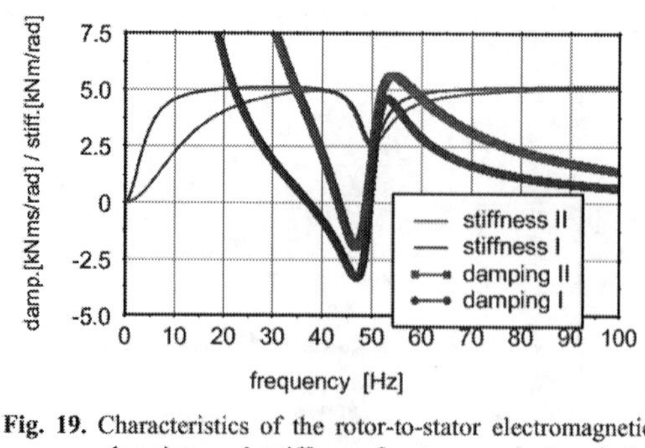

Fig. 19. Characteristics of the rotor-to-stator electromagnetic damping and stiffness for two variants of the asynchronous motor resistances.

huge transient oscillations rapidly occur with amplitudes almost 100 times exceeding the rated torque value. Here, despite of the start-up performed so mildly, such severe overloading can lead to an immediate damage of the flexible coupling. This fact can be explained by not only much slower passage through the resonance zone than in the case illustrated in Fig. 13, but by the very low, even negative electromagnetic damping generated by the asynchronous motor interacting with the torsionally vibrating mechanical system. For the considered here smaller starting motor torque Variant I the negative damping zone is significantly wider than for the greater staring torque version called Variant II. It follows from the respective characteristics of the electromagnetic

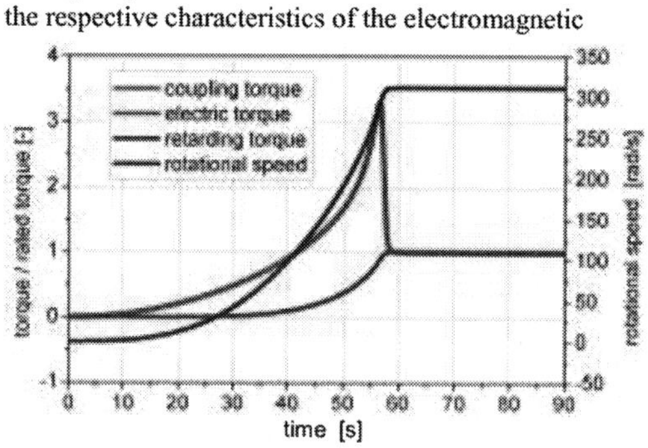

Fig. 20. Torsional transient dynamic response due to the start-up with the linearized flexible coupling characteristic for Variant II and zero initial supply voltage.

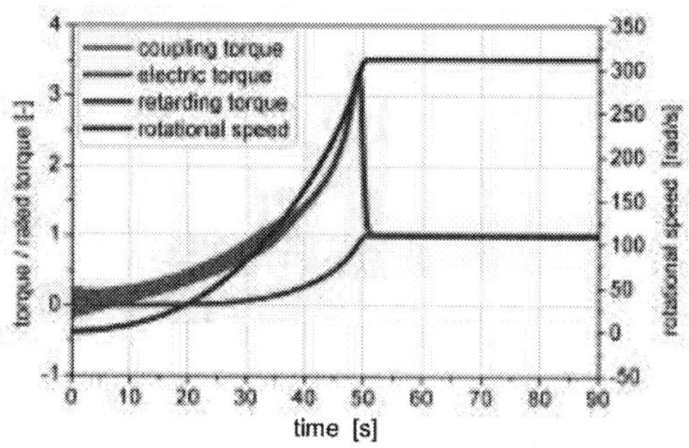

Fig. 21. Torsional transient dynamic response due to the start-up with the linearized flexible coupling characteristic for Variant II and non-zero initial supply voltage.

damping and stiffness depicted in Fig. 19 and determined for this motor using the analytical method described in [8]. Here, for Variant I this zone is contained within the range between ca. 37 and 49 Hz. Hence, the electromagnetic damping coefficient values in the vicinity of the resonant frequency 34.3 Hz are very small. Then, in the presence of so common, relatively low damping magnitude in torsionally vibrating drive systems such dangerous resonances can be easily excited. However, for Variant II corresponding to the greater starting motor torque, see Fig. 12, the negative damping zone is more narrow and placed between ca. 44 and 48 Hz. Furthermore, as shown in Fig. 19, since the electromagnetic damping coefficient values in the vicinity of 34.3 Hz are significantly greater, possibilities of resonance excitation are appropriately smaller. According to the above, when the inverter is used, the asynchronous motor with greater starting torque seems to be more convenient for start-ups of the considered rotating machine. This fact has been confirmed by the run-ups with zero- and non-zero initial voltage values. The dynamic responses induced by these start-ups are presented respectively in Figs. 20 and 21. Here, in the case of zero-initial supply voltage the completely oscillation-free response is observed, similarly as that obtained for motor Variant I and depicted in Fig. 17. However, during start-up with the initial voltage equal to 10% of the nominal value from the beginning of the process some mild transient vibrations occur, which gradually decay with time, as shown in Fig. 21.

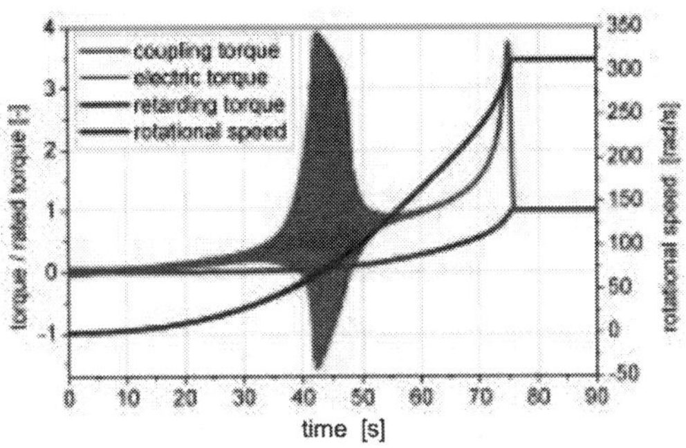

Fig. 22. Torsional transient dynamic response due to the start-up with the non-linear flexible coupling characteristic for Variant I and non-zero initial supply voltage.

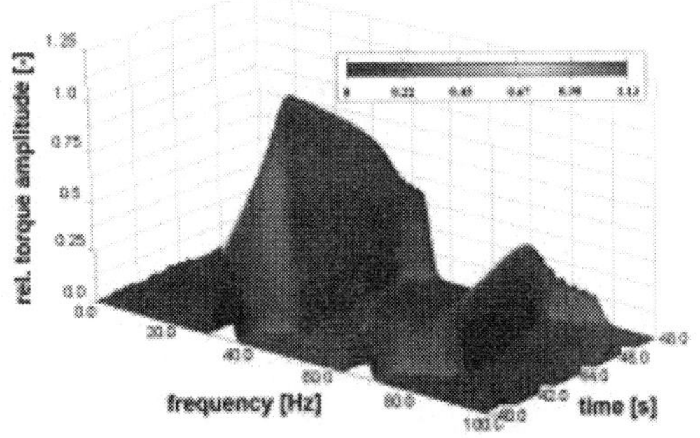

Fig. 23. Waterfall diagram for the dynamic torque amplitudes in the non-linear flexible coupling due to the start-up for Variant I and non-zero initial supply voltage.

When in the considered rotor machine drive system the flexible coupling with non-linear characteristic is assumed, for the asynchronous motor Variant I with the smaller starting torque and for the zero initial supply voltage the identical vibration-free dynamic response has been obtained as in the case of the linear coupling depicted in Fig. 17. However, during the run-up with the initial voltage equal to 10% of the nominal value remarkable transient vibrations are induced between the 30th and 55th second of the process, as illustrated in Fig. 22. These oscillations are not so severe as those resonant ones observed during start-up of the system with linear flexible coupling. Here, due to the gradual stiffening of the flexible coupling with non-linear characteristic the transient oscillations increase their frequency in a piece-wisely linear way within ca. 30-50 Hz, as demonstrated by means of the waterfall diagram in Fig. 23. Moreover, in this diagram the first super-harmonic component of the non-linear vibrations is observed.

Variant II of the asynchronous motor with the greater starting torque enables us much faster run-up. For zero initial supply voltage it also results in oscillation-free dynamic response, identical with that presented in Fig. 20 for the drive system with the linear flexible coupling. For the non-zero initial supply voltage equal to 10% of the nominal value the quicker rotational speed increase associated with the simultaneous synchronous rise of excitation frequency according to Eq. (3) makes an induction of severe transient oscillations difficult. Thus, during the start-up in such conditions the obtained

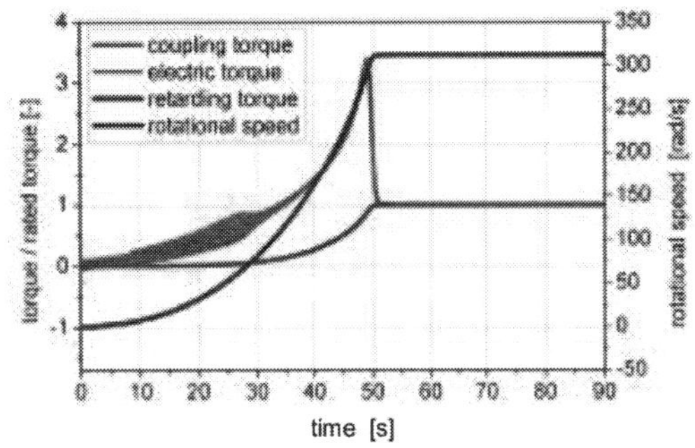

Fig. 24. Torsional transient dynamic response due to the start-up with the non-linear flexible coupling characteristic for Variant II and non-zero initial supply voltage.

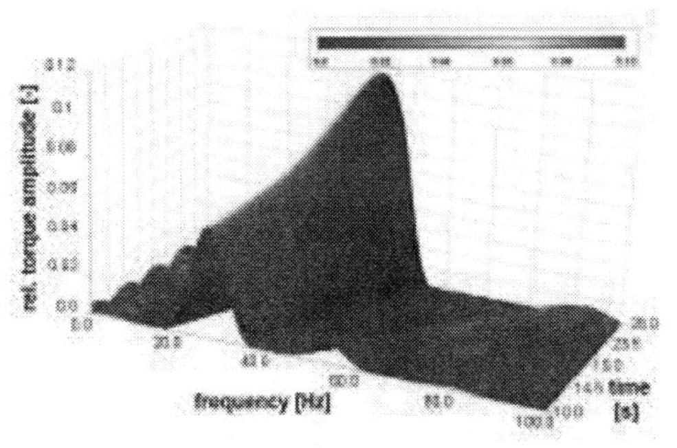

Fig. 25. Waterfall diagram for the dynamic torque amplitudes in the non-linear flexible coupling due to the start-up for Variant II and non-zero initial supply voltage.

dynamic response is characterized by very mild transient torsional vibrations transmitted by the flexible coupling, which are observed between the 10th and 28th second of the process, as shown in Fig. 24. But here, the oscillation amplitudes are significantly smaller than these observed in the case of slower run-up using Variant I of the motor starting torque. From the respective waterfall diagram depicted in Fig. 25 it follows that due to the gradual stiffening of the non-linear coupling the fundamental frequency of transient oscillations increases linearly within the abovementioned ca. 10-28 Hz as well as its first super-harmonic is also remarkable.

5 Final remarks

In the paper dynamical aspects for design and maintenance of the selected group of machines applied in the mining industry are considered. These machines and devices are driven by the asynchronous motors and they are characterized by rotational motions of their fundamental working tools. Thus, the dynamic investigations have been performed in the way typical as for the rotating machinery, i.e. in the form of bending (lateral) and torsional vibration analyses. Moreover, in these studies there was also investigated a mutual interaction between the mechanical and electrical part of the considered objects. Contrary to the traditional engineering approaches usually applied so far for design and maintenance of these machines, for the computations carried out here the advanced mechanical and electro-mechanical models and computer techniques have been used. By means of these modern computational tools the main attention was focused on qualitative aspects of dynamic behaviours typical for beater mills and crushers as well as for blowers and compressors.

From the results of bending vibrations analyses performed for the listed above machines characterized by heavy beater wheels or impellers designed in the form of overhung rotors it follows that their dynamic loadings caused by unbalances turned out to be particularly dangerous for fatigue life of their the most responsible elements. These unbalances are more dangerous than dynamic loadings associated with natural working processes, since they are caused not only by unavoidable manufacturing inaccuracies, but first of all by gradual and progressive material, thermal or erosive damages of working tool surfaces. The bending vibrations of the rotor-shafts of these machines induced by these unbalances can be a source of too fast material fatigue often leading to cracks and even to catastrophes, of dynamic over-loadings of bearing supports usually being transmitted to the environment as well as of harmful noise generation. According to the above, these detrimental and dangerous effects have to be thoroughly taken into consideration during routine design phases of these machines as well as they should be on-line monitored during regular operation in order to asses a degradation level of the working tools.

The torsional vibrations have been investigated here as entirely associated with electrical vibrations in driving motor windings in the form of electromechanical interaction analysis. From the results of performed computations it followed that steady-state operation conditions of the considered machines are not a source of essential torsional vibrations, contrary to quite frequently repeated start-ups of these devices, which can cause very severe or even dangerous transient oscillations. Here, mutual interdependencies between the electrical parameters of the driving motor and dynamic properties of the driven mechanical part significantly influence magnitudes of these vibrations. It turned out that the most important role play stiffness characteristics of the flexible couplings, an ability of negative electromagnetic damping generation by the asynchronous motors as well as a character of motor supply voltage control during start-up processes. First of all, in order to avoid severe transient torsional vibration amplitudes the asynchronous motors with inverters should be applied to control properly the supply voltage influencing runs of the driving motor torques. Then, start-ups usually last longer, but without remarkable oscillations, if system parameters had been properly selected before. Namely, the flexible couplings with non-linear characteristics seem to be more convenient, since they have an ability to mistune possible resonances. Next, the asynchronous motors with greater starting torques can assure shorter start-ups in time and thus also faster passages through eventual resonance frequency zones. Moreover, such asynchronous motors are able to generate less negative electromagnetic damping which is very dangerous from the viewpoint of an excitation of transient resonances and operational instabilities.

References

1. R. Gasch, R. Nordmann, H. Pfützner, *Rotordynamik* (Springer-Verlag, 2002)
2. G. Genta, *Dynamics of Rotating Systems* (Springer Science+Business Media, 2005)
3. M.I. Friswell, J.E.T. Penny, S.D. Garvey, A.W. Lees, *Dynamics of Rotating Machines (Cambridge Aerospace Series)* (Cambridge University Press, 2010)
4. ISO 1940/1, *Balance Quality Requirements of Rigid Rotors* (Int. Org. for Standardization, 2003)
5. ISO 11342, *Mechanical Vibration - Methods and Criteria for the Mechanical Balancing of Flexible Rotors* (Int. Org. for Standardization, 1998)
6. K. L. Shi, T. F. Chan, Y. K. Wong, L. S. Ho, Int. J. of Electrical Engineering Education, **36** (1999)
7. D.C White, H.H. Woodson, *Electromechanical Energy Conversion* (Wiley, 1959)
8. T. Szolc, R. Konowrocki, M. Michajłow, A. Pręgowska, Mech. Syst. and Signal Proc., **49** (2014)

Two-stage method for the simulation of the comminution process in a high-speed beater mill

Michał J. Marijnissen[1*], *Cezary* Graczykowski[1], *Jerzy* Rojek[1]

[1]Institute of Fundamental Technological Research, Polish Academy of Sciences, Pawińskiego 5b, 02-106, Warsaw, Poland

Abstract. The paper presents a two-stage simplified method for the simulation of comminution process which takes place in a beater mill. The first stage of the proposed method is a simulation of the flow of gas and ore particles through a mill based on a two-phase continuous-discrete model. It allows to capture the interaction between the fluid flow and embedded particles, to determine trajectories of their motion and average velocities and frequencies of their collisions against the flywheel and the mill's walls. The second stage of the proposed method is a discrete element method simulation of the process of comminution of a single ore particle. It allows to determine the size distribution of created smaller particles in terms of normal velocity and angle of impact and to estimate the global efficiency of the comminution process. The proposed simulation methodology is applied for the verification of the innovative concept of the application of high–speed beater mill for the comminution of the copper ore.

1 Introduction

Beater mills are commonly used in coal power plants for final stage comminution of coal before the combustion stage. Their operation is based on a spinning flywheel, which induces the gas flow and allows the incoming coal particles to achieve high velocity. As a result, the particles collide at high speeds against blades of the flywheel and internal walls of the mill and they are effectively crushed into smaller fractions. The process of high-speed comminution conducted in beater mills is characterised by high efficiency and relatively small energy consumption in comparison to other competitive methods.

The process of comminution in a beater mill is one of the most difficult industrial processes for numerical simulations. The reason for this is the multiplicity and complexity of diverse physical phenomena occurring simultaneously inside the mill during its operation. Amongst them, we can distinguish highly turbulent flow of the gas in the vicinity of the flywheel, the interaction of the flow and particles and, eventually, the ubiquitous process of particles comminution. All these phenomena are mutually dependent and they influence each other in an intrinsic way. Moreover, they all substantially contribute to mill operation and to the efficiency of the comminution process.

The above factors cause that direct numerical simulations of the comminution process inside beater mills are beyond the range of present computational capabilities. The literature on the holistic numerical simulation of the comminution in such a mill is rather limited. Nevertheless, various theoretical approaches and numerical models covering selected aspects of beater mill operation are extensively developed.

The first group of methods is dedicated to the precise simulation of the flow including solid particles with the use of two-phase models. This approach is based on a continuous-discrete Computational Fluid Dynamics - Discrete Element Method (CFD-DEM) coupling, which has been used in simulations and validations of particulate flow in a channel [1], fluidised beds [2] and slurry flow [3]. Additionally, it has successfully been tested against analytic and experimental data [4].

The second group of methods is dedicated to the simulation of the comminution process itself. Among these methods, the Discrete Element Method has gained the largest popularity and has been applied to the largest range of problems. The applications of DEM include impact breakage of particle agglomerates [5], impact crushing of particle-particle compounds [6] and ore breakage in laboratory scale ball mill [7]. Moreover, the mechanisms of impact fragmentation were studied in [8], the process of fragmentation in geomaterials was analysed in [9], while ultrasound assisted particle fragmentation was investigated in [10]. An extensive review of the application of DEM for the processes of fragmentation and comminution is presented in [11].

The objective of this paper is a joint application of the methods from two above groups for comprehensive simulation of the process of comminution in a beater mill. The proposed approach consists of two separate stages. The first step is a simulation of the flow of a mixture of gas and ore particles through a beater mill conducted by using two-phase continuous-discrete CFD-DEM model. The conducted simulation enables to observe the

* Corresponding author: mmarijn@ippt.pan.pl

interaction between the fluid flow and particles, determining their trajectories of motion and collision velocities. This data will be effectively used in the second stage of the method which is a DEM simulation of the comminution of a single ore particle. The conducted DEM simulation will allow to determine the mass distribution of created smaller particles in terms of normal impact velocity and to estimate the global efficiency of the comminution process. The entire proposed simulation methodology will provide deep insight into the process of comminution in beater mills. In particular, it will be applied to the verification of the innovative concept of the application of high–speed beater mills for the comminution of copper ore.

2 Models used in the computation of the flow in a beater mill

In the model used the fluid is considered a continuum, however the particles are discrete and tracked in a Langrangian reference frame. The fluid was considered compressible and the flow turbulent. The k-ω turbulence model was used to take into account for turbulent flow. Mass and heat exchange between phases were omitted.

2.1 Fluid flow

The Finite Volume Method is used to calculate the fluid flow taking into consideration the granular phase's volume fraction in given cells. A multiphase fluid is modelled with the use of conservation equations for a particular phase as given by

$$\frac{\partial \rho}{\partial t} + \nabla \cdot \{\rho \boldsymbol{u}\} = 0, \tag{1}$$

$$\frac{\partial \rho \boldsymbol{u}}{\partial t} + \{\boldsymbol{u} \cdot \nabla\}\boldsymbol{u} = \rho \boldsymbol{g} - \nabla p_r + \nabla \cdot \overline{\tau_q}, \tag{2}$$

$$\overline{\tau_q} = \mu\{\nabla \mathbf{u} + \nabla \mathbf{u}^T\} + \left\{\lambda_q - \frac{2}{3}\mu\right\}\nabla \cdot \boldsymbol{u}\mathrm{I}, \tag{3}$$

$$\frac{\partial h\rho}{\partial t} + \nabla \cdot \{\rho \boldsymbol{u} h\} = \frac{\partial p_r}{\partial t} + \overline{\tau_q} \cdot \nabla \boldsymbol{u}. \tag{4}$$

In these equations $\boldsymbol{u}$ is the velocity of the fluid, $\boldsymbol{g}$ the gravity, ρ the density, I the indentity matrix and p_r the pressure. Additionally, λ_q and μ denote the shear and bulk viscosity, $\overline{\tau_q}$ is the stress tensor and h is the specific enthalpy, while t is the time variable. Details of the k-ω turbulence model can be found in [12].

2.2 Discrete Element Method

The DEM was first introduced in [13]. In the DEM we track the position of discrete particles immersed in the fluid as given by

$$\frac{d\boldsymbol{u}_s}{dt} = F_{drag}(\boldsymbol{u} - \boldsymbol{u}_s) + \frac{g\{\rho_s - \rho\}}{\rho_s} + \boldsymbol{F}_{add.}, \tag{5}$$

$$F_{drag} = \frac{3\mu C_d Re}{4\rho_s d_s^2}, \tag{6}$$

$$Re = \frac{\rho d_s |\boldsymbol{u}_s - \boldsymbol{u}|}{\mu}, \tag{7}$$

$$C_d = \begin{cases} 0.424 & : Re > 1000 \\ \frac{24}{Re}\left\{1 + \frac{Re^{2/3}}{6}\right\} & : Re \leq 1000 \end{cases}. \tag{8}$$

Here equation (5) represents the equilibrium of forces acting upon the particles. All variables containing the subscript s consider the particles. Moreover $\boldsymbol{F}_{add.}$ represents additional accelerations acting upon the particle (ex. contact, coupling, pressure gradient) and F_{drag} can be described as shown in equations (6) and (7). Lastly remains the definition of the particle drag coefficient. We did not take into account other particles shapes than spherical, so we used the formula presented by equation (8) without any shape coefficients.

2.3 Particle contact

Contact between particles is considered. The particles are allowed to overlap for a small distance. This overlap is the deciding factor while calculating the contact force. For the overlap to be small it is crucial to use a small time step. When an overlap is detected contact forces are calculated and applied to the particles. The governing equations are given by

$$\boldsymbol{e}_{12} = \frac{x_2 - x_1}{\|x_2 - x_1\|}, \tag{9}$$

$$\delta = \|\boldsymbol{x}_2 - \boldsymbol{x}_1\| - \{r_2 + r_1\}, \tag{10}$$

$$\boldsymbol{F}_1 = \boldsymbol{F}_2 = \{\mathrm{k}\delta + \gamma \boldsymbol{u}_{12} \cdot \boldsymbol{e}_{12}\}\boldsymbol{e}_{12}, \tag{11}$$

$$\boldsymbol{F}_{tangential} = \beta \boldsymbol{F}_1. \tag{12}$$

Here $\boldsymbol{x}$ is the particle position, $\boldsymbol{r}$ is the radius of the particle, **k** and $\boldsymbol{\gamma}$ are the spring and dampening coefficients and $\boldsymbol{\beta}$ is the friction coefficient.

2.4 Coupling

Due to low particle concentration, the effect of their presence in the fluid was omitted, creating a 1-way coupling. The fluid's velocity is calculated at the particle's position and passed to equation (5). Next, the particles positions are updated after which the calculations continue for the fluid.

3 The first stage – fluid and particle flow

3.1 Geometry, mesh and boundary conditions

The simplified geometry of an existing beater mill was used for the simulation domain. The height of the mill was around 4m, with a flywheel of approximately 1m in diameter. The geometry is shown in Fig. 1. with inlets and outlets marked with coloured arrows. The mesh consists of approx. 600.000 finite volumes. A pressure inlet and outlet boundary condition were imposed on the marked surfaces. The flywheel and its nearest vicinity was set in a rotary frame and connected with the rest of the mill with an interface boundary condition. The flywheel had a constant rotational velocity of 2000 rpm. The fluid time step was set to $5 \cdot 10^{-5}$ s, while the particle time step was $5 \cdot 10^{-6}$ s. Such a small time step was required to stabilise the simulation caused by the small gap between the rotary frame and the stationary mesh near the wall.

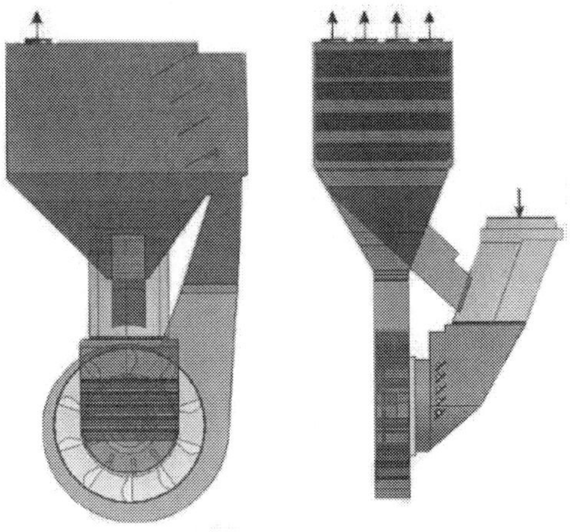

Fig. 1. Geometry of the mill, from the back (left) and side (right). Marked inlet and outlets.

3.2 Results obtained

Result obtained with the continuous-discrete method allow for detailed insight in the trajectory of every single particle. Additionally, particles of different size, density and shape can be injected inside the domain. In these calculations, we limited ourselves to spherical sandstone particles with a density of 2580 kg/m^3, based on [14]. Due to a heavily increasing computation cost with the number of particles tracked, we used a group of 360 particles of a given diameter of 40 mm. The group was evenly distributed on the inlet boundary and injected into the domain with a velocity similar to the fluid. The injection was done after the fluid flow inside the mill stabilised. The analysis of particles velocities amongst their trajectories provided information about the impact velocity and angles that occurred during collisions with the flywheel. The results for a single particle are presented in Fig.2, where the collision with the flywheel occurred at approx. 0.18s.

The results obtained from the group of particles gave a good insight into the impact statistics. The mean velocity of impact was 131.6 m/s with a variance of 17.2 m/s. An image of the particles positions just after the first collisions with the flywheel occurred was captured and is presented in Fig. 3. The image shows the chaotic trajectories after the collision with the flywheel. Particles collide not only with the wall because some of them are reflected back upon the flywheel. This leads to the conclusion that, even if the particles wouldn't break at the first collision the subsequential collision would make sure the particle would crumble. Due to the drastic differences in the behaviour of pre- and post-comminution particles further calculations of the simulation presented in Fig. 3 have not been carried out.

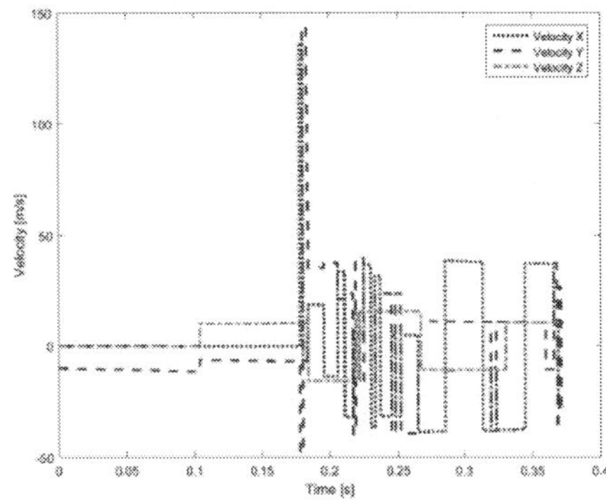

Fig. 2. The velocity of a single particle in a function of time during its transit.

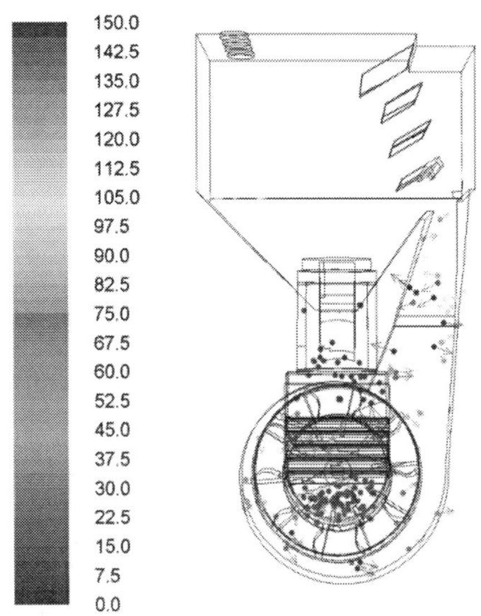

Fig. 3. Particle position in the mill after 0.2039 seconds upon injection coloured by their velocity magnitude [m/s]. Particle diameter is 40 mm.

For the simulation to have relevance with the existing counterpart, we have restarted the simulation with particles of sizes that can be found in the mill's output. For a detailed analysis of beater mills with this method, we refer the reader to [15].

4 The second stage - DEM simulation of the elementary comminution process

The elementary process of comminution is the process of crushing a single material particle during impact against a rigid surface [16]. Mechanical response of the particle during such event strongly depends on its mechanical

 DOI:10.1051/e3sconf/20171801011

properties including viscoelastic or viscoplastic characteristics and fracture durability, as well as on impact parameters such as velocity, angle and particle orientation.

The process of comminution is especially sensitive to impact velocity. Small impact velocities cause generation of micro-damages and micro-cracks which are often not visible at the particle's surface. The increase of impact velocity results in the generation of more severe fractures and local fragmentation of the particle in the vicinity of the contact area. Eventually, high impact velocities exceeding the critical value cause breakage of the entire particle into numerous pieces of diverse shapes and sizes, and generation of various material fractions

Conducting an elementary test of comminution is crucial for determining the crushing durability of the material. In particular, it enables to define the requirements for the industrial processes of crushing and grinding, to assess their efficiency and to predict the required energy consumption.

The proposed method of a comminution process simulation is based on the classical formulation of the Discrete Element Method [17,18]. In this method, the crushed material particle is modelled by a set of discrete elements, usually spheres of various sizes. At the initial state, the discrete elements are connected with each other by predefined cohesive bonds, providing the initial shape and geometrical configuration of the particle. The impact of this particle against a rigid surface causes breakage of selected cohesive bonds and resulting disintegration of the modelled ore particle into a set of smaller pieces, Fig. 4.

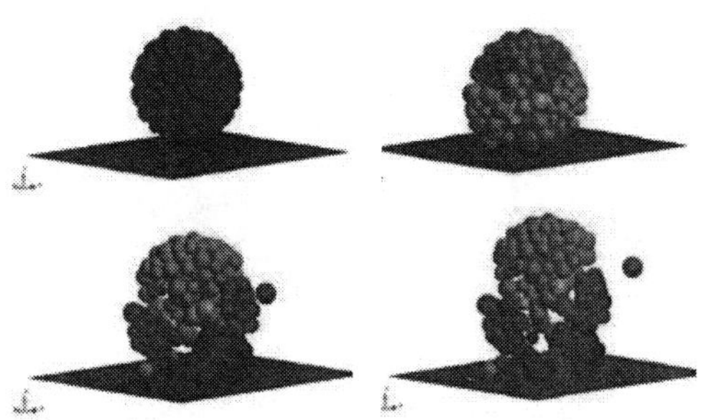

Fig. 4. Numerical simulation of the comminution of a single material particle by Discrete Element Method [18].

The mathematical formulation of the DEM model is based on equations describing the motion of each particle, which include inertial and contact forces. The inertial forces are expressed by the second derivative of the particle displacement vector **u** defined in a classical Lagrange reference frame. In turn, the contact forces transferred to the particles can be replaced by the point load $\mathbf{F}^{ext}$ and concentrated torque $\mathbf{T}^{ext}$ applied at the centre of inertia of each particle. Consequently, the equations of motion of a single particle read:

$$m\ddot{u}_C = F^{ext}, J_C\dot{\omega} = T_C^{ext}, \quad (13)$$

while matrix equilibrium equations of the whole system of particles assume the form:

$$Mr = R, J\dot{\Omega} = T, \quad (14)$$

where **M** and **J** are the matrices containing masses and moments of inertia of the discrete elements, whereas vectors **R** and **T** contain point loads and concentrated torques acting on particles. The above equations are solved by using an explicit method of integration.

The contact interactions acting between the discrete elements can be decomposed into a direction normal and tangential to the surface of contact. The contact force generated in the normal direction is defined by stiffness and damping coefficients, while the contact force generated in the tangential direction is defined by the stiffness and friction coefficients, Fig. 5.

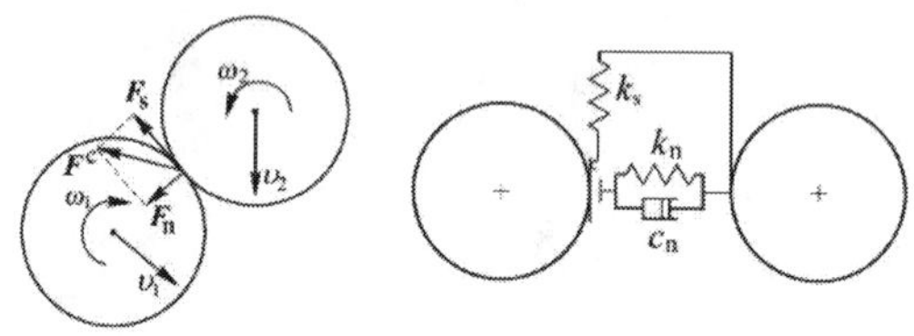

Fig. 5. Contact interactions between the discrete elements [17].

Accordingly, the cohesive bonds providing the possibility of solid material modelling are defined separately in the normal and tangential direction to the contact surface. Finally, the possibility of breakage of the cohesive bonds is provided by using stress-based criteria of breakage depicted in Fig. 6.

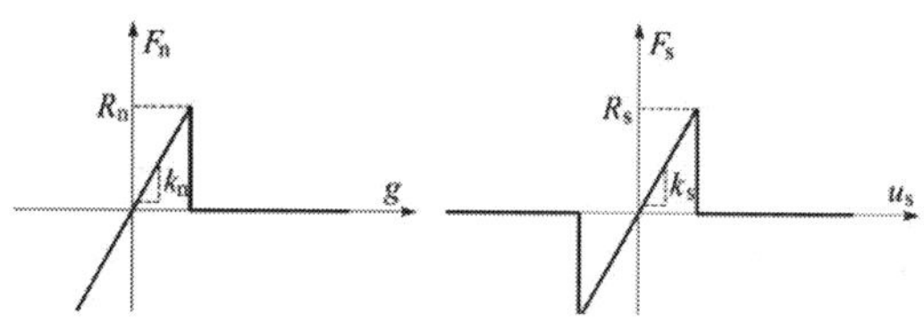

Fig. 6. Applied model of cohesive bonds with breakage [17].

All results of the elementary comminution process presented in this paper were obtained by the use of the discrete element method - based software DEM-Pack developed by the third author [17].

5 Results of DEM simulations of the comminution process

5.1 Model description

The simulations of the elementary comminution process were performed for an idealised particle of spherical shape with a diameter equal to 40 mm. The analysed model is composed of 3344 spherical elements with various radii of an average size of 1,3 mm. The average filling of the model by solid material is approximately 92%. The defined initial cohesive bonds provide that their average number corresponding to a single element equals approximately six.

The model includes a flat surface located in the vicinity of the particle, which models the rigid obstacle as well as the definition of the contact conditions between the particle and rigid surface. The initial velocity applied to the particle causes its rapid contact with the surface resulting in breaking the subsequent contact bonds and initiation of the comminution process.

The conducted numerical studies were aimed at the investigation of sandstone, being the carrier of copper ore. The macro-mechanical parameters of the sandstone were assumed as equal to: $\rho = 2580$ kg/m^3, E = 18.69 GPa, $\nu = 0.18$, $\sigma_c = 50$ MPa, $\sigma_t = 5$ MPa, based on [14]. Further, the micro-mechanical parameters of DEM model were determined by using the dependences defined in [17]. The micro-mechanical parameters were assumed to be equal for the entire comminuted particle. The possible inhomogeneity resulting from the mineralogical composition of the material was not taken into account at this stage of research.

5.2 Elementary low-speed comminution process of sandstone

The initial numerical test was a simulation of the elementary process of comminution of a sandstone particle at relatively low impact velocity of 37 m/s. The applied velocity was sufficient to cause breakage of the particle into 925 separate pieces, including 826 totally shredded particles (containing only one discrete element) and 99 partially shredded particles (containing more than one discrete element).

The initial configuration of the particle, including colouring indicating final division into pieces is presented in Fig. 7a, while the configuration of the system occurring 6 ms after the beginning of impact is presented in Fig. 7b (partially shredded particles are marked in various colours, while totally shredded particles are marked in red).

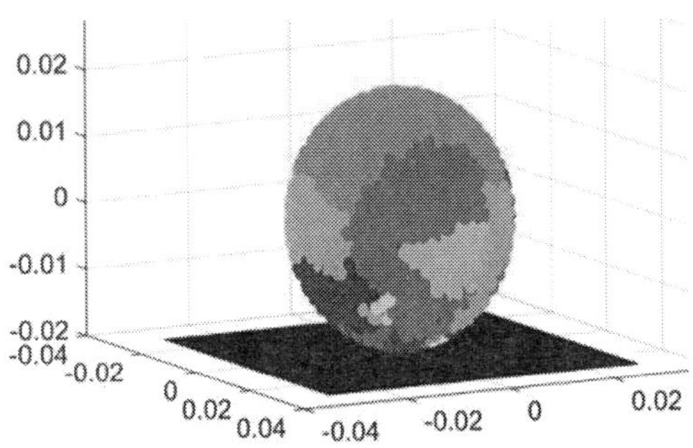

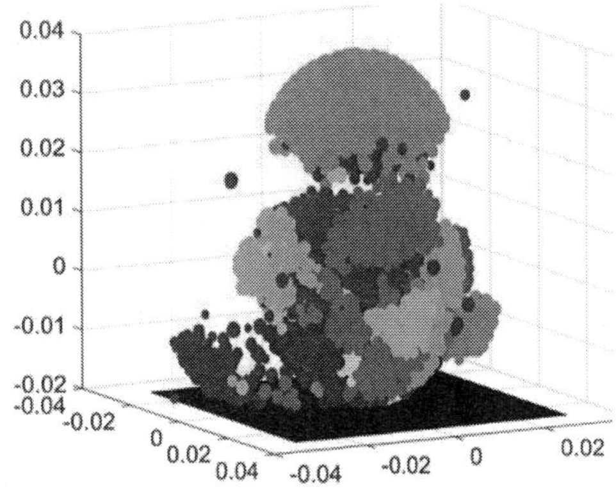

Fig. 7. Results of the numerical simulation of the elementary comminution process of a sandstone particle (impact velocity equal 37 m/s) : a) the initial state, b) state 6 ms after beginning of impact.

5.3 Sensitivity of the process of comminution on the initial impact velocity

The following objective was an investigation of impact velocity influence on the elementary process of comminution. In order to avoid a detailed analysis of the grain size distribution for each particular impact velocity, two global coefficients summarising the process of comminution were introduced:

- the number of partially shredded particles,
- the size of the largest particle.

Since both coefficients describes different properties of particle breakage they can be considered as a good global measure of the effectiveness of the comminution process. In the presented example, the comminution of a sandstone particle was analysed in a wide range of velocities from 4 to 169 m/s. Results of the conducted sensitivity study are shown in Fig. 8a-b.

The obtained results allow to assessquantitatively the intuitive effect of the intensification of the comminution process caused by an increase of the impact velocity. The critical velocity, which can be distinguished from the plot of partially shredded particles (Fig. 8a) and the plot of largest particle size (Fig. 8b) equals approximately 70 m/s. For this velocity, the particles which previously were only partially shredded become broken into smaller pieces, while the size of the largest particle decreases below 4% of the original particle size. Simultaneously, the number of fully shredded particles exceeds 2000 which allows to consider the analysed single impact comminution process as very efficient.

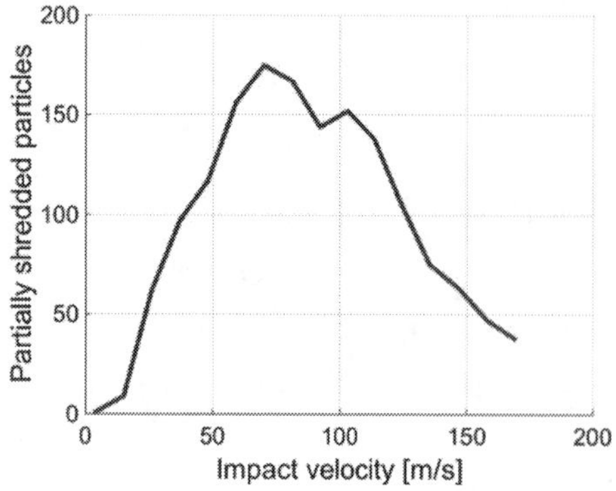

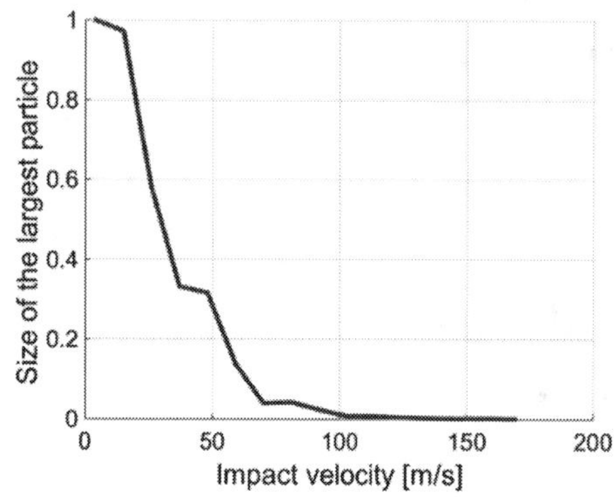

Fig. 8. Sensitivity of the sandstone particle comminution on impact velocity in the range 4-169 m/s: a) number of fully shredded particles, b) relative size of the largest particle.

5.4 High-speed comminution in a beater mill at 2000 rpm

The final analysis was simulation of the comminution of a sandstone particle under conditions which occur inside the beater mill operating at rotational velocity of 2000 rpm. The impact velocities in the corresponding DEM simulations covered the range 125–169 m/s. Five simulations of the comminution of spherical sandstone particle of diameter 40 mm were conducted and their results were collected together in order to find averaged characteristics of the process, Fig. 9.

The obtained results indicate a very high average level of comminution within the considered range of impact velocities. The products of the comminution are grouped in 12 fractions, among which the largest has the size of 0.45% of the original particle. Almost 94.6% of the total mass of the product are fully shredded elements, which belong to the fraction 0-0.03% (the entire vertical range is not shown in Fig. 9 for better visibility). Moreover, the particles smaller than 0.12% of the original size constitute almost 99% of total mass of the product.

All the above-described results indicate that the assumed range of the flywheel's rotational velocity of 2000 rpm is definitely sufficient for the purpose of sandstone comminution. Taking into account the occurrence of multiple collisions inside the mill, the assumed velocity can be regarded as too large and its decrease should be considered.

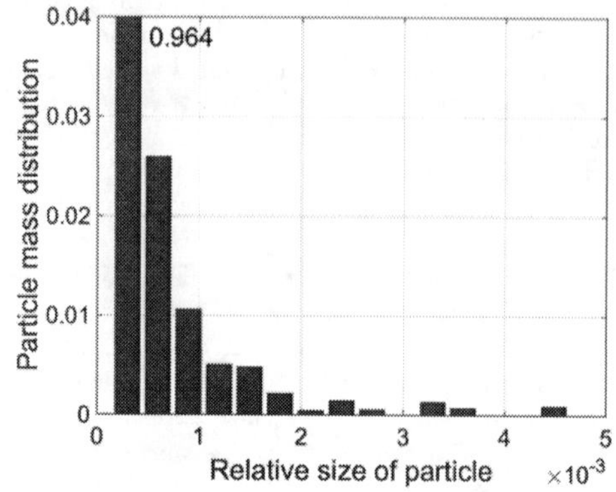

Fig. 9. Relative mass distribution resulting from comminution of a sandstone (impact velocity 125-169 m/s).

6 Conclusions

The paper introduces a simplified method for the numerical analysis of the comminution process inside a beater mill. It includes simulations of the flow of the mixture of gas and solid particles through the mill by using continuous-discrete CFD-DEM model and the following simulations of the elementary process of comminution by using the DEM model including bonds breakage.

Although the proposed method does not provide full interaction between the flow and the particle's breakage it allows to preliminarily investigate the effectiveness of the conducted comminution's process. In particular, the method was successfully applied for the verification of the innovative concept of applying beater mills for the comminution of sandstone and the extraction of copper ore. It was proved that the applied rotational velocity of the flywheel of 2000 rpm allows to achieve more than satisfactory results of sandstone comminution.

References

1. M. Azimian, M.Lichti, H.J. Bart, Open Chem. Eng. J.**8**, 1–11, (2014)
2. C. Yang, Y. Duan, Chem. Eng. Technol. **36**, 1907–1914 (2013)
3. D. Kubicki, S. LO,*Ninth International Conference on CFD in the Minerals and Process Industries.*10-12December (2012)
4. C. Kloss et al., Progr. Comp.Fluid Dyn. **12**,140–152 (2012)
5. B.K. Mishra, C. Thornton, Int. J. Miner. Process. **61**, 225–239 (2001)
6. W. Schubert, M. Khanal, J. Tomas, Int. J. Miner. Process. **75**, 41– 52(2005)
7. P.W. Cleary, R.D. Morrison, Miner. Eng. **24**, 352–366(2011)

8. F. K. Wittem et al., Int. J.Fract. **154,** 105–117 (2008)
9. G.A. D'Addettaet al., Lect. Not. Phys., **568**, 231-258 (2001)
10. F. K. Wittel, Gran. Matt. **12**,447–455 (2010)
11. H.A. Carmona, F.K. Wittel, F.Kun,Eur. Phys. J. Special Topics **223**, 2369–2382 (2014)
12. D.C. Wilcox, DCW Indust. Inc. La Canada, (1998)
13. P.A. Cundall, O.D.Stark, Géotechnique, **29**, 47-65 (1979)
14. N. Kunysz, Rudy i Metale, (1972)
15. M.J. Marijnissen, J. Rojek, J. Phys. Conf. Ser. **760**, 1-8 (2016)
16. M.Brozek,W. Maczka,T. Tumidajski, *Modele matematyczne procesów rozdrabniania* (1995)
17. J.Rojek, *Modelowanie i symulacja komputerowa zlozonych zagadnien mechaniki nieliniowej metodami elementów skonczonych i dyskretnych* (2007)
18. J.Rojek, C.Graczykowski, M.J.Marijnissen, T.Szolc, I.Marczewska, *Proceedings of the XIth International Conference on Non-ferrous Ore Processing* (2015)

Analysis of a grinding efficiency in the electromagnetic mill for variable process and feed parameters

Marta Wołosiewicz-Głąb[1,*], *Dariusz* Foszcz[1], *Daniel* Saramak[1], *Tomasz* Gawenda[1], and *Damian* Krawczykowski[1]

[1]AGH University of Science and Technology, Faculty of Mining and Geoengineering, Department of Environmental Engineering and Mineral Processing, Mickiewicza 30 Av., 30-059 Krakow, Poland

Abstract. Article describes issues related to the assessment of raw materials grinding efficiency in an electromagnetic mill. Application of such type of mill in grinding processes can result in measurable benefits of more efficient raw material grinding by reducing time needed to produce a product of a suitable particle size compared to conventional mills. The expected effect of application of a system based on an electromagnetic mill is to achieve a lower energy consumption of grinding process for raw material processing compared to the existing equipment used for the fine grinding. A key element of correct design of comminution circuits in processing of particular feed mineral is to identify the impact of key factors affecting the efficiency of the grinding process. The most important factors determining the efficiency of the grinding process are the time, the construction parameters of the machine, the type of grinding media used in grinding process, as well as the parameters of the feed, among others the crushing resistance or particle size. The paper presents the results of kinetic study of copper ore grinding in electromagnetic mill for variable feed particle size distribution of and different sizes of grinding media.

1 Introduction

Fine grinding operations are part of processing in case of many mineral raw materials. Depending on the required particle size distribution and maximum grain size of final product, grinding process is conducted in appropriate types of mills. Such devices work with different efficiency, determined by the energy consumption, performance and especially comminution degree of the final product. In ores processing, the required particle size is primarily conditioned by mineral particle size, the applied beneficiation technology and the grinding costs. Due to constant decrease of useful component content in the currently used ores, in order to obtain a comparable amount of metal in the concentrate, manufacturers are striving to increase production which means an increase in operating costs. Such situation stimulates further research on technical and technological solutions enabling the process of ore treatment to be run at the lowest possible cost. Due to the fact that comminution processes account for about 50-60% of operating costs, the main efforts are directed to these operations, and since the finer is a comminution product, the more energy has to be delivered, and the newer comminution technologies and devices are applied [8]. One of the new grinding devices is an electromagnetic mill, which, due to its grinding capabilities has a high potential for efficient comminution of raw materials at low energy costs. An important issue in application of electromagnetic mill is determination of the feed parameters which determines the efficiency of the mill's operation [4].

Feed preparation for beneficiation of fine-grained products (limestone meal, cement clinker) is a complex circuit of comminution (crushing, grinding) and classification (with recycle flows). Grinding is the basic operation, depending on the conditions and accepted technological assumptions, can take place in drum mills (eg rods, ball or cylpeps), Semi-Autogenous mills (SAG) and Autogenous ones (AG). In subsequent stages of grinding, vertical, vibratory, and other rotor mills are used. However, grinding the material (feed, ore) to a size below 0.1 mm is a very energy-intensive process. The amount of energy required for proper grinding depends primarily on the lithological composition of the raw material but also on the type of device [1,2,3].

The grinding performance of the mill is determined by its work parameters, such as grinding time, roll speed, load, grinding media fulfilment, grinding media composition and size, physical and mechanical properties of the feed (particle size distribution, including maximum particle size and fine particles participation; susceptibility to crushing). By changing these parameters it is possible to reduce or increase the processing capacity of the mill. Properly selected parameters, taking into account their mutual impact on the efficiency of the operation of comminution equipment, are crucial in determination of the working conditions and can lead to improvement in the comminution process course and cost reduction [1,2,3].

* Corresponding author: wolosiewiczm@gmail.com

 DOI:10.1051/e3sconf/20171801012

2 Comminution technology in electromagnetic mill

The electromagnetic mill is one of the devices, in which it is possible to achieve high technological efficiency of the process, measured by comminution ratio value, at maintaining high productivity and relatively low energy consumption.

Investigations in this field indicate that application of a circuit of simultaneous grinding and drying with the electromagnetic mill, offers the potential for achieving the desired fineness of the grinding product with favourable energy consumption ratios, as well as the ability to effectively control the process [5]. The electromagnetic mill is a device in which ferromagnetic grinding media are deliberately moved by a rotating electromagnetic field, which works as an energy carrier. The basic elements of the mill are: rotary magnetic field inductor and a tube in its axis, making the working chamber.

The working chamber is filled with ferromagnetic grinding media in approximately 20%. They should be selected accordingly to the particle size of the feed [6,7]. In case of large material differentiation, a mixture of grinding media of different sizes is used. Possible changes in the size of the mills are connected to the physical properties of the feed and the final particle size distribution of grinding product.

The ability to control the feed-to-grinding media ratio, real-time particle size analysis and the operating condition of the mill, as well as flow volume of recycle allow for much more effective mechanical activation of the produced particles with strictly defined physicochemical parameters (size, shape, specific surface, surface properties). They also increase the grinding capacity and reduce the energy consumption [6,7]

Fig. 1. Electromagnetic field inductor (by D. Foszcz).

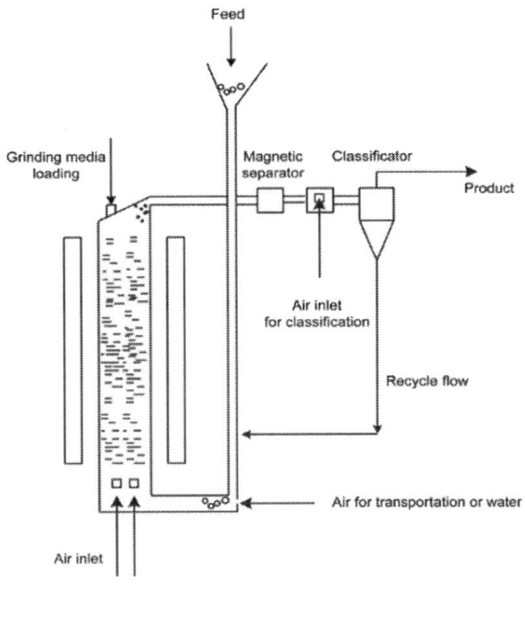

Fig. 2. Scheme of material flow through the electromagnetic mill [5].

The concept assumes the vertical position of the electromagnetic mill's working chamber, which is loaded from the bottom by screw feeder with the feed, and the stream of transport air, which is suitably moistened, is being provided [7]. Above the working chamber, there is a preliminary classifier integrated with a mill, which forms the inner recycle stream. The material is received from the working chamber of the mill from the top and goes to the pre-classifier. Course (not grinded) material returns along with the feed from the top to the working chamber and the material with suitable particle size is directed upwards towards the second (exact) classifier, which is designed as an inertial-impingement classifier. To ensure adequate transportation air flow speed and accurate classification, there is an adjustable additional air intake between the initial and the exact classifier. Exact classifier separates the stream of ground material to the end product and the recycle stream, providing the closed loop cycle of the mill operation [7].

However, due to the small size of the particles in recycle, to force the next passage through the working chamber, the recycle stream is fed from the bottom of the working chamber and its flow is controlled by a variable recycle air stream flow. The final product, coming out of the exact classifier, is separated from the transport air in the cyclone and goes to the final product tank. In addition, the installation is equipped with a cooling system of the working chamber in the form of a set of variable-capacity fans [7].

3 Material and methods

The experiments were carried out in a laboratory electromagnetic mill with a diameter of 200 mm – Fig. 3. The feed consisted of copper ores with variable particle sizes 0-0.5 mm, 0-1 mm and 1-2 mm.

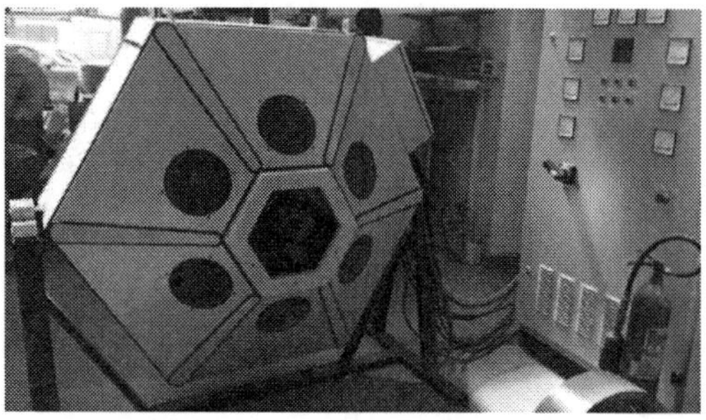

Fig. 3. Laboratory class electromagnetic mill with a diameter of 200 mm used for investigations (by D. Foszcz).

The tests involved the grinding of ore with three parameters of particle size distribution and four sets of grinding media of different sizes. Approximately 500 grams of material and 1500 grams of grinding media were used in each grinding test. Grinding times for the ore 0-0.5 mm and 0-1 mm were 5, 10, 15, 20, 25 and 30 seconds; while for the ore 1-2 mm: 5, 10 and 20 seconds.
For each grinding product, the sieve analysis was performed and the increments of the analysed particle size classes 0-0,071 mm and 0-0,2 mm were determined.
The aim of the study was to evaluate the efficiency of the grinding process in an electromagnetic mill, depending on the size of the grinding media and the particle size distribution of the feed. The measure of effectiveness in this case was the yield of fine products – below 0.071 mm and 0-0.2 mm. The objective of the research was to try to select the appropriate size of the grinding media in order to obtain optimum technological results in the fine particles yield and the obtained fineness in grinding products for variable particle size distribution of the feed.

4 Results of investigations

The obtained particle size of the ore from the electromagnetic mill milling products is shown in Tables 1-3.

Fig. 4. Material and grinding media after grinding in electromagnetic mill [by M. Wołosiewicz-Głąb).

Table 1. Particle size distribution of the ore grinding products in the electromagnetic mill for the feed size 1-2 mm.

d [mm]	Grinding media 10/1				Grinding media 12/1,5				Grinding media 12/2				Mixture of grinding media			
	Feed	t=5 sec	t=10 sec	t=20 sec	Feed	t=5 sec	t=10 sec	t=20 sec	Feed	t=5 sec	t=10 sec	t=20 sec	Feed	t=5 sec	t=10 sec	t=20 sec
	ϕ(d) [%]	ϕ(d) [%]	ϕ(d) [%]	ϕ(d) [%]	ϕ(d) [%]	ϕ(d) [%]	ϕ(d) [%]	ϕ(d) [%]	ϕ(d) [%]	ϕ(d) [%]	ϕ(d) [%]	ϕ(d) [%]	ϕ(d) [%]	ϕ(d) [%]	ϕ(d) [%]	ϕ(d) [%]
0	0	0	0	0	0	0	0	0	0	0	0	0	0	0	0	0
0.071	0	20.3	31.6	46.1	0	16.7	35.0	54.6	0	17.1	33.2	56.1	0	20.5	39.9	53.3
0.16	0	22.5	33.7	48.1	0	21.2	40.0	59.7	0	21.5	39.0	64.6	0	24.3	43.6	59.2
0.2	0	23.3	34.5	48.7	0	22.4	41.7	61.1	0	23.3	41.5	67.1	0	25.6	45.2	60.4
0.4	0	26.4	37.3	50.6	0	27.6	48.2	65.5	0	29.9	50.5	74.2	0	30.5	50.0	64.0
0.63	0	29.3	40.4	52.7	0	33.5	54.2	68.7	0	36.8	59.0	78.9	0	35.9	55.5	67.3
1	0	39.9	48.8	58.5	0	44.3	64.2	75.3	0	50.3	69.9	83.9	0	47.4	62.9	73.1
2	100	100	100	100	100	100	100	100	100	100	100	100	100	100	100	100

 DOI:10.1051/e3sconf/20171801012

Table 2. Particle size distribution of the ore grinding products in the electromagnetic mill for the feed size 0-1 mm.

d [mm]	Grinding media 10/1							Grinding media 12/1,5						
	Feed	t=5 sec	t=10 sec	t=15 sec	t=20 sec	t=25 sec	t=30 sec	Feed	t=5 sec	t=10 sec	t=15 sec	t=20 sec	t=25 sec	t=30 sec
	ϕ(d) [%]	ϕ(d) [%]	ϕ(d) [%]	ϕ(d) [%]	ϕ(d) [%]	ϕ(d) [%]	ϕ(d) [%]	ϕ(d) [%]	ϕ(d) [%]	ϕ(d) [%]	ϕ(d) [%]	ϕ(d) [%]	ϕ(d) [%]	ϕ(d) [%]
0	0	0	0	0	0	0	0	0	0	0	0	0	0	0
0.071	52.5	83.5	96.2	99.8	100.0	100.0	100.0	54.9	77.0	81.5	95.5	99.5	99.9	100.0
0.12	58.0	88.1	98.8	100.0	100.0	100.0	100.0	59.3	82.5	88.4	99.8	100.0	100.0	100.0
0.2	64.1	94.5	99.9	100.0	100.0	100.0	100.0	67.4	89.2	95.2	100.0	100.0	100.0	100.0
0.4	80.2	100.0	100.0	100.0	100.0	100.0	100.0	97.8	99.4	100.0	100.0	100.0	100.0	100.0
0.63	96.8	100.0	100.0	100.0	100.0	100.0	100.0	100.0	100.0	100.0	100.0	100.0	100.0	100.0
1	100.0	100.0	100.0	100.0	100.0	100.0	100.0	100.0	100.0	100.0	100.0	100.0	100.0	100.0
d [mm]	Grinding media 12/2							Mixture of grinding media						
	Feed	t=5 sec	t=10 sec	t=15 sec	t=20 sec	t=25 sec	t=30 sec	Feed	t=5 sec	t=10 sec	t=15 sec	t=20 sec	t=25 sec	t=30 sec
	ϕ(d) [%]	ϕ(d) [%]	ϕ(d) [%]	ϕ(d) [%]	ϕ(d) [%]	ϕ(d) [%]	ϕ(d) [%]	ϕ(d) [%]	ϕ(d) [%]	ϕ(d) [%]	ϕ(d) [%]	ϕ(d) [%]	ϕ(d) [%]	ϕ(d) [%]
0	0	0	0	0	0	0	0	0	0	0	0	0	0	0
0.071	54.9	67.3	77.5	94.2	98.8	99.5	99.5	54.9	79.2	88.5	98.5	99.9	100.0	100.0
0.12	59.3	73.3	84.2	98.8	100.0	100.0	100.0	59.3	84.2	94.9	100.0	100.0	100.0	100.0
0.2	67.4	82.0	91.8	99.9	100.0	100.0	100.0	67.4	91.5	99.3	100.0	100.0	100.0	100.0
0.4	97.8	99.1	99.8	100.0	100.0	100.0	100.0	97.8	99.7	100.0	100.0	100.0	100.0	100.0
0.63	100.0	100.0	100.0	100.0	100.0	100.0	100.0	100.0	100.0	100.0	100.0	100.0	100.0	100.0
1	100.0	100.0	100.0	100.0	100.0	100.0	100.0	100.0	100.0	100.0	100.0	100.0	100.0	100.0

Table 3. Particle size distribution of the ore grinding products in the electromagnetic mill for the feed size 0-0,5 mm.

d [mm]	Grinding media 10/1						
	Feed	t=5 sec	t=10 sec	t=15 sec	t=20 sec	t=25 sec	t=30 sec
	ϕ(d) [%]	ϕ(d) [%]	ϕ(d) [%]	ϕ(d) [%]	ϕ(d) [%]	ϕ(d) [%]	ϕ(d) [%]
0	60.1	89.9	98.6	99.9	100.0	100.0	100.0
0,071	66.1	95.8	100.0	100.0	100.0	100.0	100.0
0,16	75.7	99.1	100.0	100.0	100.0	100.0	100.0
0,2	97.6	100.0	100.0	100.0	100.0	100.0	100.0
0,4	100.0	100.0	100.0	100.0	100.0	100.0	100.0
0,63	100.0	100.0	100.0	100.0	100.0	100.0	100.0
1	60.1	89.9	98.6	99.9	100.0	100.0	100.0
2	66.1	95.8	100.0	100.0	100.0	100.0	100.0

In order to evaluate the grinding efficiency of the electromagnetic mill for the analysed variable sizes of grinding media and particle size distribution, the increments of the size fraction 0-0.071 mm and 0-0.2 mm were calculated. The results are shown in Figures 5-7.

The results of ore grinding in the electromagnetic mill confirmed the high efficiency of this device in the process of ore comminution – the expected fineness was reached after about 10-15 seconds. Analysis of selected particle size fractions indicates that for fine grained feed with particle size less than 1 mm, the most suitable choice of grinding media are the ones with dimensions 10/1 mm or a mixture of grinding media. For feed with particle size 1-2 mm, the 0-0.071 mm particle size fraction increment for individual grinding media sizes and mixtures was not so unequivocal and depends on the grinding time. For 5 seconds time the most advantageous for obtaining a high increment of this particle size fraction are 10/1 mm grinding media. In case of 0-0,200 mm particle size fraction, the most suitable were the ones with 12/2 mm dimension. The high efficiency of grinding in the electromagnetic mill causes that some of the increments of the analysed particle size fractions reach a maximum value of 100% after 5-15 seconds. This implies the need to control the process in terms of material residence time in working chamber because further grinding results in material fineness below the expected dimension, which is undesirable.

The obtained results indicate the need for optimization of the size of the grinding media or their mixtures, as well as the grinding time depending on the particle size distribution of the feed in order to obtain the desired product characteristics. It is particularly important to determine the residence time of the material in the mill in the context of product comminution and operating costs requirements.

DOI:10.1051/e3sconf/20171801012

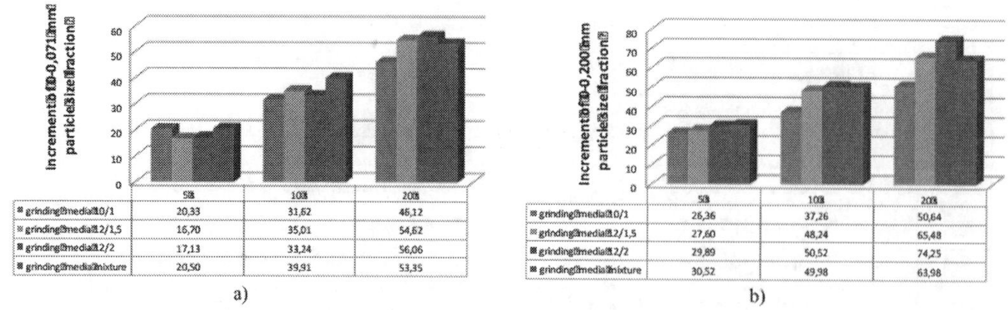

Fig. 5. Increment of 0-0,071 mm and 0-0,2 mm particle size fraction during grinding in electromagnetic grinding mill for the feed particle size 1-2 mm.

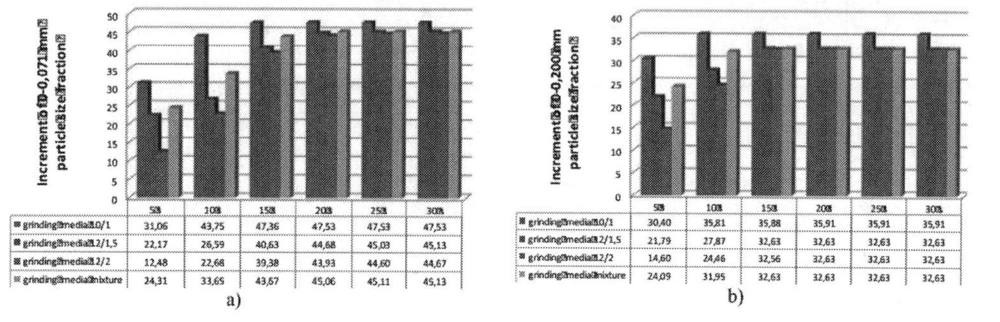

Fig. 6. Increment of 0-0,071 mm and 0-0,2 mm particle size fraction during grinding in electromagnetic grinding mill for the feed particle size 0-1 mm.

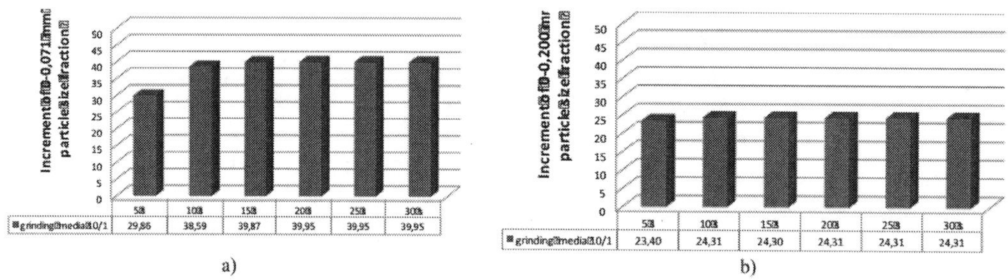

Fig. 7. Increment of 0-0,071 mm and 0-0,2 mm particle size fraction during grinding in electromagnetic grinding mill for the feed particle size 0-0,5 mm.

5 Conclusions

Proper selection of equipment according to feed parameters is crucial in achieving the intended aims of the grinding process. Grinding circuit based on electromagnetic mill offers great opportunities for the process, comminution performance and control of product parameters. The research shows possibilities of additional control of the particle size of products by changing the grinding time and the size of the grinding media. The results of the investigations presented in the article indicate that small grinding media should be used for grinding of fine feed. For feed with particle size less than 1 mm, grinding media should not be greater than 1 x 10 mm.

The tests carried out in this article have been conducted in a laboratory unit with periodic operation and should not be directly translated into an industrial size installation. The device described in this article will allow for a more complex control of the grinding process in order to obtain a product with specific parameters.

The article has been written within the framework of the project: PBS3/B3/28/2015.

References

1. D. Foszcz, T. Gawenda, AGH Journal of Mining and Geoingineering, (AGH University of Science and Technology Press, **36**, 2012)
2. D. Saramak, T. Tumidajski, B. Brożek, T. Gawenda, Z. Naziemiec, Gospodarka Surowcami Mineralnymi Polska Akademia Nauk, **26** (2010)

DOI:10.1051/e3sconf/20171801012

3. T. Tumidajski, E. Kasińska-Pilut, T. Gawenda, Z. Naziemiec, R. Pilut, Gospodarka Surowcami Mineralnymi, Polska Akademia Nauk, **26** (2010)
4. T. Gawenda, *Monograph No. 304* (Wydawnictwo AGH, 2015)
5. M. Wołosiewicz-Głąb, D. Foszcz, T. Gawenda, Mineral Engineering Conference, (Szczwanica, 212–227, 2016)
6. M. Wołosiewicz-Głąb, Sz. Ogonowski, D. Foszcz, *IFAC Symposium On Control, Optimization And Automation In Mining, Mineral And Metal Processing* (Vienna, Austria, 2016)
7. Sz. Ogonowski, Z. Ogonowski, M. Pawełczyk, *21st International Conference MMAR* (2016)
8. D. Saramak, Gospodarka Surowcami Mineralnymi Mineral Resources Management, **28**, 4 (2012)

DOI:10.1051/e3sconf/20171801013

Effect of some operating variables on the performance of a 150 mm Heavy Medium Cyclone treating high ash Indian Coking Coal

Thonangi Gouri Charan[1,*], *G.S.* Jha[1], *K.M.K.* Sinha[1], *U.S.* Chattopadhyay[1], and *Kalyan* Sen[2]

[1]Scientist, CSIR –Central Institute of Mining and Fuel Research, Dhanbad – India
[2]Retried Director, Central Fuel Research Institute, Dhanbad

Abstract. Tests were carried out on a 150 mm heavy medium cyclone test rig treating coal in the size range - 13 +0.5 mm. A total of twenty seven experiments were carried out by varying the parameters like vortex finder diameter, spigot diameter and inlet feed pressure. The effect of these variables on the performance of the 150 mm Heavy Medium Cyclone was investigated. The results indicate that it is possible to achieve about 30.1% clean coal at an ash content of 18.5% at a feed pressure of 10 lb/in2, vortex diameter of 63.75mm and spigot diameter of 31.5mm. The tromp curves obtained from operating the cyclone gave a probable error of 0.04, indicating good separation efficiency.

1 Introduction

The Indian Coking coal constitute about 14% of the total reserves and the domestic availability for steel industry is limited and therefore the Indian Steel industry has to depend heavily on imported coking coal to meet its needs. Currently, domestic steel makers meet 80% of their coking coal requirement through imports. The quantum of imports may go up significantly in the near future as steel production in a large number of new projects is likely to be through the BF-BOF route and also to meet the requirement of existing steel plants.

Coking coal is an essential input for production of Iron & Steel through blast furnace route. To save steel industry facing acute dependence on imported coking coal, domestic availability of coking coal in desired quality has become imperative. The good quality coking coals of the upper seams are fast depleting leaving behind the inferior quality lower seam coal or Low Volatile Coking (LVC) coals. The cleaning of the Indian coals requires crushing to a reasonable size for liberation of ash forming minerals and suitable technology adopted for washing this size fraction is processing them through Heavy Medium Cyclone. Further for processing of the LVC (Low Volatile Coking) coals or lower seam coal, the coal is to be first deshaled and crushed to 13 mm and for processing the coarser fractions (- 13 +0.5 mm),HM Cyclones are the only efficient washers. It has been established worldwide that coal washing of intermediate sizes (- 13 +0.5 mm), HM cyclone is the best separator. (1-4)

Around seventy percent of the existing Indian coal washeries use heavy medium cyclones (HMC) as the main unit operation in the washing circuit. Most of these washeries are old and feed characteristics had drastically changed over the period, as a result it has become difficult to optimize the process which ultimately reduced efficiency. With this view, tests were carried out to study the performance of 150 mm heavy medium cyclone test rig treating LVC coal in the size range -13+0.5 mm.

2 Experimentation

The LVC coal from operating coking coal mines of Jharia Coalfields was taken for the study. For beneficiation by HMC scheme, the "as received" sample was crushed to below 75 mm in a double roll crusher and a representative portion of crushed coal was screened at 13 mm. The screened fraction of -75+13 mm was deshaled at density of 1.80 .The deshaled product of -75+13 mm size fraction was crushed to below 13 mm and crushed product was then mixed with untreated minus 13 mm fraction to form the sample of -13+0 mm.

Fig. 1. H.M. Cyclone Test Rig.

* Corresponding author: gouricharan@yahoo.com

The -13+0 mm size fraction was screened at 0.5 mm. Thereafter, Tests were carried out on a 150 mm heavy medium cyclone test rig treating coal in the size range - 13+0.5 mm. Detailed washability studies were carried out for size - 13 +0.5 mm .The experimental set up of 150 mm diameter H M Cyclone test rig is shown in Figure 1.

The medium of desired specific gravity is prepared in the slurry tank. Finely ground magnetite (95 percent passing through 44 micron) was used to prepare the heavy medium. The medium is fed to the cylindrical vessel and Coal is fed from the top at the rate of desired quantity to maintain pulp density and suitable media to coal ratio. The media and the coal particles are mixed in the slurry tank and the mixture is fed to cyclone. There is a by-pass arrangement also to find out the specific gravity as well as ratio of media and coal fed to the cyclone. The by-pass line in the slurry tank is adjusted to feed the cyclone at definite pressure. The products of the cyclone are passed over launder divided into two parts for cleans and sinks.

The coal particles (cleans and sinks) coated with magnetite are water sprayed, cleaned and collected. The dilute media is collected in a separate tank where the media settles and is reused. The cleans and rejects weights were recorded and analyzed for its ash content and also to study the efficiency of the cyclone float & sink tests on the products were carried out.

Twenty seven experiments were carried out to study the effects of Vortex Finder Diameter (VFD), Spigot diameter (SPD) and Feed Pressure (P) on the quality and quantity of the cyclone products (5). The level of the parameters studies is shown in Table-1, while Table-2 shows the ash content of the cleans and rejects and efficiency data with respect to organic efficiency, d50 and Ep is shown in Table -3.

Table 1. Parameters for operating the cyclone.

Sl.No.	Feed Pressure, lb/in^2	Vortex Finder Diameter, mm	Spigot Diameter, mm
1	5	52.5	31.5
2	7	63.75	42
3	10	75	52.5

3 Results and discussion

From the washability data of coal size 13-0.5 mm the theoretical yield at 18 % ash level is 35.1%, where the corresponding rejects being at ash of 45.2 %.The cut density is 1.59 and the NGM at this gravity is 31 %, which categorize the coal as difficult-to-wash coal.

Table 2. Experimental data for 150 mm diameter H.M. Cyclone treating coal of size - 13 +0.5 mm.

Test	Pres sure	Vortex	Spig ot	Cleans		Rejects	
No.	lb/in 2	mm	mm	Wt %	Ash %	Wt %	Ash %
1	5	52.5	31.5	59.7	34.3	40.3	32.5
2	7	52.5	31.5	46.0	30.4	54.0	36.8
3	10	52.5	31.5	62.0	31.3	38.0	37.7
4	5	52.5	42	8.4	19.0	91.6	35.6
5	7	52.5	42	7.9	17.9	92.1	36.0
6	10	52.5	42	19.8	25.8	80.2	36.5
7	5	52.5	52.5	3.5	21.1	96.5	34.7
8	7	52.5	52.5	4.6	19.1	95.4	35.0
9	10	52.5	52.5	4.0	19.0	96.0	35.3
10	5	63.75	31.5	19.1	21.2	80.9	38.0
11	7	63.75	31.5	19.1	22.7	80.9	36.4
12	10	63.75	31.5	30.1	18.5	69.9	40.5
13	5	63.75	42	6.4	15.4	93.6	35.0
14	7	63.75	42	5.3	14.3	94.7	35.0
15	10	63.75	42	3.1	11.8	96.9	34.6
16	5	63.75	52.5	3.8	12.5	96.2	37.0
17	7	63.75	52.5	3.9	12.8	96.1	34.9
18	10	63.75	52.5	3.0	12.5	97.0	37.0
19	5	75	31.5	31.1	24.6	68.9	39.5
20	7	75	31.5	43.9	27.1	56.1	38.9
21	10	75	31.5	74.7	31.6	25.3	41.2
22	5	75	42	23.3	22.2	76.7	38.0
23	7	75	42	25.9	22.2	74.1	38.0
24	10	75	42	22.6	23.2	77.4	36.5
25	5	75	52.5	13.8	21.8	86.2	36.5
26	7	75	52.5	11.9	18.8	88.1	36.0
27	10	75	52.5	12.9	19.3	87.1	36.1

Based on the findings the cyclone parameters in terms of yield and ash content of the cyclone cleans and sinks at different operating conditions and also considering the efficiency parameters the H. M. Cyclone was standardized and the final parameters which meet the required quality is shown in Table 2.

Table 3. Standardization of H. M. Cyclone Parameters.

Cyclone diameter:	Pressure, lb/in^2(g)	Vortex diameter, mm	Spigot diameter, mm	Cleans wt.%	Cleans ash%	Rejects wt.%	Rejects ash%
150 mm	10	63.75	31.5	30.1	18.5	69.9	40.5

 DOI:10.1051/e3sconf/20171801013

Table 4. Efficiency data for 150 mm diameter H.M. Cyclone treating coal of size - 13 +0.5 mm.

Test No.	Pressure lb/in2	Vortex mm	Spigot mm	Organic Efficiency	d50	Ep
1	5	52.5	31.5	76.70	1.69	0.095
2	7	52.5	31.5	60.00	1.63	0.181
3	10	52.5	31.5	72.10	1.73	0.158
4	5	52.5	42	46.00	1.44	0.035
5	7	52.5	42	55.30	1.45	0.035
6	10	52.5	42	42.30	1.46	0.079
7	5	52.5	52.5	18.40	1.33	0.043
8	7	52.5	52.5	20.90	1.32	0.025
9	10	52.5	52.5	21.30	1.31	0.019
10	5	63.75	31.5	64.30	1.52	0.065
11	7	63.75	31.5	48.80	1.47	0.113
12	**10**	**63.75**	**31.5**	**56.40**	**1.59**	**0.044**
13	5	63.75	42	48.70	1.41	0.057
14	7	63.75	42	48.20	1.40	0.055
15	10	63.75	42	45.30	1.37	0.046
16	5	63.75	52.5	64.80	1.42	0.041
17	7	63.75	52.5	41.80	1.37	0.054
18	10	63.75	52.5	24.70	1.33	0.023
19	5	75	31.5	24.70	1.33	0.023
20	7	75	31.5	70.70	1.59	0.021
21	10	75	31.5	70.70	1.59	0.021
22	5	75	42	65.80	1.54	0.062
23	7	75	42	72.50	1.55	0.048
24	10	75	42	54.30	1.50	0.119
25	5	75	52.5	43.70	1.41	0.087
26	7	75	52.5	54.60	1.44	0.085
27	10	75	52.5	49.00	1.42	0.099

4 Conclusions

Hence, it may be concluded that for recovery of clean coal at desired ash level, the lower seam coal may be deshaled and the deshaled cleans may be crushed to 13 mm, followed by washing of 13 – 0.5mm fraction in 150 mm dia HM Cyclone for optimum recovery. The effect of these variables on the performance of the 150 mm Heavy Medium Cyclone (HMC or HM Cyclone) was investigated. The results indicate that it is possible to achieve about 30.1% clean coal at an ash content of 18.5% at a feed pressure of 10 lb/in^2, vortex diameter of 63.75mm and spigot diameter of 31.5mm. The tromp curves obtained from operating the cyclone gave a probable error of 0.044, indicating good separation efficiency. Since, the NGM of the lower seam coal is very high, it is always beneficial to wash the coal in Heavy medium cyclone. The rejects which is as high as 69.9% at 40.5% ash level can be used for power generation.

References

1. Geological Survey of India (GSI) Report, Government of India. Inventory of Indian Coal Resources, 2014.
2. K. Sen, S.G. Chaudhuri, K.S. Narasimhan, Asia Energy Vision-2020, **15-18**, 137-74, (1991)
3. K. Sen et al "*Multi product beneficiation of inferior coals to user specific products*", XIV International Coal Preparation Congress, (2002)
4. T.Gouri Charan et al " *Washability and Pilot Plant Studies to Generate Bulk Cleans at Desired Qualities from Low Volatile Coking Coal of Jharia Coalfields, BCCL, CIL India*", International Seminar on Coking Coals and Coke Making: Challenges and Opportunities, 219-229 (2009)
5. Project No. MLP-31/2012-13, Report on "Studies on the Effects of design and operating variables of 150 mm & 150 mm diameter Heavy Medium Cyclone for achieving maximum yield at desired quality from difficult-to-wash Indian Coking Coals"

Reuse of Cement Kiln Dust for backfilling and CO_2 carbonation

Marcin Lutyński[1,*], and *Piotr* Pierzyna[1]

[1]Faculty of Mining and Geology, Silesian University of Technology, Gliwice, Poland

Abstract. The study aims to investigate possible alternative paths of reusing Cement Kiln Dust in mining technologies or as mineral sorbent for CO_2 capture. Properties of CKD and bottom slag slurry were assessed and these were ia.: chemical composition, compressive strength and excess water content. Results show that CKD/bottom slag slurry mixed in the proportion of 25%/75% can be used as a backfill material if concentration of contaminants in the leaching tests is at the acceptable level. Second part of the study was devoted to the assessment of CKD as a sorbent in Calcium looping technologies or for mineral carbonation. TGA and DSC study shows that the rate of CO_2 capture (carbonation) is determined by the free CaO content. The highest carbonation rate was within the temperature range of 600-800°C.

1 Introduction

Cement industry generates various waste both in the form of solids (dust) as well as gases (CO_2). This industry accounts for over 5% of global CO_2 emissions [1]. One of the by-products of cement manufacturing is the Cement Kiln Dust (CKD). CKD is created in the kiln during the production of cement clinker. The dust is a particulate mixture of partially calcined and unreacted raw feed, clinker dust and ash, enriched with alkali sulfates, halides and other volatiles. Particle size distribution is irregular as in case of fly ash. Particulates of CKD are captured by the exhaust gases and collected in dust control devices such as cyclones, baghouses and electrostatic precipitators. Each Mg of manufactured cement creates 41 kg of CKD [2]. Taking into consideration global production of cement which was approximately 4200 million Mg of cement in 2016 [3] the CKD accounts for a considerably large share of generated solid waste. Therefore, there is a need to utilize CKD in order to reduce its impact on the environment. In general, CKD consists primarily of calcium carbonate and silicon dioxide which is similar to the cement kiln raw feed, but the amount of alkalies, chloride and sulfate is usually considerably higher in the dust.

High content of calcium oxide is of particular importance when considering CKD as a mineral binder of CO_2. In most of the cases the CKD is reversed back into the kiln however there is still a considerably large amount which is either deposited or beneficially used.

Common uses of CKD include soil stabilization, waste stabilization or solidification, filling of voids in mine reclamation operations, etc. Summary of technologies where CKD can be used is shown in Table 1. The problem with CKD is its variation in chemical composition which depends on the type of operation, dust collection method, type of fuel used and other factors. Nevertheless, in this study an attempt is made to assess the possibility of using CKD in mine reclamation purpose or void filling and to assess the potential of CO_2 mineral carbonation of CKD. This is particularly important given the fact the high energy intensity of cement industry and as a consequence - CO_2 emissions. Recovering of waste such as CKD in underground mine technologies (backfilling, void filling) helps to reduce the environmental impact both as a result of removing fine particulate from the atmosphere but also by reducing the impact of mining (surface subsidence reduction). The same refers to mineral carbonation where gaseous waste (CO_2) is combined with a solid waste, in this case CKD, forming non- reactive solid which can be also used for backfilling [4].

Table 1. Beneficial uses of Cement Kiln Dust (CKD) [5, 6].

Current uses	Potential uses
Soil stabilization	Permeable Reactive Barrier filling
Waste stabilization/solidification	Heavy metal adsorption
Cement additive / blending	Fertilizer production
Landfill liner	CO_2 mineral carbonation
Wastewater neutralization	
Road foundation	
Concrete products	
Mine reclamation	

Since CKD contains high content of unreacted CaO under high temperature the CaO should react with CO_2 in carbonation-calcination thermochemical reaction:

$$CaO\ (g) + CO_2(g) = CaCO_3\ (s) \qquad (1)$$

This reaction has been thoroughly studied mostly for the CO_2 capture processes [7,8] and in the calcium looping systems (CaL) [9].

* Corresponding author: marcin.lutynski@polsl.pl

 DOI:10.1051/e3sconf/20171801014

2 Materials and methods

In this study a sample of Cement Kiln Dust from one of the cement plants from the East of Poland which is representative for a typical cement manufacturing process in Poland was selected.

Two separate analysis were carried out, first the CKD was assessed in terms of its suitability in reusing it as a backfill material for underground void filling. For this case CKD has to be mixed with other silicate material which increases the strength of a mixture. CKD acts as an activator and a CaO supplier. For this purpose a bottom slag (or sand from fluidized bed) was chosen. Fluidized bed sands are a waste in accordance with the European Union waste catalogue with a code no. 10 01 24.

Chemical composition of bottom slag and CKD used in the study is shown in Fig. 2.

Table 2. Chemical composition of CKD and bottom slag.

Constituents	CKD	Bottom slag 10 01 24
	(% m/m)	(% m/m)
SiO_2	13.7	52.4
Al_2O_3	3.4	18.5
Fe_2O_3	1.1	5.9
CaO	48.7	11.2
(including free) CaO	14.3	4.2
MgO	0.4	0.9
Na_2O	2.5	0.7
K_2O	16.0	1.6
SO_3	8.6	4.9
TiO_2	0.1	0.8
P_2O_5	0.2	0.1
SrO	0.1	1.1
ZnO	0.1	0.2
Chloride	0.4	0.01
Ignition losses	6.3	3.7

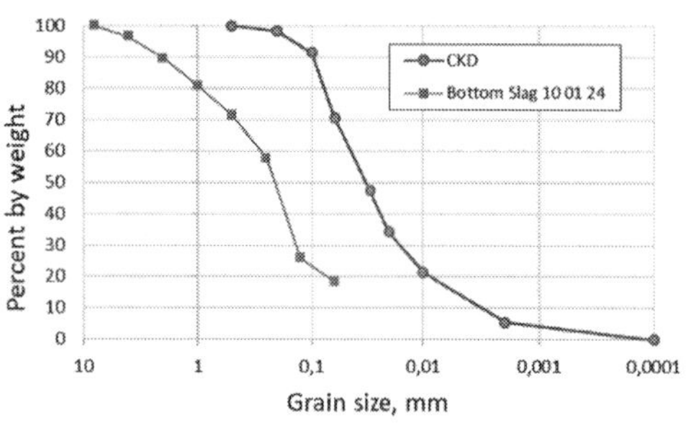

Fig. 1. Particle size analysis of materials used in the study.

Since the mixture of CKD and bottom slag is to be used for the purpose of backfilling the content of trace metal elements has to be analyzed. In Fig. 3 metal trace elements content both for the CKD and bottom slag is shown.

Particle size distribution of both materials is shown in Fig. 1. Particle size analysis of CKD were carried out with light scattering particle size analyser Micromeritics Saturn DigiSizer II Mo. 5205, for the bottom slag the analysis were done with the use of a conventional sieve analysis.

Table 3. Metal trace elements content in CKD and bottom slag.

Element	CKD	Bottom slag 10 01 24
	(ppm)	(ppm)
Ag	<2	< 2
As	<2	< 2
Ba	97	284
Cd	241	< 2
Cr	14	73
Cu	186	521
Mo	<2	5
Mn	85	287
Ni	16	29
Pb	2174	98
Rb	1042	87
Sb	44	< 2
Sr	829	256
V	9	117
Zn	1423	203

Particle size distribution curve shows that the particle size distribution in CKD is similar to that of a raw cement. This was also observed by Wang et al. [10]. Particle size analysis show that the distribution is irregular and is comprised mostly of sizes 0,0001 – 0,1. The uneven distribution of grains and low sphericity of grains as proven by Ahmari and Zhang [11] (see Fig. 2) causes high water demand when preparing the slurry.

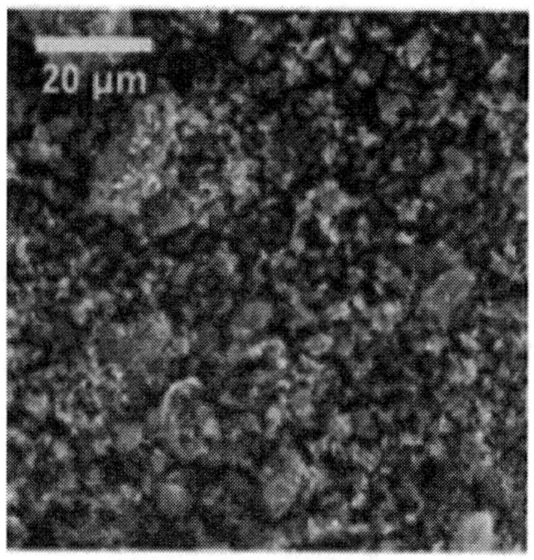

Fig. 2. SEM image of Cement Kiln Dust [11].

CKD and bottom slag were mixed in the proportion 25%/75% and tested in accordance with the polish standard PN-G-11011:1998 "Materials for self-solidifying backfill and gob grouting. Materials and testing". The following properties were tested:

- Spread of a slurry – selection of W/S ratio
- Excess water

- Compressive strength
- Soaking resistance

Other properties such as compressibility, permeability and curing time were not included in the tests since these parameters are determined for a specific application. The compressive strength tests were conducted in accordance with the standard PN-EN 196 1:2006. „Methods of testing cement - Part 1: Determination of strength". Compressive strength was tested on prismatic test specimens 40x40x160 mm in size. Example specimen is shown in Fig. 3. For comparison same test was carried out for bottom slag without the addition of CKD. The W/S ratio of a slurry was chosen to match the spread of 180 mm. The spread of 180 mm of a slurry provides proper hydraulic transportability with minimum excess water [12]. Specimens were cured for 28 days in a climatic chamber in order to simulate in-situ conditions of a mine i.e. 90% humidity and temperature of 22°.

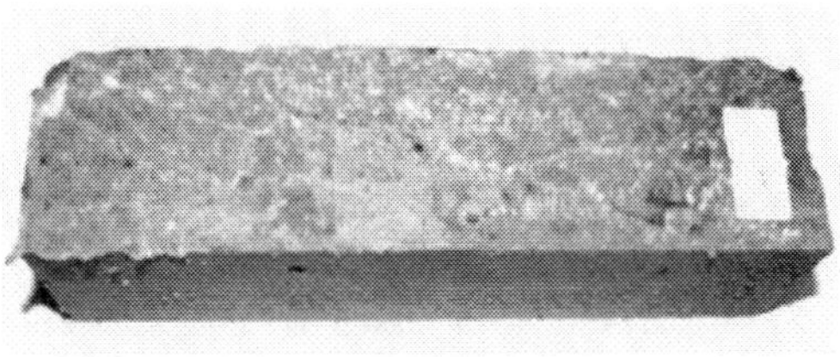

Fig. 3. Specimen prepared for a compressive strength test of CKD and bottom slag mixture after 28 days of curing in climatic chamber.

The CO_2 mineral carbonation potential was assessed with the use of thermal analyzer NETZSCH 449 F3 Jupiter. Approximately 20g of sample was used for the tests. The heating rate during the experiment was 10°C/min up to the temperature of 1100°C with consecutive cooling phase down to 40°C with the same rate. During the heating phase a CO_2 flow of 25 ml/min was applied whereas during the cooling stage – an inert gas (N_2) was flowing with the same rate as CO_2. Two analytical methods were used ie. Differential Scanning Calorimetry and Thermogravimetry which allowed to record changes in the weight of the sample.

3 Results

3.1 Backfill material

Results of the test of CKD/bottom slag slurry for the purpose of backfilling are shown in Table 4.

Compressive strength of CKD/bottom slag slurry after 28 days of curing is 4,7 MPa and is much above the minimum requirement for self-solidifying backfill which is 0,5 MPa (according to the PN-G-11011:1998 standard). The soaking resistance (12,4%) is also below the threshold which is specified in the same standard as 20%. The excess water observed after the curing was also considerably low (0,6%) and is much below the standard value for self-solidifying backfill (7%) as well as for gob-grouting (15%). The water to solid ratio corresponds to the spread of the slurry which in accordance with the standard for both self-solidyfing backfill and gob-grouting should be above 90 mm. Since metal trace elements were measured for solid materials (see Tab. 4) in order to transpose these results into real conditions, i.e. after solidifying, there is a need to conduct a leaching test. Therefore, it was decided to conduct an additional leaching test of solidified sample in accordance with the same standard. Result of the leaching test is shown in Table 5.

Table 4. Results of tests of CKD/bottom slag slurry for the purpose of backfilling.

Water/solid ratio (-)	0.32
Spread (mm)	175
Excess water (%)	0.6
Compressive strength after 28 days (MPa)	4.7 (2.1 for bottom slag)
Soaking resistance (%)	12.4

Leaching test shows that the majority of leached out elements (contaminants) is below the threshold for the wastewaters introduced into the soil. The only parameter above the threshold is Potassium (max. 80 mg/dm^3) and Lead (max. 0,5 mg/dm^3) [13]. High contents of Pb and K_2O are observed in CKD rather than bottom slag which implies that other slurry proportion eg. 10% CKD and 90% of bottom slag would probably dilute and immobilize sufficiently these elements. Since CKD acts as an activator and increases overall strength of the m ixture the mechanical strength is much above the minimum value of 0,5 MPa and most probably will be enough with the addition of 90% of bottom slag. This indicates that CKD/bottom slag slurries could be successfully used as the self-solidifying backfill and/or as a gob-grout. This would be a safe and efficient method of reusing this waste.

Table 5. Results of CKD/bottom slag slurry leaching test.

Parameter	Value
Cd (mg/dm^3)	< 0.005
Cr (mg/dm^3)	< 0.02
Zn (mg/dm^3)	0.08
Mg (mg/dm^3)	< 0.5
Mn (mg/dm^3)	< 0.02
Cu (mg/dm^3)	< 0.02
Ni (mg/dm^3)	< 0.02
Pb (mg/dm^3)	0.66
K (mg/dm^3)	1 380
Na (mg/dm^3)	143
Sr (mg/dm^3)	1.9
Ca (mg/dm^3)	204
Fe (mg/dm^3)	0.5
Chlorides (mg/dm^3)	20
Sulphates (mg/dm^3)	215
pH	12

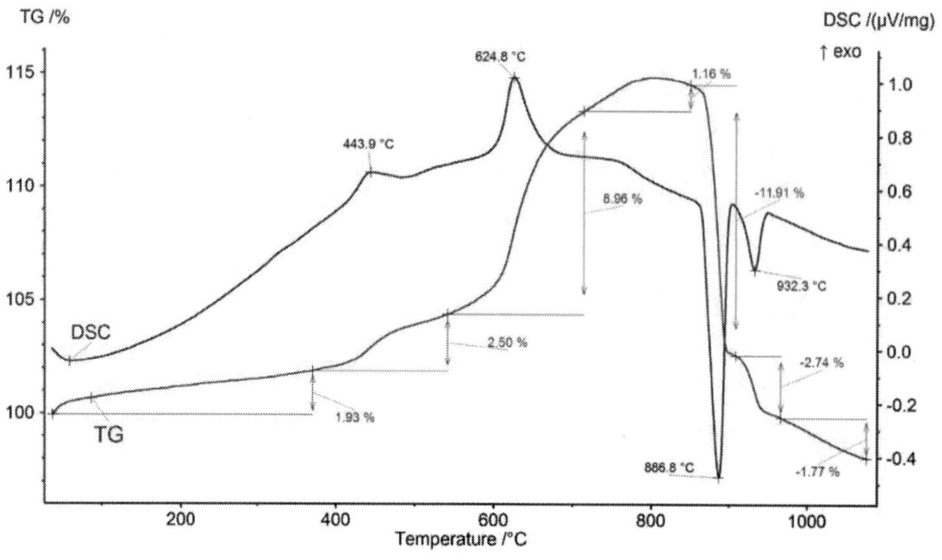

Fig. 4. TGA and DSC analysis of CKD sample.

3.2 CO_2 mineral carbonation

Thermogravimetric and DSC analysis results are shown in Fig. 4. As it was mentioned, the maximum temperature of the experiment was 1100°C and it was sufficient to see the decomposition of carbonates. The TGA curve increases by 1,93% up to the temperature of approximately 480°C which probably indicates physical sorption of CO_2 on CKD particles (slight increase in DSC curve). The second and third stage of mass growth up to the temperature of approximately 850°C indicates carbonation with two maxima at 443,9°C and 624,8°C. Above 800°C a rapid two stage decomposition of calcium carbonate (calcination) is observed with two maxima at the temperature of 886,8°C and 932,3°C. The total mass uptake by the sorbent is 14,55% which corresponds to the free CaO content in the CKD (see Tab. 2). This indicates that CKD could be potentially used for carbon looping technologies or as a mineral sorbent where CO_2 forms $CaCO_3$ from free CaO making the CKD less reactive and inert to the environment. Temperature of carbonation and calcination observed in the experiment are similar to those in other studies on limestone or fly ash [14–16]. Further tests need to be conducted in order to obtain more information on the regeneration of the sorbent or its properties after carbonation.

4 Conclusions

In this study and attempt was made to show alternative paths of Cement Kiln Dust reuse in mining technologies (as a backfill) and as a mineral adsorbent of CO_2 in capture technologies. The following conclusions can be drawn:

- CKD has a relatively high content of CaO which acts as an activator when mixed with silicate materials increasing their compressive strength,
- CKD/bottom slag slurry can be used as a backfill or gob grouting material in underground mining if the chemical leaching test results are positive. Both compressive strength and excess water content allow to use the tested slurry as a backfill material. The concentration of contaminants in the leachate is at the acceptable level except Pb and K.
- The rate of CO_2 capture (carbonation) is determined by the free CaO content. The highest carbonation rate is within the temperature range of 600-800°C.
- Additional studies are necessary to observe the regeneration rate of the sorbent and decrease in the carbonation capacity

References

1 E. Worrell, L. Price, N. Martin, C. Hendriks, L.O. Meida, Annu. Rev. Energy Environ. **26,** 303–329 (2001)

2 A.H. Sulaymon, A.A. Faisal, Q.M. Khaliefa, J. Hazard. Mater. **297,** 160–172 (2015)

3 www.statista.com

4 P. Buryan, P. Donat, S. Buryan, Inż. Miner. **16** (2015)

5 W.S. Adaska, D.H. Taubert, others, in:, Cem. Ind. Tech. Conf. Rec. 2008 IEEE, IEEE, 210–228 (2008)

6 M.-J. Kim, S.Y. Pak, D. Kim, S. Jung, KSCE J. Civ. Eng. **21,** 629–635 (2017)

7 J.C. Abanades, E.J. Anthony, J. Wang, J.E. Oakey, Environ. Sci. Technol. **39,** 2861–2866 (2005)

8 V. Nikulshina, C. Gebald, A. Steinfeld, Chem. Eng. J. **146**, 244–248 (2009)

9 A. Perejón, L.M. Romeo, Y. Lara, P. Lisbona, A. Martínez, J.M. Valverde, Appl. Energy **162**, 787–807 (2016)

10 K. Wang, A. Mishulovich, S.P. Shah, J. Mater. Civ. Eng. **19**, 112–119 (2007)

11 S. Ahmari, L. Zhang, Constr. Build. Mater. **40**, 1002–1011 (2013)

12 P. Mazurkiewicz, Z. Piotrowski, A. Tajduś, *Lokowanie Odpadów w Kopalniach Podziemnych*, (Kraków, 1997)

13 Regulation of the Minister of Environment of 24 July 2006 on Conditions to Be Met for the Introduction of Sewage into the Water or Soil, and on Substances Particularly Harmful to the Aquatic Environment (Journal of Laws No 137, Item 984, as Amended)

14 V. Nikulshina, C. Gebald, A. Steinfeld, Chem. Eng. J. **146**, 244–248 (2009)

15 D.N. Huntzinger, J.S. Gierke, L.L. Sutter, S.K. Kawatra, T.C. Eisele, J. Hazard. Mater. **168**, 31–37 (2009)

16 M. Erans, V. Manovic, E.J. Anthony, Appl. Energy **180**, 722–742 (2016)

 DOI:10.1051/e3sconf/20171801015

An update on possibilities of metals recovery from Polish copper ores by biotechnology. Project Ecometals

Agnieszka Szubert[1*], *Anne-Gwénaëlle* Guezennec[2], *Françoise* Bodénan[2], *Stefan* Dirlich[3], *Agnieszka* Pawłowska[4], *Andrzej* Grotowski[1], *Zygmunt* Sadowski[4], and *Kajetan* Witecki[1]

[1]KGHM Cuprum Sp. z o.o. Research and Development Center, Environmental Protection and Processing Department, 53-659 Wroclaw, Poland
[2]Bureau de Recherches Géologiques et Minières - BRGM, 45060 Orléans Cedex, France
[3]Helmholtz-Zentrum Dresden-Rossendorf, Helmholtz-Institut Freiberg für Ressourcentechnologie, 09599 Freiberg, Germany
[4]Wrocław University of Science and Technology, Faculty of Chemistry, Department of Chemical Engineering, 50-373 Wroclaw, Poland

Abstract. The possibilities of metals recovery from copper ores with the biotechnological methods are widely known. The methods consist in bioleaching of copper ores, copper concentrates and byproducts of their production, as well as metals recovery from leaching solutions. Biohydrometallurgical methods were tested for years to be applied at KGHM Polska Miedź S.A., in order to improve efficiency of copper production. Several different research units worked on the topic, and the most significant and wide range initiatives in this area are mentioned in this article. One of the initiatives is an ongoing German and French Ecometals project. KGHM Polska Miedź S.A. and KGHM Cuprum Ltd. Research and Development Centre are this project Partners. In the frame of the project different metals bearing materials (ores, concentrates and tailings) are tested. Among them three lithological types of the copper ore from Rudna mine and the copper concentrate from Lubin concentrator are used for studies. The article gives a general overview of these activities, with the main focus on results of bioleaching studies of selected materials, conducted by KGHM Cuprum. In these studies sandstone and shale, as well as so called "shale concentrate" (containing 39% of the shale) were used for experiments, and possibilities of their bioleaching were evaluated.

1 Possibilities of biohydrometallugical methods application for treatment of Polish copper ores. A state of the art

The methods of metals recovery from copper ores with the biotechnological methods are widely known. The methods consist in bioleaching of coper ores, concentrates or by-products of the concentrates production processes, followed by metals recovery from leaching solutions.

Biohydrometallurgical methods, and especially bioleaching, were tested for years to be applied for the Polish copper ores, as well as for by-products and concentrates produced at KGHM Polska Miedź S.A., in order to improve efficiency of copper production. Several different research units have been working on the topic.

A special focus was given to ores and to the low grade concentrate from the area of the Lubin mine. At the end of the seventies, a test of an underground bacterial leaching of a sandstone ore was tried in the Lubin mine. The principle of the underground test was to leach the ore in an underground installation located in a working of the mine, without transportation on surface [1]. The principle of the underground test did not consist in leaching the rock directly into the deposit, but in its disassembling from the body of the rock and placing it in a tight, enclosed dam built in the designated ramp in the working area for that purpose. The ore in the bacterial leaching dam was immersed in a circulating bath of sulfuric acid solution, flowing from the dam to a settler. The copper ore leached at that time was mainly sandstone of 1.8% copper grade. No detailed information about the type of bacteria used is available.

In the nineties, an extensive work of possibilities of application of hydrometallurgical methods at KGHM, aiming to increase the efficiency of the mining and processing was carried out [2], considering both acidic and alkaline conditions of the process. It was not the first work which pointed out that due to the nature of Polish copper ore, the use of leaching in an acidic environment may be difficult to be justified, both from technical and economical point of view. This applies not only to the dolomitic or the shale ores, but also to the sandstone ore, in which the content of carbonate minerals is still very high. Therefore, other solutions of leaching under alkaline conditions were also analyzed. An example was a cyanide leaching which allows not only for copper, but also for silver recovery. However, because of a toxicity of this agent and necessity of its regeneration after the leaching – contrary to gold - its application was not feasible. A detailed analysis in this matter was confirmed by a concept of the leaching method of flotation tailing from

* Corresponding author: aszubert@cuprum.wroc.pl

the old "Konrad" mine, elaborated at ZBiPM Cuprum [3]. As another example, an ammonia leaching was given. At the end of seventies, and then at nineteens, the ammonia leaching was proposed by the Institute of Nonferrous Metals (IMN) in Gliwice, and discussed as a method of treatment of a by-product of the copper ore flotation process at KGHM. The method allowed for copper recovery from the selected by-products, without an influence on copper recovery, but without possibility of silver recovery.

Also ferric iron leaching methods, both in a sulfuric acid and in a hydrochloric acid environment, were analysed and studied (like Clear and Cymmet methods, or the concept proposed by Wroclaw University of Technology), including bioleaching method, which was a rising technology at that time. As an interesting method, also leaching in nitric with an addition of the sulfuric acids was analysed.

Studies on possibilities of application of the hydrometallurgical and biohydrometallugical methods have been also carried out over the last fifteen years. A possibility of an application of atmospheric and pressure leaching methods for the copper concentrate processing have been investigated at the Wroclaw University of Technology with using of the copper concentrates and by-products from the Lubin concentrator (Project HYDRO supported by National Centre for Research and Development, Poland) [4].

In 2007, a three years international *BIOSHALE* project has been ended. It was funded by European Commisision in the frame of 6th Framework Programme. In the frame of this project, different black shale bearing raw materials were tested. As the main one, the Lubin copper concentrate was analysed to be processed by the bioleaching method [5]. The method was developed by BRGM (France) and the metals recovery (Cu, Ag, Pb) method from PLS (pregnant leaching solution), by Technicas Reunidas (Spain), with cooperation of the project partners. Preliminary results indicated that after six days more than 90% of copper and more than 85% of silver can be extracted from this concentrate using a continuous stirred tank reactors technology (CSTR) [5]. The solid/liquid ratio applied in the bioleaching circuit was 15%. As one of the main results of this project, a general concept of the bioleaching technology for the Lubin copper concentrate was proposed. Within this concept, the bioleaching of the Lubin concentrate in cascade of the stirred tanks bioreactor was followed by copper recovery from the leaching solution via solvent extraction and electrowinning (SX-EW), and then by hydrometallurgical silver recovery from a solid residue after the process.

The research studies on Lubin concentrate CSTR technology were continued in the next European project *PROMINE*, carried out in the frame of the 7th Framework Programme of the UE. As an effect, improvements of the process in terms of S/L ratio, as well as in PLS treatment methods were proposed. The solid/liquid ratio was increased to 25% without an effect on the process efficiency after the same residence time. It was shown that after 4.7 days of a residence time of the solids in the bioreactors, the copper extraction rate reached 87% [6,7]. The increase of the S/L ratio allowed for reducing the capacity of the bioreactors. As a result, higher copper content in the leaching solution could be obtained. This allowed for a direct use of the electrowinning process for copper recovery from the leaching solution, without necessity of application of the solvent extraction unit. A new flowsheet of the PLS treatment was proposed by Institute of Nonferrous Metals (IMN) (Gliwice, Poland) [7].

Based on the analyses of results of the above-mentioned attempts, it is possible to remark that a range of tools and agents possible to be applied for leaching process did not change over the last decades. It is also difficult to update a list of the potential substrates selected at KGHM's mine-concentrator circuits of in the past years as the most perspective for application of the hydrometallurgical methods, i.e. Lubin copper concentrate, the by-products of the flotation process or the sandstone ores. Although results of the leaching tests were promising, a complex concept of the biohydrometallugical technology, including full economic analysis, of the selected materials has not been elaborated so far. As results of the biohydrometallugical projects showed the feasibility of bioleaching by autotrophic microorganism at small and pilot scale, it encourages for further studies on possibilities of application of the methods, and for studies on possibilities of application of the methods, not only for Polish copper ores but taking into account possibilities of development of prospective European deposits.

Currently, a challenging project aiming at evaluation of application possibilities of the bioleaching method for in-situ leaching of copper ores is carried out in the frame of an European BioMOre project (Horizon2020 programme) [8]. The new method has a potential to be implemented, especially for deposits which cannot be mined with traditional mining technologies because of technical and/or economic reasons. This potential will be evaluated as the final result of the project, also based on results of a pilot test, being prepared underground, in Rudna mine (KGHM Polska Miedź S.A.).

The second example of the currently ongoing project focused on application of biohydrometallurgical methods for European copper resources is the *ECOMETALS* project. The main challenges, current results and goals of the Ecometals project together with examples of laboratory results of bioleaching of the selected raw materials obtained from KGHM Polska Miedź S.A., are presented in this article.

2. Ecometals project

The Ecometals project is a German and French project, with participation of KGHM Polska Miedź S.A. and KGHM Cuprum Ltd. Research and Development Centre. In the frame of the project different metals bearing materials (ores, concentrates and mineral waste) are tested to evaluate possibilities of metals recovery with using biotechnological methods. The main focus put on the copper shale – like raw materials, originating from a sediment hosted copper deposit (known as

Kupferschiefer), which previously was exploited in Germany. Currently the shale copper ore are mined in Poland but a potential for its exploration and their future exploitation in this area still exits. The main German areas of the deposit considered in the project were the Mansfelder Land, Spremberg and the region of Weißwasser [9]. Apart from mineral samples from Germany, the main samples tested in the project were the copper ore samples from Rudna mine (KGHM Polska Miedź S.A) and the Lubin copper concentrate.

The project is structured into six work packages and involves 17 Partners. The first three WP's deal with the identified scientific bottlenecks and case studies on primary and secondary resources. WP4 supports the three aforementioned WP's with respect to analyses, characterization of the resources as well as with the development of appropriate methodologies. WP5 aims at an integration of the processes as well as the assessment of the sustainability of the new methods proposed within the project [9].

In the frame of WP1 the above-mentioned potential areas of a future interest were characterized. Also studies on the pretreatment of the copper ores in in order to prepare potential substrates for bioleaching were carried out. The studies used traditional concentration methods (flotation) but also other methods being developed over last years, like an optical sorting of the organic rich material were applied. In WP2, a further progress in studies on efficiency and mechanisms of the bioleaching process was done. This is valuable especially in case of the bioleaching with the CSTR method, which industrial use for treatment of the copper ores is still very limited. In these terms, a multi-scale approach from molecular techniques [10] to bench-scale small pilot tests was used to demonstrate bioleaching feasibility for Cu recovery from the *Kupferschiefer* ores. Heterotrophic strains were studied by HZDR (Germany) while BRGM (France) and BGR (Germany) focused on autotrophic bacteria and optimization of operating parameters. It was shown that sparkling the bioreactors with oxygen enriched gas instead of air offers an interesting alternative process to provide an adequate oxygen supply in order to satisfy the oxygen demand, which allow operating at higher solid load [11]. Besides O_2 enrichment, this study demonstrated that CO_2 supplementation is required to avoid limitation of the leaching kinetics [12]. Major achievements were also carried out regarding the optimization of the temperature: temperature increase in successive bioleaching reactors improves significantly copper recovery compared to conventional bioleaching operated at constant temperature [13]. Finally, autotrophic bioleaching of the copper concentrate from Lubin concentrator was performed in an innovative pilot bioreactor (2 m^3) at BRGM (France) (see Figure 1.) to produce pregnant leach solution (PLS) for further metal recovery testing at GEOS (Germany).

Results are currently analysed to propose an integrated concept of bio-treatment for the copper resources (WP3). Such a preliminary concept was already proposed and a first environmental evaluation of the concept was carried out by Cuprum (WP5), indicating the main environmental aspects of the process i.e. a storage of the solid waste from the process, or an energy consumption, as the major ones. This will be verified in the final evaluation of the feasibility of the process, at the end of the project.

Fig. 1. 2m^3 tank for the bioleaching of the Lubin Cu concentrate pilot test at BRGM. Source: BRGM, France.

3. Bioleaching experiments

One of KGHM Cuprum's tasks in the Ecometals project consisted in bioleaching of selected copper bearing materials from KGHM. The main focus was given to a so called "shale concentrate" (containing about 39% of shale), which was obtained from the optical sorting pilot tests of the copper ore from Rudna concentrator. This pilot test was carried out at Tomra company (Germany). The studies on possibilities of the optical sorting of the feed for the Rudna concentrator aiming in separation of the shale fraction from other lithological types of the copper ore mined at Rudna mine were planned and carried out by KGHM Cuprum within WP1.

Currently, a method of separation of the copper shale from other lithological types of the copper ore mined at KGHM does not exist. The sorting test showed that a separation of the shale material from other types of rocks is possible but the method requires further improvements to obtain a satisfying product. The idea of the separation of shale was considered as attractive because of the possibility to recover easily base and other valuable metals concentrated in this type of the ore, and because of a potential possibility of improving the enrichment factors at existing concentrator by separating this refractory type of ore from the circuit. A separation of shale ore from the feed was also supposed to separate the high organic content material from the technological circuit, and thus limiting a necessity of its removal in the smelter, prior to its processing in the flash furnaces. In addition, the separation would also limit the amount of concentrates processed in the smelter and thus reduce SO_2 and dust emissions to the atmosphere. However, if a concentrate of 100% shale content is obtained from the sorting, none of the methods currently used at KGHM would be suitable for its further processing. Therefore, the (bio-) hydrometallurgical methods may be the only possible ones to be applied for metals recovery from this concentrate. Depending on the shale fraction, copper and valuable metals contents in the shale concentrate, two possible ways of its treatment can be considered, i.e. bioleaching with the CSTR method or on a heap. In the case of a significant silver content, like for Polish ores, a bioleaching in tank reactors seems to be more suitable, allowing for this metal recovery from the bioleaching residue. If the content of metals, which are not leachable in sulfuric acid environment is not significant, a heap leaching option could be considered. Such option is also preferred for the raw material characterized by coarse particles, which is the case for the product of the optical sorting process. Depending on characteristics of product, only a preliminary comminution of the ore can be required, for better performance of the heap (see Figure 2).

In the view of the possibility of obtaining a suitable shale concentrate in the future, examples of results of the bioleaching laboratory experiments carried out both with using the shale the mine, and with the shale concentrate, are shortly presented.

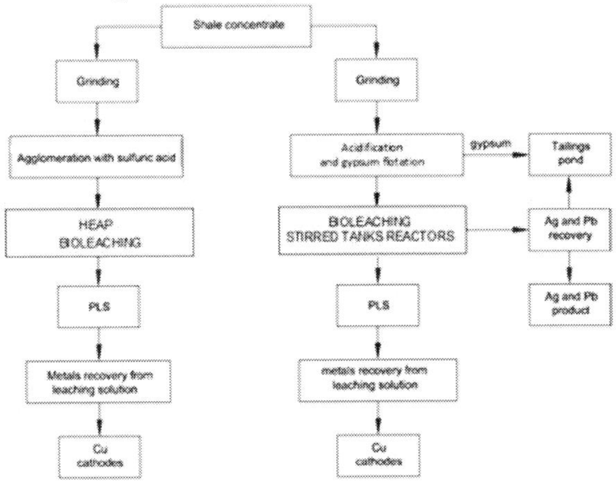

Fig. 2. Possibilities of processing of the shale concentrate.

3.1. Bioleaching of the shale ore

Column and bath bioleaching tests of the shale ore were carried out. The bath tests were done in 500 ml Erlenmeyer flasks, with using ground samples, at at solid concentration of 20% (by weight). A control sample (without microorganisms) was prepared as well: in the controls, the pH was maintained at <2. The process was monitored by pH and redox potential measurements. Also, iron(II), iron(III) and copper(II) concentration were analyzed. For column tests the shale (10-40 mm) from Rudna mine was used. The content of shale used for the column test is presented in Table 1.

Table 1. Chemical composition of the shale ore sample.

Cu	C_{total}	C_{org}	Fe	Cl	CaO	MgO	Co	Ni	As
%							g/Mg		
6.96	12.10	8.23	1.68	0.110	11.6	4.4	194	337	133

The material was placed in a column with dimensions 10/60 cm. In the first step, washing of the feed was carried out with sulfuric acid solution until a stable pH conditions (i.e. <2.0) were obtained. During this time, chlorides content in the solution was also checked, as too chlorides concentration is harmful for the bacteria. The solution flowrate at this stage was higher than during bioleaching (11.3 l/h). The washing of the shale was performed during 26 days. After that period the solution was partially removed from the circuit and the bacterial culture was added to the system. During bioleaching, the flow rates of the solution were smaller, to simulate an assumed flowrates applied on heaps (8 -12 l/m^2/h): during the last ten days of the test the flowrate was twice increased, as the wetting efficiency appeared to be too low.

The leaching solution containing microorganisms was heated (37.5 °C) and aerated in the collection tank, below the column. Aeration of the fixed bed in the column adapted at the beginning caused too intensive evaporation of the solution, so it was given up. During the process concentrations of chloride ions and ions of other metals (Ni, Co, As) were also measured, apart from iron, pH and Eh measurments.

The microorganisms used in the bioleaching experiments were obtained from the project partner: Federal Institute for Geosciences and Natural Resources (BGR), Germany. The medium for the bacteria growth contained basic mineral salts, like magnesium, sodium or ammonia sulfates, nitrate, and phosphate, small amounts of specific elements like a selected metals ions, as well as copper concentrate. After adjusting the pH of the solution to 2.0 the culture was maintained at 40°C.

In the bath tests of the shale control sample, the copper recovery after 27 days was 46.0% compared to the 60.0% of the copper recovery in the samples with the presence of the bacteria. In all flasks containing bacteria, the redox potential increased during the process from about 400 mV to more than 650 mV. In the control samples its value was lower than 450 – 500 mV.

Bioleaching of the shale in the column was carried out for a period of 95 days. For the acidification phase of the shale sample, only 200 ml of concentrated sulfuric acid (approx. 2.5 kg /t) was used for neutralization of carbonates being present in the ore. Such low acid consumption indicated the reaction occurred only on the surface of the ore particles, which was proved by very low MgO and CaO leaching efficiency at the end of the process (17 % and 4,5% respectively). Low efficiency of carbonates dissolution at the acidification stage resulted in difficulties with pH control in the system. This led also to local increases in pH in the column and consequently caused the precipitation of iron (as hydroxides or jarosite) on the surface of the shale particles. Among the precipitates, a presence of elemental sulfur was also identified in the biomass, that accumulated in the collection tank. Copper recovery was very low during such long period of time (6.8% after acidification and 33.4% after bioleaching). It indicated that the solid has to be pre-treated before starting bioleaching in a fixed bed (column/heap), in order to deal with the rock alkalinity and gypsum issues. An efficiency of dissolution of arsenic was similar for copper whereas nickel and cobalt were not leached at all.

3.2. Bioleaching of the shale concentrate

The shale concentrate with low copper grade (slightly higher than 2.0%), organic carbon content of 2.0% (which is approx. 3x higher than in the feed to the concentrator), and a total carbon content of 8.4%. The content of other metals was similar or slightly higher than in the feed being currently directed to the concentrators. The content of the shale in this concentrate was 39.1%: unfortunately the content of dolomitic fraction in this concentrate was very high (>50%).

The shale concentrate was ground to -1.0 mm and bioleached in an aerated, single stirred tank reactor of the working volume of 7 L, at temperature of 35°C and S/L ratio of 25%. The pH of the suspension was maintained at 1.6 – 1.8. At the acidification stage 545 ml (311.0 kg acid /ton of ore) of concentrated sulfuric acid was added to the system to get the appropriate pH for the bioleaching process. After acidification stage, the inoculum was introduced to the system (10% vol). During the process the solution was analyzed for iron (II) and iron (III), copper (II) and if necessary, also for other metals. The pH and redox potential were also measured. The microorganisms used in the bioleaching experiments were obtained from the project partner, BGR (Germany), and prepared in the same way as for the tests with the shale ore.

In the bath tests with using the shale concentrate (-1.0 mm), the copper recovery in the control sample was 63%, compared to the 83% in the samples where the bacteria were added, after 27 days. The redox potential and pH observations were similar for this case as for the tests with the shale ore.

It was found that the copper recovery after the acidification stage was 15%, and after the addition of microbial culture, the recovery increased to almost 50 % after 15 days of the experiment, and then raised to more than 90%. After 41 days (96% of cu recovery) the experiments was stopped.

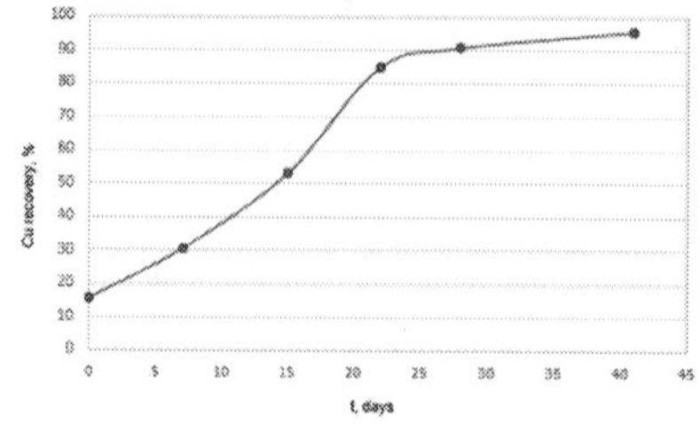

Figure 2. Bioleaching of the shale concentrate in the singe stirred tank reactor.

The bath tests experiments with the shale materials carried out in the frame of the project by Partners differed, depending on specific conditions of the process.

4. Summary

All the studies carried out so far indicate, the biohydrometallurgy can be an option for processing of Polish copper ores and products of their processing, as well as other European copper ores/concentrates from the sediment hosted copper deposit (of the *Kupferschiefer* deposit area). Justification of a future application of the method will depend on a cost-effectiveness and a technological feasibility evaluation of the treatment of a selected feed. The final results of the Ecometals project, based on data from the bioleaching pilot test of the Lublin concentrate will allow to conclude about such possibility. However the alkalinity of these rocks, as well as a high content of salts (NaCl) remain still a challenge with regards to the processing of these raw materials with biohydrometallurgy. From this point of view, the effective, cheap and selective ways to remove the carbonates matrix from the ores prior to bioleaching need be proposed and developed.

Ecometals project is co-funded by the German Federal Ministry of Education and Research (BMBF under the Grant ID 033RF001) and the French National Research Agency (ANR-13-RMNP-0006). KGHM Cuprum sp z o.o. CBR activities were co-funded by the Ecomentals project (HZDR) and internal funds of KGHM Polska Miedź S.A. The authors would like to thank to Prof. Axel Schippers and Dr. Sabrina Hedrich from the Federal Institute for Geosciences and Natural Resources (BGR), Germany, for delivery of the bacterial cultures used for the bioleaching experiments.

References

1. Zakłady Badawcze i Projektowe Miedzi (ZBPM) „Cuprum", *Projekt techniczny dołowego stanowiska prób bakteryjnego ługowania miedzi* (Wrocław, 1977)

2. A. Grotowski, Z. Sadecki, J. Lehman, M. Kalisz, W. Charewicz, T. Chmielewski et al., *Rozpoznanie możliwości zastosowania niekonwencjonalnych metod pozyskiwania miedzi i innych metali w złożach o trudnych warunkach geologiczno -górniczych*, report of ZBPM Cuprum sp. z o.o., in Polish, (Wrocław, 1998)
3. A. Grotowski, P. Kijewski, S. Lasko, T. Wiślicki, K. Krzysik et al., *Study of Selected Problems of Reprocessing of Flotation Tailings from the no 3 Tailings Pond (Wartowice) in ZG „Konrad"*, CBPM Cuprum, (Wrocław, 1995)
4. T. Chmielewski, Physicochem. Probl. Miner. Process. **51(1),** 335–350 (2015)
5. P. d'Hugues, P.R Norris., K.B., Hallberg F. Sanchez, J. Langwaldt, A. Grotowski, T. Chmielewski, S. Groudev, Minerals Engineering **21,** 111–120 (2008)
6. S. Kutschke, A.G. Guézennec, S. Hedrich et al., Minerals Engineering **75,** 116–125 (2015)
7. A-G. Guezennec, M. Hanke., A. Chmielarz, C. Joulian, Y. Menard, P. d'Hugues, Hydrometallurgy, **10** (2014)
8. www.biomore.info
9. http:// www.ecometals.org
10. S. Hedrich, A-G Guézennec, M. Charron, A. Schippers, C. Joulian, Front. Microbiol. (2016)
11. A-G Guezennec, C. Joulian, J. Jacob, A. Archane, D. Ibarra, R. de Buyer, F. Bodénan, P. d'Hugues, Minerals Engineering **106**, 64–70 (2017a)
12. A-G Guezennec, C. Joulian, J. Jacob, F. Bodénan, P. d'Hugues, S. Hedrich, Influence of CO_2 supplementation on the bioleaching of a Cu concentrate from Kupferschiefer ore, International Bioleaching Symposium (2017b)
13. S. Hedrich, A. Schippers, C. Joulian, A-G Guezennec, Effect of temperature ramping on stirred tank bioleaching of a copper concentrate. International Bioleaching Symposium (2017)

DOI:10.1051/e3sconf/20171801016

Studies of the aromatic β-diketones as extractant of copper ions

Katarzyna Witt[1,*], and *Elzbieta* Radzyminska-Lenarcik[1]

[1]Faculty of Chemical Technology and Engineering, UTP University of Sciences and Technology, Seminaryjna 3, PL-85326 Bydgoszcz, Poland

Abstract. Benzoylacetone 2, dibenzoylmethane 3 and acetylacetone 1 were studied as extractants of copper ions by liquid-liquid partition and spectrophotometric methods. Extraction ability increase in order 1 < 2 < 3. The values of $pH_{1/2}$, corresponding to 50% of the extraction of copper, are equal to 10.75, 10.08 and 9.85 for 1, 2, 3, respectively. The extraction process is the most effective when 3 was used as extractant, the maximum amount of copper ions passed to the organic phase is 50% and 99% for 1 and 3, respectively. The spectra have shown that in the aqueous and in the organic phase β-diketone form two chelated complexes $[CuL]^+$ and $[CuL_2]$. Partition ratios (P_2) of $[CuL_2]$ complexes have very high values, much higher than P_1. That complexes determine the efficiency of extraction process of Cu(II) ions, what confirm the electron absorption spectra.

1 Introduction

Hydrometallurgy is concerned with aqueous solutions, and thus metal ions are the species of interest. Extraction of metal ions directly into a water-immiscible oil phase is energetically unfavorable, and this process can only be accomplished by arranging that the metal ion be transferred to the organic phase as a neutral species. Such species can be produced by aqueous complexing or by association with anions whereby, by careful choice of ligand concentration, a neutral species extractable by an organic solvent can be produced [1,2].

Copper is also presented in a number of effluents or waste waters of various nature, being also its elimination a primary target from an environmental point of view, with solvent extraction and membrane technologies gaining an important role in this approach [3].

β-Diketones are successfully used as extractants in solvent extraction. The most known β-diketone extractants are the commercial compounds of the LIX group. These include, e.g. LIX 54-100, which is a composition of six β-diketone isomers: 1-phenyldecane-1,3-diones, heptane-8,10-dione and 1,3-diphenylpropane-1,3-dione. Ochromowicz et al. [4] and Kyuchoukov et al. [5] conducted extraction of copper(II) from ammoniacal solutions by among others a LIX 54-100. Also other metals, for example nickel, can be separated from ammoniacal media by means of LIX 54 [6]. Authors proved that nickel extraction is very sensitive to equilibrium pH and the extraction decreases beyond pH 9.0.

In the literature one can find articles on other commercial β-diketones used in solvent extraction. For example Weng et al. [7] carried out separation and recovery of copper from ammoniacal chloride solutions by solvent extraction using 1-phenyl-3-heptyl-1,3-propanedione, 1-phenyl-4-ethyl-1,3-octanedione and 1-(4′-dodecyl) phenyl-3-tert-butyl-1,3-propanedione in kerosene. The results indicated that the last demonstrated the best selectivity for the separation of copper from nickel and zinc.

In our previous paper [8] we described a liquid-liquid partition and spectrophotometric methods, which were used to examine the process of complexes formation of acetylacetone, 3-butyl-acetylacetone and 3-allyl-acetylacetone with Cu(II) ions during extraction process. We proved that the highest efficiency of extraction of copper(II) complexes was observed in the case of 3-allyl-acetylacetone.

The process of extraction of Th(IV), U(VI) and Mo(VI) ions was introduced by Monhanty et al. [9]. For separation U(VI) ions a mixture of 10% (v/v) LIX 84 and 0.1M dibenzoylmethane at pH 4.2 and mixture of 10% LIX 84 and 0.05M thenoyl-trifluoro-acetone in the pH range 5.5–7.3 was used. The Mo(VI) ions was extracted from chloride media by 10% LIX 84 at pH 1.5. It was shown that extraction decreases very rapidly with increase in the concentration of HCl as compared to that from H_2SO_4, HNO_3 and $HClO_4$ acid solutions. Inconsiderable extraction has been observed in case of Th(IV) by LIX 84 or its mixtures.

Virnig et al. [10] synthesed novel sterically hindered β-diketones and proposed method of their use in the extraction of copper from aqueous ammoniacal solutions containing copper resulting from commercial processes, including leaching of copper from ores, such as sulfidic ores, or concentrates resulting from flotation of such sulfidic ores.

Aromatic diketones, such as benzoylmethane and dibenzoylmethane, are also commonly known as very good complexing agents. This made them also very effective extractant in the separation of metal ions using

* Corresponding author: Katarzyna.Witt@utp.edu.pl

solvent extraction. El-Naggar, Abdel-Gawad, and Someda were proposed process of extraction of Co(III) by benzoylacetone from acetateacetic acid solutions [11]. Also Nezhadali et al. investigated competitive metal ion extraction of Cu(II), Ni(II), Zn(II) and Mn(II) ions by benzoylacetone with addition of acetylacetone. The experiments were conducted at pH 3–7. Results shown that in case of Cu(II) ions both extractants have a similar effect. Copper ions have been also more effectively extracted compared with the other tested cations like Ni(II), Zn(II) and Mn(II) [12].

Whereas dibenzoylmethane was used as extractant of copper ions by Kondo et al. [13]. They research were mainly concentrated on kinetic studies of the extraction of copper.

The aims of this work were investigated acid-base and extraction properties of two derivatives of β-diketones: benzoylacetone (2) and dibenzoylmethane (3) in comparison with unsubstituted acetylacetone (1) (Fig. 1). The extraction process was conducted in ammonia solutions at 25°C. Methylene chloride was used as a solvent.

2 Experimental

2.1 Reagents

β-Diketones used for research were commercial compounds (Fig. 1). Acetylacetone 1 was purchased from Avantor (Gliwice, Poland), dibenzoylacetone 2 and benzylacetone 3 were purchased from Sigma Aldrich (Poznan, Poland).

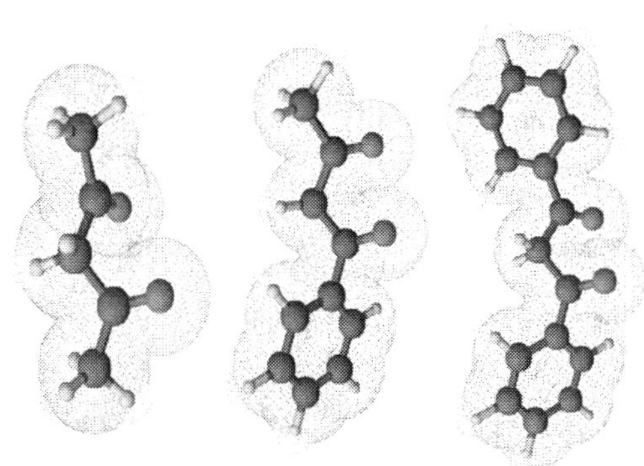

acetylacetone	benzoylacetone	dibenzoylmethane
CH_3-CO-CH_2-CO-CH_3	CH_3-CO-CH_2-CO-C_6H_5	C_6H_5-CO-CH_2-CO-C_6H_5
1	2	3

Fig. 1. Structure of the studied extrahents.

The potassium and copper(II) nitrates, both of analytical reagent grade (Avantor, Gliwice, Poland) were recrystallized from distilled water. The concentration of potassium nitrate was determined gravimetrically as potassium sulphate. Ammonia buffer was prepared from ammonia and ammonium nitrate (both analytical reagent grade, Avantor, Gliwice, Poland). Carbonate free potassium hydroxide (the titrant, Avantor, Gliwice, Poland) was prepared and standardized against a standard potassium hydrogenphthalate solution. Methylene chloride and methanol (analytical reagent grade, Avantor, Gliwice, Poland) were used as received. The pH-meter was calibrated using commercial buffer solutions (Radiometer, Warsaw, Poland) having a pH of 7.00 ± 0.01 and 9.21 ± 0.01. The pH was also checked against hydrochloric acid according to IUPAC recommendations [14].

2.2 Equipment

Potentiometric measurements were carried out using a multifunctional computer-aided pH-meter (PHM 250, Radiometer, Warsaw, Poland) equipped with a glass-calomel combination electrode C 2401-8 (Radiometer, Copenhagen, Denmark). The atomic absorption spectrophotometer 210 VGP (BUCK Scientific, East Norwalk, USA) was used for determination of Cu(II) concentrations. A Cary 50 (Varian, Melbourne, Victoria, Australia) spectrophotometer was used for recording the absorption spectra of the Cu(II) complexes in the aqueous and organic phase over the range 400nm-1000nm.

2.3 Determination of dissociation constants (pK_a)

The dissociation constants (pK_a) of β-diketones (2, 3) were determined by potentiometric titration. Measuring cell was thermostated at 25°C ± 0.5°C. Two water-methanol solutions were prepared. One contained KOH (1 mmol/dm^3), chosen β-diketone, and a proper quantity of potassium nitrate to ensure a constant ionic strength (1 mol/dm^3). Another solution was prepared in a similar manner but without β-diketone. Measurements were taken for two β-diketone concentrations (0.06 mol/dm^3 and 0.05 mol/dm^3) in the first solution. A 50 cm^3 volume of first solution was placed in the measuring cell, and the solution was titrated with standard solution of 5 mmol/dm^3 KOH. Then the second solution was titrated in the same way. Three series of measurements were done.

2.4 Extraction Procedure

The measurements were run at 25°C, at a fixed ionic strength (0.5 mol/dm^3) maintained in the aqueous phase with KNO_3. A constant pH was maintained using ammonia buffer (NH_3 + NH_4NO_3, 1:1). Before an extraction, concentrations of Cu(II) ions in the aqueous phase were held constant (0.01 mol/dm^3) whereas the ligand (1-3) concentrations in organic phase (methylene chloride) were varied (from 0.01 to 0.03 mol/dm^3). A volume of 5 cm^3 of the aqueous phase was placed in a graduated test tube and an equal volume of a β-diketone (1-3) solution in the organic solvent was added. The test tubes were then shaken for 30 min. The equilibrium was established after a few minutes; however, never longer than 30 min. Afterwards, it was checked if any changes in the phase volumes had occurred, then the phases were separated and the pH of the aqueous phase was measured. The Cu(II) concentration in the aqueous phase was determined by atomic absorption spectrophotometry. The

DOI:10.1051/e3sconf/20171801016

detection limit of the method was found to be 0.2 μg/dm³ and the relative standard deviation was about 6%.

2.5 Spectrophotomeric examination

UV-Vis spectra of the aqueous and organic phase were recorded in the wavelength range of 400 nm to 1000 nm. Water solution of 0.5 mol/dm³ KNO_3 was used as reference solution while testing a aqueous phase, whereas the organic phase was tested against methylene chloride. The spectra of the samples were recorded with spectral slits of 1 nm and with 1 nm steps using curettes having 1 cm path lengths.

3 Results and discussion

3.1 Dissociation constants of aromatic β-diketones

Aromatic β-diketones dissociate according to the following reaction:

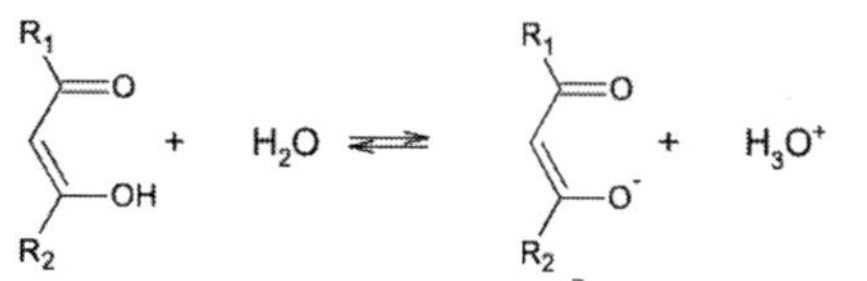

(1)

β-diketone β-diketone⁻

Dissociation constants (pK$_a$) of investigated ligands (2 - 3) were determined by potentiometric method using equilibrium constant of the above reaction:

$$K_a = \frac{[H_3O^+][L^-]}{[HL]} = \frac{[H_3O^+](C_{KOH} + [H_3O^+])}{(C_{HL} - C_{KOH}) - [H_3O^+]} \quad (2)$$

where $[H_3O^+]$, $[L^-]$, and $[HL]$ are equilibrium concentrations of the hydronium ion, the free ligand (β-diketone anion) and its undissociated form, respectively, and C_{L^-}, C_{KOH} are analytical concentrations of the ligand, HL and KOH in aqueous solution, respectively.

The values of determined dissociation constants are summarized in Table 1.

Lower values of pK$_a$ provide stronger acid properties of molecules. pK$_a$ value of 2 indicate that compound have slightly lower value than 1 and 3 (Table 1). On the other hand dibenzoylmethane is as strong basis as acetylacetone.

Table 1. The dissociation constants pK$_a$ of investigated β-diketone derivatives in aqueous-methanol (1:1) solution (I=1 mol/dm³ KNO_3) at 25°C.

β-diketone	pK$_a$	ref.
acetylacetone 1	9.65	[15]
benzoylacetone 2	9.61	[this work]
dibenzoylmethane 3	9.65	

The percentage extractions of the test metal ions in the test systems were calculated for each measured pH of the aqueous phase using Formula (3):

$$E\% = \frac{C^0_{Cu} - C_{Cu}}{C_{Cu}} \cdot 100\% \quad (3)$$

where: C^0_{Cu} and C_{Cu} denote analytical Cu(II) concentrations in the aqueous phase before and after attaining a partition equilibrium, respectively.

Changes in the percentage extraction vs. pH of the aqueous phase is presented in the Figure 2.

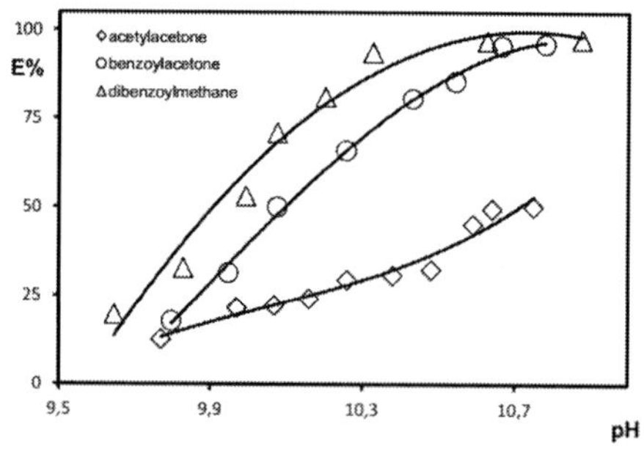

Fig. 2 Percentage extraction vs. pH for Cu(II) complexes with acac ◊, benzoylacetone o, dibenzoylmethane Δ in water-methylene chloride phase systems.

β-Diketones containing the aromatic ring in the molecule (2 and 3) are more effective extractants than acac (1). The extraction process is the most effective when 3 was used as extractant, the maximum amount of copper ions passed to the organic phase is 50% and 99% for 1 and 3, respectively; although its pK$_a$ values are the same. Differences in the efficiency of the extraction process have impact on the ability to form complexes. β-Diketones form chelate complexes with Cu(II) ions according to equation (4):

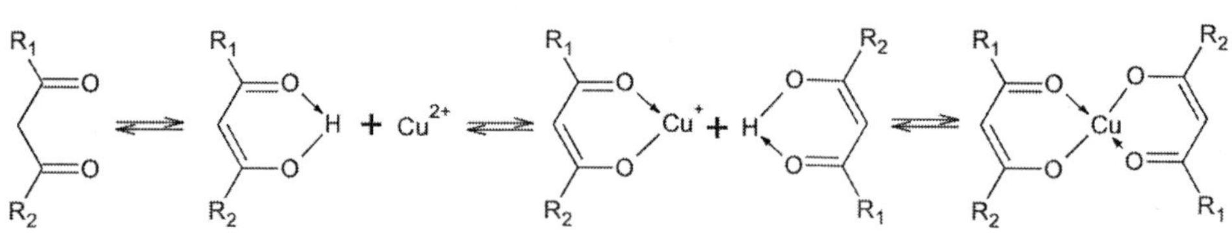

(4)

DOI:10.1051/e3sconf/20171801016

The logarithms of the reaction equilibrium constant (4) for the copper complexes (β_n) with **1**, **2** and **3** are 11.85, 12.05 and 12.98 respectively [16]. The insertion of the aromatic ring into the acetylacetone molecule increases the complexion properties of that extractants.

For **1**, **2**, **3** the values of $pH_{1/2}$, corresponding to 50% of the extraction of copper, are equal to 10.75, 10.08 and 9.85, respectively. This ability to form the complexes also increases their susceptibility to extraction to the organic phase. This is evidenced by the absorption spectra of both phases, recorded after the fix a equilibrium of distribution of the organic phase/aqueous phase, as shown in Figures 3 and 4.

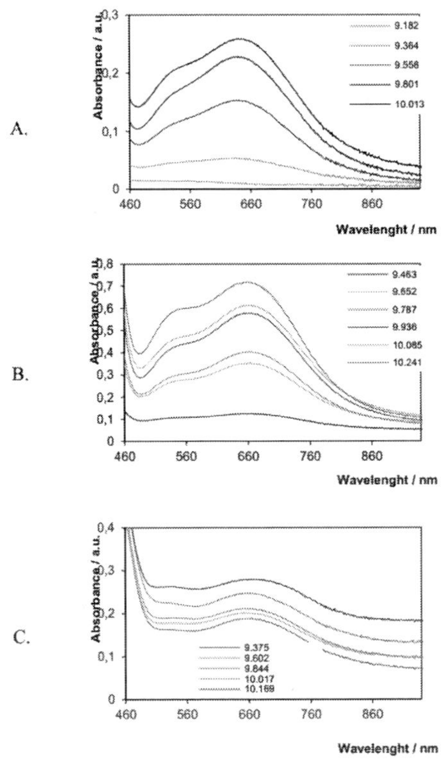

Fig. 3 Absorption spectra of organic phase after extraction of the Cu(II) complexes with A. acetylacetone, B. benzoyl-acetone, C. dibenzoylmethane in methylene chloride together with the corresponding pH of the aqueous phase.

To the organic phase in the extraction process (Fig. 3) pass the green copper complexes. Upon raising the ligand concentration, absorption maxima become slightly displaced towards longer wavelengths. The absorption spectra reveal both octahedral (absorption max. around 520 nm) and tetrahedral (660 nm) complexes. The tetrahedral complexes appear at the higher ligand concentrations in the aqueous and organic phases.

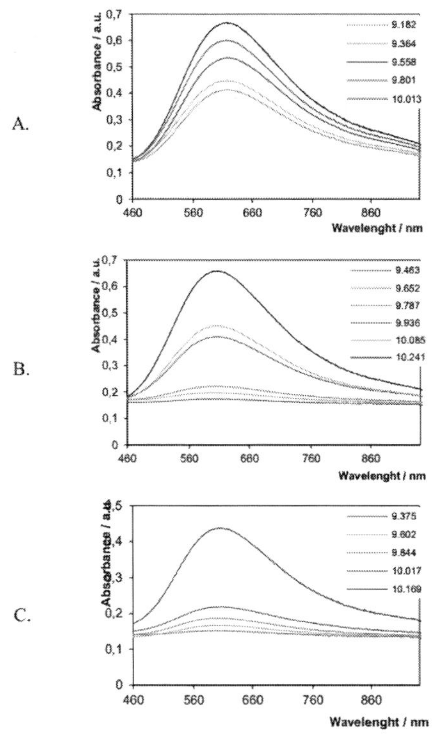

Fig. 4 Absorption spectra of aqueous phase after extraction of the Cu(II) complexes with A. acetylacetone, B. benzoyl-acetone, C. dibenzoylmethane depending on various pH.

In aqueous phase the Cu(II) complexes with β-diketones are blue. Copper complexes with **2** and **3** are being created in higher values of pH of water phase (Fig. 4), which is advantageous in case of copper recovery after leaching by ammonia.

The spectra have shown that in both the aqueous and organic phases β-diketone form two chelated complexes $[CuL]^+$ and $[CuL_2]$. The first of them can contain two molecules of water in coordination sphere $[CuL(H_2O)_2]^+$. The second one is completely anhydrous.

Their partition ratios (P_n) were calculated from formula (5):

$$\frac{[L^-]^n}{E\%/100} = \frac{\sum_{n=0}^{n=N} \beta_n [L^-]^n}{P_n \beta_n} \quad (5)$$

and summarized in the Table 2.

Table 2. Partition ratios of studied β-diketones

β-diketone	P_1	P_2
acetylacetone **1**	$7.2*10^4$	$4.9*10^7$
benzoylacetone **2**	$1.8*10^5$	$5.6*10^9$
dibenzoylmethane **3**	$2.5*10^6$	$8.3*10^9$

* Corresponding author: Katarzyna.Witt@utp.edu.pl

The values of the partition ratios of the investigated complexes separated between the aqueous and organic phases (methylene chloride) for the studied β-diketones increase grow in a order **1** > **2** > **3**. The values of the partition ratios of the second complexes are at least three orders of magnitude higher than the first one ($P_2 >> P_1$). Thus, the extraction process is determined by the possibility of forming a second complexes, which takes place at higher pH values of the aqueous phase. It is preferable to use aromatic β-diketones as copper extractants from solutions after leaching of ores or waste using ammonia or basis.

4 Conclusion

The studied β-diketones have all qualities to regard them as selective extractants of Cu(II) ions. It depends on the plasticity of the copper ions coordination sphere, which can easily change shape and adjust to the geometric imposed by β-diketone. As a result of this interaction a stable copper chelate complexes are formed. In the aqueous and in the organic phase β-diketone form two chelated complexes $[CuL]^+$ and $[CuL_2]$. The values of extraction percentage are the highest for dibenzoylmethane. The maximum amount of copper ions passed to the organic phase is 50% and 99% for 1 and 3, respectively and extraction ability increase in order 1 < 2 < 3.

The new developments presented above were carried out within the 2007-2013 Innovative Economy Operational Programme, Sub-action 1.3.2., Support of the protection of industrial property generated by scientific entities as result of R&D works within project no. UDA-POIG.01.03.02-04-077/12-01, financed by the European Regional Development Fund (ERDF) (85% of co-financing) and from a designated subsidy (15% of co-financing).

References

1. J. Aromaa, L. Rintala, M. Kahari, O. Forsen, Physicochem. Probl. Miner. Process. **51(1)**, 269 (2015)
2. K. Ochromowicz, T. Chmielewski, Physicochem. Probl. Miner. Process. **46**, 207 (2011)
3. T. Chmielewski Physicochem. Probl. Miner. Process. **51(1)**, 335 (2015)
4. K. Ochromowicz, M. Jeziorek, K. Wejman, Physicochem. Probl. Miner. Process. **50(1)**, 327 (2014)
5. G. Kyuchoukov, M.B. Bogacki, J. Szymanowski, Ind. Eng. Chem. Res. **37(10)**, 4084 (1998)
6. F.J Alguacil, A Cobo, Hydrometallurgy **48**, 291 (1998)
7. F. Weng, Ch. Qiyuan, H. Huiping, N. Chunlin, Z. Qinqin.. Sep. Purif. Technol. **80(1)**, 52 (2011)
8. E. Radzyminska-Lenarcik, K. Witt, Sep. Sci. Technol., DOI: 10.1080/01496395.2017.1329838 (2017)
9. R.N. Mohanty, S. Singh, V. K. Chakravortty, C. Dash, J. Radioanal. Nucl. Chem. **132(2)**, 359 (1989)
10. M.J. Virnig, G.A. Kordosky, S.I. Kang, K.V. Martin, P.L. Mattison, US 6107523 A (1997)
11. H.A. El-Naggar, A.S. Abdel-Gawad,. H.H. Someda, J. Radioanal. Nucl. Chem. Lett. **128**, 233 (1988)
12. A. Nezhadali, A. Sadeghi, M. Roigar, Arab. J. Chem. **8(2)**, 164 (2015)
13. K. Kondo, T. Tsuneyuki, K. Hiramatsu, F. Nakashio, Sep. Sci. Technol., **25(11-12)**, 1213 (1990)
14. A. Braibanti, G. Ostacoli, P. Paoletti, L.D. Pettit, S. Sammartano, Pure Appl. Chem. **59**, 1721 (1987)
15 K. Witt, E. Radzyminska-Lenarcik, W. Urbaniak, Sep. Sci. Technol. **51(15-16)**, 2620 (2016)
16. L.G. Van Uitert, W.C. Fernelius, B.E. Douglas, J. Am. Chem. Soc. **75**, 457 (1953)

Studies on the separation of some transition metals using trialkylimidazole as selective extractant

Elzbieta Radzyminska-Lenarcik[1,*], and *Katarzyna* Witt[1]

[1]Faculty of Chemical Technology and Engineering, UTP University of Sciences and Technology, Seminaryjna 3, PL-85326 Bydgoszcz, Poland

Abstract. The possibility of Cu(II) ion separation from an equimolar Cu-Cd-Zn tertiary mixture by solvent extraction was investigated. The process was based on the use of 1,2,4-trimethylimidazole (TMI) and 1-decyl-2,4-dimethylimidazole (DDMI) as extractant. Dichloromethane was used as the solvent in the extraction process. The dissociation constant of extractants were found potentiometrically. Extraction data were used for finding stability constants and partition constants for the complexes being formed in the aqueous solution. For TMI, separation coefficients for Cu(II) with regard to zinc and cadmium are 2.9 and 1.3, respectively. For DDMI separation coefficients for Cu(II) with regard to zinc and cadmium are 1.4 and 1.6, respectively. Sparingly soluble in water 1-decyl-2,4-dimethylimidazole is a more useful extractant.

1 Introduction

Industrial wastewater containing heavy metals (eq. zinc, copper, cadmium) belongs to hazardous wastes. Wastewater discharge is a primary source of this metal release into the environment [1]. On the other hand, these metals are very important due to wide application in the manufacturing of alloys, batteries, pigments and metal plating [2]. Therefore, the recovery and recycling of heavy metals from various waste materials is necessary and very important to gain the metal values and protect the environment [2,3]. Moreover, the increasingly stringent regulations regarding the discharge of metals into the environment, and the increasing stress upon the recycling of effluents have focused the great interest on the development of new approaches for the recovery of metals from industrial wastes [4]. Cote scored copper, zinc and cadmium for strategic metals [5].

A survey of literature indicates that separation or removal of heavy metal ions from different aqueous solutions can be achieved by solvent extraction [6-9]. Several extractants have been proposed for separation of Cu(II), Zn(II), and Cd(II) from aqueous solutions, e.g., carboxylic acids [10], derivatives of phosphinic acid [11], derivatives of acid amides, trioctylamine [11], alkylphenylamines [12], complex oximes [13], crown ethers [14-16] derivatives of ketones [17-20], derivatives of pyridine [21, 22], as well as alkyl derivatives of imidazole [6, 23-25] and quaternary ammonium salt [26].

The possibility of using alkylimidazoles for solvent extraction of metals has been noted for a long time [27-29]. They are now a fairly well-known group of basic ligands (extractants) whose properties can be easily modified. The pK_a value of the imidazole molecule is 6.98 at 20°C [30]. The stability of its complexes with the cations of the 1st row of transition metals is higher than ammonia due to the binding of $\pi_{M\rightarrow L}$ [31].

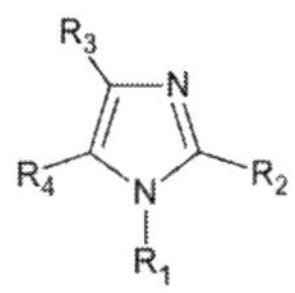

Fig. 1. General formula of alkylimidazole.

The alkyl group at position 1 (R_1 Fig. 1) slightly increases the basicity of the donor nitrogen atom as well as the stability of the complexes. Even small alkyl group at position 2 (R_2 Fig. 1) or 4 (R_3 Fig. 1) constitute a steric hindrance making it difficult to form of complexes with a rigid, hardly deformable coordination sphere with octahedral or flat symmetry. Although 2 or 4 substituted alkylimidazoles are stronger bases than 1-alkylimidazole their complexes have lower values of stability constants. The introduction of alkyl groups into the imidazole ring, irrespective of their location, reduces the water solubility in all homologous series of alkylimidazoles. Compounds containing a total of more than 5 carbon atoms in substituents are practically insoluble in water. They can be used for metal separation as extractants in hydrometallurgy or as a carrier in membrane techniques. So far, the extraction properties of 1-alkylimidazoles [6, 32, 33], 1-alkyl-2-methylimidazoles [7, 34, 35], 1-alkyl-2-ethylimidazoles [36], 1,2-dialkylimidazoles [23, 25] and 1-alkyl-4-methylimidazoles [25, 37] have been studied. On the other hand, our knowledge of donor and extraction properties of tri-substituted alkylimidazoles is very small.

The purpose of the present study was to determine the basicity, complexity and extraction properties of

* Corresponding author: elaradz@utp.edu.pl

 DOI:10.1051/e3sconf/20171801017

1,2,4-trimethylimidazole (TMI) and 1-decyl-2,3-dimethylimidazole (DDMI) and to compare their effectiveness in copper(II) ion separation from an equimolar Cu-Cd-Zn mixture. TMI and DDMI molecules differ in their hydrophobicity. Methylene chloride was used as the dilutent in the solvent extraction process.

2 Experimental

2.1 Reagents

Inorganic chemicals: copper(II), zinc(II), cadmium(II), and potassium nitrates were of analytical grade and were supplied by POCh (Gliwice, Poland). Nitric acid was of analytical grade and were supplied by MERCK. Aqueous solutions were prepared with double distilled water (conductivity of 5 μS/m).

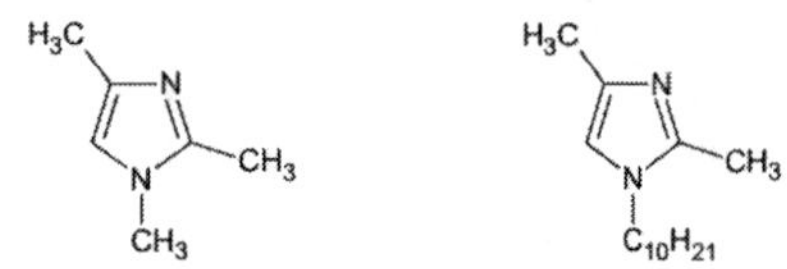

Fig. 2. Structure of the extrahents: 1,2,4-trimethylimidazole (TMI) and 1-decyl-2,4-dimethylimidazole (DDMI).

1,2,4-Trimethylimidazole (TMI) and 1-decyl-2,4-dimethylimidazole (DDMI) (Fig. 2) were synthesized at Medical University of Lodz, Poland.

2.2 Determination of dissociation constants (pK_a)

The dissociation constants of TMI was determined by potentiometric titration. Measurements of pH were made using PHM 250 pH-meter (Radiometer) equipped with a glass-calomel combination electrode C 2401-8 (Radiometr). Measuring cell was thermostated at 25°C ± 0.5°C. Two solution were prepared. One contained nitric acid (0.01 mol/dm^3), ligand (TMI or DDMI), and a proper quantity of potassium nitrate to ensure a constant ionic strength (0.5 mol/dm^3). Another solution was prepared in a similar manner but without ligand. Measurements were taken for two ligand (TMI or DDMI) concentrations (0.06 mol/dm^3 and 0.05 mol/dm^3) in the first solution. A 20 cm^3 volume of first solution was placed in the measuring cell, and the solution was titrated with the second. Three series of measurements were run.

2.3 Solvent extraction

Equal volumes of organic and aqueous phases (phase volume ratio O/A = 1) were mechanically shaken for 20 min at 25 ± 2°C. 0.05-0.25 mol/dm^3 ligand (TMI or DDMI) in dichloromethane was used as the organic phase. After establishing equilibrium, the aqueous phase was separated from the organic phase. Equilibrium pH of aqueous phases was measured with a Radiometr PHM 250 pH-meter equipped with a glass-calomel combination electrode C 2401-8 (Radiometr). Metal ions concentrations in aqueous phases were analyzed with BUCK Scientific 210 VGP spectrophotometer (AAS — Atomic Absorption Spectrometry). The concentrations of metal ions in aqueous phases were determined before and after solvent extraction for calculating extraction efficiency. The concentrations of metals in organic phases were deduced by mass balance. Each experiment was carried out three times and the error did not exceed 3%.

3 Results and discussion

3.1 Dissociation constant of ligands (TMI and DDMI)

The acid dissociation constants of the protonated TMI and DDMI molecules were calculated from the equation:

$$K_a = \frac{[H_3O^+][L]}{[HL^+]} = \frac{[H_3O^+](C_L - C_{HNO3} + [H_3O^+])}{C_{HNO_3} - [H_3O^+]} \quad (1)$$

where $[H_3O^+]$, [L], and $[HL^+]$ are equilibrium concentrations of the hydronium ion, the free ligand (TMI or DDMI) and its protonated form, respectively, where C_L and C_{HNO3}, are analitycal concentrations of the ligand and HNO_3 in aqueous solution, respectively.

Table 1 presented the dissociation constants of the protonated TMI and DDMI together with the previously determined the dissociation constants of imidazole derivatives.

Table 1. Comparison of the dissociation constants pK_a of chosen alkylimidazole derivatives in aqueous solution (I=0.5 mol/dm^3 K(H)NO$_3$) at 25°C.

	pK_a	ref.
1,2,4-trimethylimidzole (TMI)	8.49	[this work]
1-decyl-2,4-dimethylimidazole (DDMI)	8.62	[this work]
imidazole	6.99	[30]
1-methylimidazole	7.06	[31]
2-methylimidazole	8.10	[38]
4-methylimidzole	7.80	[39]
1,2-dimethylimidzole (1,2-DMI)	8.21	[31]
1,4-dimethylimidzole (1,4-DMI)	7.72	[40]
2,4-dimethylimidazole (2,4-DMI)	8.57	[38]
4,5-dimethylimidzole (4,5-DMI)	8.19	[41]

The introduction of even a small alkyl group in positions 2 or 4 of the imidazole ring results in the molecule having a much higher basicity (Table 1). 2-Alkylimidazole and 4-alkylimidazole are harder bases in comparison with imidazole and 1-alkylimidazoles. 1,2,4-Trimethylimidazole (TMI) is a stronger base than the 1,2-dimethylimidazole (1,2-DMI), but 1-decyl-2,4-dimethylimidazole (DDMI) is the strongest base.

DOI:10.1051/e3sconf/20171801017

3.2 Solvent extraction

In the case where there is no phase change in volume, distribution ratio (D_M) was defined as follows:

$$D_M = \frac{C_{M(org)}}{C_{M(aq)}} = \frac{C_M^0 - C_M}{C_M} \quad (2)$$

where: C_M^0 and C_M denote analytical metal concentrations in the aqueous phase before and after attaining a partition equilibrium, respectively.
Based on analytical calculations for every equilibrium pH value for the aqueous phase, the distribution ratios (D_M) were calculated from formula (2) for each tested metal and then extraction curves were plotted (log D_M = f(pH)) (Figs. 3 and 4).

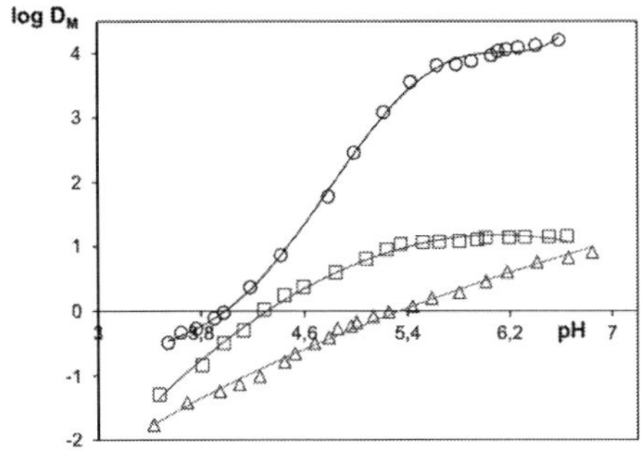

Fig. 3. Extraction curves of Cu(II) (○), Cd(II) (□), and Zn(II) (△) complexes with 1,2,4-trimethylimidazole (TMI) into dichloromethane.

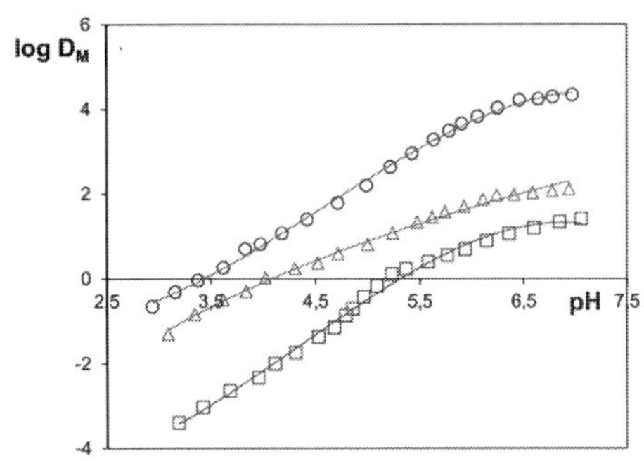

Fig. 4. Extraction curves of Cu(II) (○), Cd(II) (□), and Zn(II) (△) complexes with 1-decyl-2,4-dimethylimidazole (DDMI) into dichloromethane.

Fig. 3 and Fig 4 indicate that partition ratios for the test metal complexes with TMI and DDMI tend to increase with an increase in the pH values of the aqueous phase. For TMI (Fig.3), the order of extracted ions is Cu(II) > Cd(II) > Zn(II), but for DDMI (Fig.4) Cu(II) > Zn(II) > Cd(II).

In the case of Cu(II) and Cd(II) complexes with TMI (Fig.3) for pH > 5.8 (for Cu(II)) and 5.4 (for Cd(II)) the values of D_M are constant and they do not depend on the increase in the extractant's concentration in the aqueous and organic phases. Best extracted were Cu(II) complexes with TMI, worst extracted were Zn(II).

In the case of DDMI, the extraction of Zn(II) and Cu(II) complexes is almost the same at lower pH values (to about pH=5), while for Cd(II) is much worse. Above pH=5.5, the values of D_M increase rapidly for Cu (II), but in case of Cd(II) and Zn(II) these values are approaching each other.

For TMI the values of $pH_{1/2}$, corresponding to 50% of the extraction of copper, cadmium and zinc, are equal to 4.0, 4.3 and 5.3, respectively. This means, they can be separated by partition in the liquid-liquid system, conducting a multistep extraction at a pH value which is optimum for a given metal (Fig. 3).

For DDMI (Fig. 4), 50% of the extraction of copper, cadmium and zinc, occurs at pH equal to 3.45, 5.4 and 4.05, respectively.

Distribution ratio (D_M) is related to stability constants of the complexes and their partition ratios by the following equation:

$$D_M = \frac{P_n\beta_n[L]^n + P_{n+1}\beta_{n+1}[L]^{n+1} + ... + P_N\beta_N[L]^N}{\sum_{n=0}^{n=N}\beta_n[L]^n} \quad (3)$$

where: β_n are cumulative stability constants of the complexes in the aqueous phase, P_n are organic solvent/water partition ratios of the complexes, ($P_n=[ML_n]_{(org)}/[ML_n]_{(aq)}$), [L] is the free ligand (TMI or DDMI) concentration (mol/dm^3) in the aqueous phase and n is the number of ligands molecules in the first metal(II) complex that is so hydrophobic that it freely passes into the organic phase.

Cumulative stability constants are described by the equations:

$$\beta_1 = \frac{[ML]}{[M][L]} \quad \beta_2 = \frac{[ML_2]}{[M][L]^2} \quad \ldots \quad \beta_n = \frac{[ML_n]}{[M][L]^n} \quad (4)$$

The pooled D_M values and the corresponding sets of the independent variable $[L]$ were used for determination of the stability constants β_n and partition ratios P_n of the metal(II) complexes by using the equation proposed by Rydberg [42]:

$$\frac{[L]^n}{D_M} = \frac{\sum_{n=0}^{n=N}\beta_n[L]^n}{P_n\beta_n} \quad (5)$$

Equation (5) can be used for determination of the composition of the first complex being extracted from the aqueous phase and the stability constant, β_1, of the first complex formed in the aqueous phase. To do this, the shape of the function $[L]^n/D_M = f[L]$ is monitored when varying the exponent n between 1 and 6. In this way, a bundle of curves were obtained of which only one was a straight line with a positive slope and b values. It is assumed that the n value is equal to the number of ligand molecules attached to the central ion in the first complex being extracted. The method was used previously [6, 7, 23-25, 32-37]. The stability constants, β_n, obtained in this way are collected in Table 2 together with stability constants for 1,2-dimethylimidazole [31], and 1,4-dimethylimidazole [40].

Table 2. Comparison of the stability constants β_n of Cu(II), Zn(II), and Cd(II) complexes with 1,2,4-trimethylimidazole (TMI), 1-decyl-2,4-dimethylimidazole (DDMI), 1,2-dimethyl-imidazole (1,2-DMI), and 1,4-dimethylimidazole (1,4-DMI) at 25°C, ionic strength 0.5 mol/dm^3 (KNO_3).

ligand/pK_a	metal ion	log β_1	log β_2	log β_3	log β_4	ref.
TMI 8.49	Zn(II)	2.03	4.17	6.27	9.51	[this work]
	Cu(II)	3.75	6.83	9.26	11.38	
	Cd(II)	2.34	3.79	5.32	6.95	
1,2-DMI 8.21	Zn(II)	1.92	4.32	7.11	9.00	[31]
	Cu(II)	3.70	6.80	9.18	10.80	
	Cd(II)	2.20	3.45	5.05	6.34	
1,4-DMI 7.72	Zn(II)	1.70	4.48	5.40	8.90	[40]
	Cu(II)	3.54	6.53	9.17	10.47	
	Cd(II)	2.05	3.93	5.11	5.81	
DDMI 8.62	Zn(II)	2.17	4.48	6.39	9.87	[this work]
	Cu(II)	3.95	7.18	9.42	11.69	
	Cd(II)	2.53	4.21	5.68	7.56	

A comparison of the values of stability constants in Table 2 indicates that stability for all complexes tends to increase with ligand's alkalinity and is the highest for Cu(II) complexes. In the case of TMI, stability constants for Zn(II) complexes are higher than those for Cd(II) complexes.

In aqueous solutions, cations of Zn(II), Cd(II) and Cu(II) exist in the form of octahedral aquacomplexes $[M(H_2O)_6]^{2+}$. Owing to the effect of ligands (extractants), octahedral aquacomplexes of certain cations tend to change their coordination number (c.n.) from 6 to 4 and they change their coordination sphere into flat or tetrahedral, depending on the structure of their d-electron layer. The process is illustrated by the equation.

$$[M(H_2O)_{6-n+1}L_{n-1}]^{2+} + L \leftrightarrow [M(H_2O)_{4-n}L_n]^{2+} + (n+2)H_2O \quad (6)$$

Therefore, tetrahedral complex (c.n.=4) are formed in the solution in addition to octahedral ones (c.n.=6). Tetrahedral complexes are specific to Zn(II), and Cd(II). Also Cu(II) ions tend to change the symmetry of their coordination sphere, forming distorted tetrahedral or flat structures [43].

Changes in the coordination sphere lead to higher sums of complex stability constants (stability constants are the sum of the stability constants for the complexes with c.n.=4 and with c.n.=6) and higher susceptibility to extraction in the liquid-liquid system, because complexes with the c.n.=4, being less hydrated, pass more readily into the organic phase after at least part of the molecules of water, which is bound in the metal coordination sphere, is replaced by the molecules of the more hydrophobic ligand. For the fourth stage of complexation (n=4) this can be illustrated by the following equations:

$$[M(H_2O)_6]^{2+} + 4\,L \leftrightarrow [ML_4]^{2+} + 6\,H_2O \quad (7)$$

$$[M(H_2O)_6]^{2+} + 4\,L \leftrightarrow [ML_4(H_2O)_2]^{2+} + 2\,H_2O \quad (8)$$

wherein L is the extractant (TMI or DDMI) molecules.

Table 3. Comparison of the partition ratios P_n of Cu(II), Zn(II), and Cd(II) complexes with 1,2,4-trimethylimidazole (TMI), 1-decyl-2,4-dimethylimidazole (DDMI) and 1,4-dimethyl-imidazole (1,4-DMI) between the aqueous and the organic phases, at 25°C, ionic strength 0.5 mol/dm^3 (KNO_3).

ligand/ solvent	metal ion	P_1	P_2	P_3	P_4	ref.
TMI methylene chloride	Zn(II)		0.84	1.23	29.21	[this work]
	Cu(II)		1.26	1.87	84.72	
	Cd(II)		0.3	1.56	65.46	
1,4-DMI benzyl alcohol	Zn(II)				8.72	[40]
	Cu(II)		0.1	1.65	20.75	
	Cd(II)			1.54	110.7	
DDMI methylene chloride	Zn(II)		1.12	2.46	76.43	[this work]
	Cu(II)		0.98	2.15	106.8	
	Cd(II)		0.15	1.71	67.39	

The partition ratios between the aqueous and organic phases for the investigated metal complexes with TMI and DDMI, were found from Equation (5), and they are collected in the Table 3 together with partition ratios for the Cu(II), Zn(II) and Cd(II) complexes with 1,4-dimethylimidazole (1,4-DMI) [40].

The data in Table 3 indicate that in the process of extraction of the tested metal complexes with TMI and DDMI as well as 1,4-DMI [40], mainly the fourth complexes tend to pass into the organic phase. The partition constants P_4 (Table 3) well reflect susceptibility to extraction of the tested metal complexes with TMI and DDMI. Their high values are proof that, most probably, they are 4-coordination complexes of the formula $[ML_4](NO_3)_2$. Extraction order for depends on extractant, for TMI extraction order was Cu(II) > Cd(II) > Zn(II), but for DDMI – Cu(II) > Zn(II) > Cd(II).

For TMI separation coefficients Cu(II) with regard to Zn(II) and Cd(II), calculated as $S = P_{4(Cu)} / P_{4(M)}$, are 2.9 and 1.3, respectively. For DDMI separation coefficients SCu(II)/Zn(II) decreased (1.39), but $S_{Cu(II)/Cd(II)}$ increased (1.58).

4 Conclusion

The results of the presented investigation prove that 1,2,4-trimethylimidazole and 1-decyl-2,4-dimethyl-imidazole in dichloromethane can be successfully used as the selective extractant for copper(II) from the Cu-Zn-Cd mixture. For TMI, separation coefficients for Cu(II) with regard to zinc and cadmium are 2.9 and 1.3, respectively. For DDMI separation coefficients for Cu(II) with regard to zinc and cadmium are 1.4 and 1.6, respectively. Sparingly soluble in water 1-decyl-2,4-dimethylimidazole is a more useful extractant. It can be used to separate Zn (II) from Cd (II) at a pH of 3 - 4.5.

Moreover, the investigation indicates that trialkylimidazole with all of the test metal cations tend to form 4-coordination complexes in addition to 6-coordination ones in the solution. This is possible because, in the case of Cu, Cd and Zn ions, a change of the coordination number takes place at various stages of complexation. Moreover, the configurational equilibrium tetrahedron ↔ octahedron for Zn(II) is clearly shifted more to the right than for Cd(II). This leads to an increase in all total stability constants β_n and an increase in the values of their partition ratios P_n.

The new developments presented above were carried out within the 2007-2013 Innovative Economy Operational Programme, Sub-action 1.3.2., Support of the protection of industrial property generated by scientific entities as result of R&D works within project no. UDA-POIG.01.03.02-04-077/12-01, financed by the European Regional Development Fund (ERDF) (85% of co-financing) and from a designated subsidy (15% of co-financing).

References

1. C. Sgarlata, G. Arena, E. Longo, D. Zhang, Y. Yang, Bartsch R.A., J. Membr. Sci. **323**, 444 (2008)
2. M.K. Jha, V. Kumar, J. Jeong, J. Lee, Hydrometallurgy **111-112**, 1 (2012)
3. K.A. Schimmel, S. Ilias, S. Akella, Sep. Sci. Technol. **36**, 805 (2001)
4. Agrawal, K.K. Sahu, J. Hazard. Mater. 171, 61 (2009)
5. G. Cote, Solv. Ext. Ion Exch. **18(4)**, 703 (2000)
6. E. Radzyminska-Lenarcik, Solv. Ext. Ion Exch. **25**, 53 (2007)
7. E. Radzyminska-Lenarcik, Sep. Sci. Technol. **42**, 2661 (2007)
8. F. Xie, D. Dreisinger, Solv. Ext. Ion Exch. **27**, 459 (2009)
9. A. Kumari, M.K. Sinha, S.K. Sahu, B.D. Pandey, Hydrometallurgy **165**, 159 (2016)
10. W. Apostoluk, Hydrometallurgy **21**, 345 (1989)
11. K. Wejman-Gibas, M. Pilsniak-Rabiega, K. Ochromowicz, Physicochem. Probl. Miner. Process. **53**, 202 (2017)
12. J.G.H. du Preez, Solv. Ext. Ion Exch. **18**, 679 (2000)
13. J. Szymanowski, Hydrooximes and Copper Hydrometallurgy (CRC Press, Boca Raton, USA, 1993)
14. M. Ulewicz, K. Sadowska, J.F. Biernat, Desalination **214**, 352 (2007)
15. M. Ulewicz, J. Szczygelska-Tao, J.F. Biernat, J. Membr. Sci. **344**, 32 (2009)
16. S. Chun, S.V. Dzyuba, R.A. Bartsch, Anal. Chem. **73(15)**, 3737 (2001)
17. M.L.F. Gameiro, M.R.C. Ismael, M.T.A. Reis, J.M.R. Caravalaho, Sep. Purif. Technol. **63**, 287 (2008)
18. G. Goetz, J. Arichi, Z. Lakkis, , J.P. Brunette, M. Hebrant, Solv. Ext. Ion Exch. **28**, 793 (2010)
19. S.S. Sombhatla, A. Kumar, S. Mushruwala, K. Rokkam, A. Shukla, Hydrometall. **166**, 94 (2016)
20. E. Radzyminska-Lenarcik, K. Witt. Sep. Sci. Technol.,DOI:10.1080/01496395.2017.1329838 (2017)
21. A. Borowiak-Resterna, H. Chlebowska, M. Giezek, Hydrometall. **103**, 158 (2010)
22. M. Tomaszewska, A. Borowiak-Resterna, A. Olszanowski, Hydrometall. **85**, 116 (2007)
23. E. Radzyminska-Lenarcik, Sep. Sci. Technol. **44**, 954 (2009)
24. E. Radzyminska-Lenarcik, Solv. Ext. Ion Exch. **28**, 636 (2010)
25. E. Radzyminska-Lenarcik, Chem. Pap. **65**, 226 (2011)
26. B. Wassink, D. Dreisinger, J. Howard, Hydrometall. 57, 235 (2000)
27. M.E. Cuprey, US Patent 3,843,667. 22 October (1974)
28. J.G.H. du Preez, T.I.A. Gerber, W. Egde, V.L.V. Mtotywa, J.A.M. van Brecht, Solv. Ext. Ion Exch. **19**, 143 (2001)
29. J.M. Schaekers, J.G.H. du Preez, US Patent US 2004/0208808 A1. 21 October (2004)
30. R.J. Sandberg, R.B. Martin, Chem Rev. **74**, 471 (1974)
31. B. Lenarcik, B. Barszcz, Dalton Transactions, 24 (DOI:1039/DT9800000024) (1980)
32. B. Lenarcik, P. Ojczenasz, Sep. Sci. Technol. **39**, 199 (2004)
33. B. Lenarcik, A. Kierzkowska, Solv. Ext. Ion Exch. **22**, 449 (2004)
34. B. Lenarcik, A. Kierzkowska, Solv. Ext. Ion Exch. **24**, 433 (2006)
35. E. Radzyminska-Lenarcik, K. Witt, Sep. Sci. Technol. **50**, 676 (2015)
36. E. Radzyminska-Lenarcik, Sep. Sci. Technol. **43**, 794 (2008)
37. B. Lenarcik, A. Kierzkowska, Sep. Sci. Technol. **39**, 3485 (2004)
38. S. Nakatsuji, R. Nakajima, T. Hara, Bull. Che. Soc. Jpn. **42**, 3598 (1969)
39. B. Lenarcik, K. Oblak, Roczniki Chemii **51**, 2079 (1977)
40. B. Lenarcik, K. Kurdziel, R. Czopek, Solv. Ext. Ion Exch. **4**, 165 (1986)
41. B. Lenarcik, K. Kurdziel, Polish J. Chem. **55**, 737 (1981)
42. J. Rydberg, C. Musakis, G.R. Choppin, Principles and Practices of Solvent Extraction, vol.1 (M. Dekker, Inc., New York, 1992)
43. J. Gazo, I. B. Bersuker, J. Garaj, M. Kabesova, J. Kohout, H. Langfelderowa, M. Melnik, M. Serator, F. Valack, Coord. Chem. Rev. **19**, 253 (1976)

The uptake of silver(I) from chloride solutions by amine extractants

Katarzyna Wejman[1,*], and *Magdalena* Pilśniak-Rabiega[1]

[1]Wroclaw University of Science and Technology, Faculty of Chemistry, Division of Analytical Chemistry and Chemical Metallurgy, 50-370 Wrocław, Poland

Abstract. The amine extractants, bis(2-ethylhexyl)amine, N,N-dimethylethanolamine, and trioctylamine were used to recover silver(I) ions from chloride solutions. The effect of the pH, contact time, extractant concentration and reextraction were studied. It was found that extraction of silver(I) depended on the pH, extractant concentration and strongly on the contact time. Reextraction of Ag(I) ions from the loaded organic phase showed that the metal can be removed in over 50% for the three extractant using sodium hydroxide. The recovery of silver from the chloride leaching solutions were above 85% for bis(2-ethylhexyl)amine, above 58% for N,N-dimethylethanolamine, and above 70% for trioctylamine.

1 Introduction

Use Sulphurous copper ores have polymetallic character, which means that they contain a series of other chemical elements and some of them may be significant for reclamation process. In practice, there are none other recognised copper ore deposits in the world which contains so many supporting metals than polymetallic and polymineral copper ore deposits in Legnicko – Głogowski Copper Zone. The existence of supporting elements has a big influence on technological process; it could be the reason of serious disturbances as well as the source of additional profit. The supporting elements, which concentrate in different production streams, constitute useful ingredients, such as silver, nickel, cobalt, gold, platinum, or undesirable pollutants whitch have to be removed from processing systems (Pb, As, Cd, Hg, F, Sb) [1, 2].

According to various sources, the demand for silver is still increasing – it is mainly used in the production of kitchen vessels, photographic industry, production of electronic, electric and cooling devices, as well as manufacturing jewellery and coins. When there are exploited, the majority of those products may become the potential raw material for the silver reclamation. It is now being estimated that about a half of worldwide production of silver is a result of the reclamation of its various raw materials and wastes, which is a phenomenon of growing tendency [3-5].

The polish national leader of silver manufacture is Lower Silesia's KGHM Poland, which is now collecting silver using pyrometallurgical techniques. However, the diversity, dynamically changing characteristic and type of raw materials, which contain a great part of valuable metals resources, led to the growing interest in hydrometallurgical methods. Those methods are more flexible concerning scale and technical capabilities; moreover they guarantee at the same time a big changeability of the material and a high effectiveness of the processes. Taking silver raw materials into consideration, it is possible to treat them in an effective way using hydrometallurgical methods [2]. The basic process is leaching a solid chase that contains silver mostly in cyanide solutions, as well as chloride, thiosulfate, thiourea, ammonia, or by using concentrated sulphuric acid in increased temperature (ca. 200 °C) [6, 7]. Even though in this case it is cyanide leaching that is most commonly used, chloride leaching is much more beneficial after all, because it is less toxic and the process itself is more efficient and less time – consuming, while technological costs remain significantly lower [6].

Selective and effective extraction of the metal from the solutions after leaching is mostly achieved with solvent extraction, or polymeric resin sorption [8, 9]. They may be used in hydrometallurgical methods of metals recovery, which allows to concentrate and extract metals from the solutions that contains only trace amounts of a particular metal (even to the concentration of a few ppm), sometimes directly in the production plants where e. g. waste solutions are produced. In comparison with pyro-metallurgical methods, hydrometallurgical ones do not require high temperatures and do not produce harmful gases (carbon monoxide), particulates (lead and arsenic), while investment costs are significantly low. Moreover, solvent extraction process is highly selective and effective; it is also not time – consuming and the apparatus used is not complicated.

Solvent extraction, also called liquid-liquid extraction is one of the most important, economical and practical process in hydrometallurgical industry for separation and purification of several metals - nonferrous, rare earths, the platinum group metals (PGMs). This method have

* Corresponding author: katarzyna.wejman@pwr.edu.pl

a number of advantages like: high selectivity and metal purity [10-12]. Despite widespread use in industrial hydrometallurgical processes, phosphorus-containing compounds are considered less environmentally friendly than extractants which have hetero atoms such as oxygen, nitrogen or sulphur in their structures [13]. For this reason, this study tested the extraction of Ag(I) chloride solutions with amines.

Extracting silver in solvent extraction processes was widely examined and described in scientific literature. However, the usage of solvent extrication for recovering silver from solutions after chloride leaching, where the feed was a remain from atmospheric leaching of the concentrate from ZWR Lublin, was previously described in a scientific work only with the usage of three amines [14]. In order to broaden the perspective of the process of recovering silver from chloride solutions containing over 4 M chloride ions, the below examination was presented. The objective of this study was to examine the process of silver(I) ions extraction from chloride solutions. Effect of the initial pH, contact time, extractant concentration and stripping on silver(I) recovery was investigated. Additionally, the process of silver(I) ions extraction was tested from chloride pregnant leach solutions (PLS), coming from an atmosphere leaching of the copper flotation concentrate from Lubin Concentrator (KGHM Polska Miedź S.A.).

2 Experimental

2.1 Reagents

Amine extractants were used as organic solvents without further purification, detailed information was included in Table 1. Ksylene (Avantor) were the diluent.

Table 1. Characteristics of the amine extractants.

Amine	Sign	Concentration [%]	Producer
N,N-Dimethylethanolamine	DMEA	99.5	Sigma-Aldrich
Bis(2-ethylhexyl)amine	BEHA	99.0	Sigma-Aldrich
Trioctylamine	TOA	98.0	Sigma-Aldrich

2.2 General extraction procedure

Synthetic solutions of silver(I) ions were prepared by dissolving a known amount of silver nitrate (AgNO3) in distilled water with hydrochloric acid and sodium chloride (3.7 g/L of HCl and 234 g/L of NaCl). The initial Ag(I) ion concentration for synthetic solutions was 50 mg/L. Chloride leach solutions were generated in atmosphere chloride leaching of a solid residue after sulphate leaching of the commercial flotation sulphide concentrate produced at the Lubin Concentrator (KGHM Polska Miedź S.A.). Solutions after chloride leaching were contained silver(I) and other metals, such as Cu(II), Pb(II), Fe(III), Zn(II), Co(II), Ni(II). The chemical composition of "Lubin" solutions after chloride leaching was presented in Table 2. The organic phase was contained one of three extractants and ksylene as a diluent. Different solutions were used as stripping agent: NaOH, NaNO3, NH4NO3 and NH3•H2O. Aqueous and organic phases were shaken at 325 rpm and ambient temperature (23 ± 1°C) using a mechanical shaker type ELPAN 357. The concentration of Ag(I) ion in the aqueous phase before and after the extraction and the reextraction was determined by means of flame atomic absorption spectrometry (FAAS) using a Varian SpectrAA 20 Plus instrument.diluent.

Table 2. Composition of solutions after chloride leaching.

Part	Metal concentration [mg/L]						
	Ag(I)	Pb(II)	Cu(II)	Zn(II)	Ni(II)	Co(II)	Fe(III)
a	44.8	197.9	25.8	30.6	25.8	26.4	-
b	46.2	501.2	364.2	12.4	15.3	23.7	501.2

3 Results and discussions

3.1 The effect of the initial pH

It was expected that the efficiency of the extraction process of Ag by amine extractants, bis(2-ethylhexyl)amine (BEHA), N,N-dimethylethanolamine (DMEA), and trioctylamine (TOA) should mainly depend on the metal-extractant complex formation and the concentration of the extractable Ag(I) species (closely related to pH of the aqueous phase). Additionally, it was presumed that in case of the amine extractant use in this experiment nitrogen could positively influence on extraction of Ag(I), according to theory hard and soft acids and bases (HSAB) [15]. In these conditions, it could enable to coordinate Ag species via the anion exchange mechanism. Therefore, the effect of the initial pH on the extraction of Ag(I) was examined. As shown in Fig. 1, the extraction of Ag(I), regardless of the extractant used, highly depends on the acidity of the aqueous phase. The extraction efficiency at pH ~1.0 yielded values of 76, 80 and 87% for TOA, DMEA, and BEHA, respectively. A decrease in the solution acidity resulted in a reduction of the extraction efficiency for all studied extractants, reaching the level below 30% at pH values above 6. This pointed a high effectivity of examined amine (TOA). Furthermore, the separation of phases was very fast (< 15 s), and no emulsion formation was noticed.

DOI:10.1051/e3sconf/20171801018

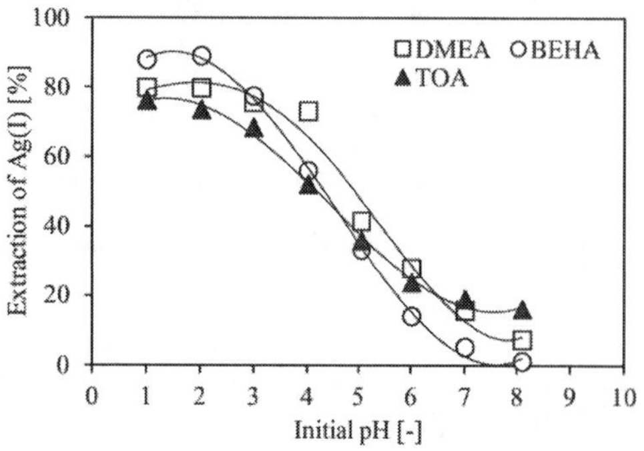

Fig. 1. Influence of initial pH on silver(I) extraction using BEHA, DMEA and TOA as extractants from chloride solution ([extractant] = 10% (v/v), [time] = 60 s).

3.2 Effect of contact time on silver(I) extraction

The effect of contact time on silver uptake is presented in Fig. 2, where the extraction efficiency of silver(I) was plotted vs. time. The process was carried out at ambient temperature and the contact time was varied in the range of 5 – 120 minutes. The concentration of the extractants (BEHA, DMEA, TOA) was equal to 10% (v/v). Phase ratio A/O (aqueous/organic) equal to 1. For two extractants (BEHA, DMEA), results show that the silver(I) ions were removed after 60 min of the shaking (Fig. 2). The extraction efficiency equals 89 and 82% for BEHA and DMEA, respectively. Results for TOA show that the majority of Ag(I) ions (77%) was removed after a 30 min of shaking. A further increase of the extraction time up to 60 min was found to slightly increase the extraction efficiency. Therefore, contact time of 60 min was recommended as the time, which enables to achieve of an equilibrium of the silver extraction for all extractants. emulsion formation was noticed.

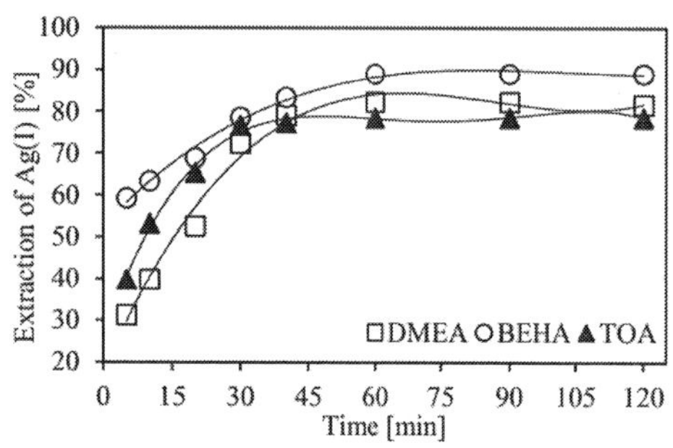

Fig. 2. Effect of contact time on silver(I) extraction using BEHA, DMEA and TOA as extractants from chloride solution ([extractant] = 10% (v/v), [time] = 5 – 120 min.).

3.3 Effect of extractant concentration

To study the effect of extractant concentration the content of BEHA, DMEA and TOA in organic phase was varied in the range of 2.5 – 20% (v/v) (Table 3 and Fig. 3). Aqueous (A) and organic phases (O) were contacted for 60 minutes at phase ratio A/O equal to 1. Extraction was performed at ambient temperature. It was observed that extraction efficiency increases with increase in extractant concentration. The maximum extraction efficiency of silver(I) (89.3, 87.6 and 75.5%, respectively) was obtained at the highest concentration of extractant (BEHA, DMEA and TOA, respectively), 20% (v/v). Additionally, Table 3 presents the effect of extractant concentration on the distribution coefficient (O/A) on the studied extractants. It appears that extraction of Ag for all extractants increases with increasing extractant concentration, same as the extraction efficiency. This result can be explained considering increasing of number of extractant molecules which leads to increasing the uptake. Applying too high concentrations of extractant in organic phase while silver concentrations in solutions are low may be unprofitable. Therefore, it was assumed that optimal extractant concentration in described experiments would be 10% (v/v).

Table 3. Silver(I) extraction using amine extractants (BEHA, DMEA and TOA) from chloride solutions.

Extractant concentration [% (v/v)]	DMEA		BEHA		TOA	
	E [%]	D	E [%]	D	E [%]	D
2.5	7.5	0.1	4.3	0.05	3.6	0.04
5	21.1	0.3	54.2	1.2	10.1	0.1
7.5	68.1	2.1	82.2	4.6	45.1	0.8
10	80.5	4.1	87.2	6,8	75.4	3.1
15	85.5	5.9	87.9	7.2	75.4	3.1
20	87.6	7.1	89.3	8.3	75.5	3.1

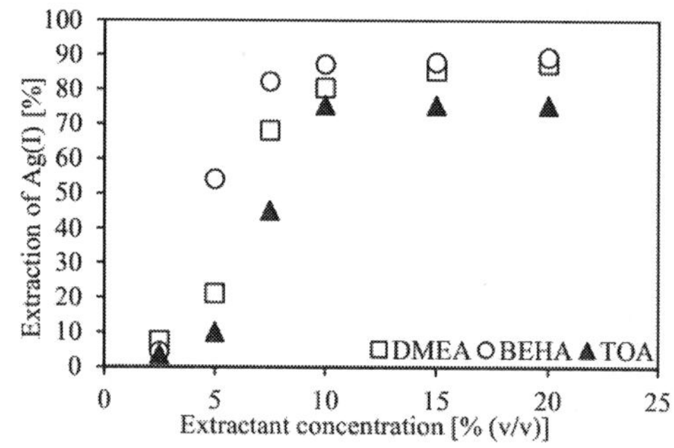

Fig. 3. Effect of extractant concentration on silver(I) extraction using BEHA, DMEA and TOA as extractants from chloride solution ([extractant] = 2.5 - 20 % (v/v), [time] = 60 min).

3.4 Reextraction

Stripping test was performed using freshly loaded and pre-filtered organic phases containing 77.4 - 89.3 mg/L of Ag(I) ions and 1.0 M of stripping agents ($NH_3 \cdot H_2O$, NaOH, NH_4NO_3, and $NaNO_3$) in the aqueous phase. Stripping of loaded organic was carried out at one to one volume ratios of loaded organic and stripping solutions. After the 60 min contact, the phases were allowed to disengage. The aqueous phases were separated and analysed. The results presented in Fig. 4 show that for all tested extractants were obtained low reextraction efficiency of Ag(I) using as a stripping agents: $NH_3 \cdot H_2O$, NH_4NO_3, and $NaNO_3$. Yield of the reextraction process was not exceeded 23%. The best stripping of Ag(I) ions from loaded organic phase in the case of all extractants (DMEA, BEHA, and TOA) was observed by using of 1.0 M NaOH solution and reextraction efficiency was consecutively 54, 57 and 77%, respectively.

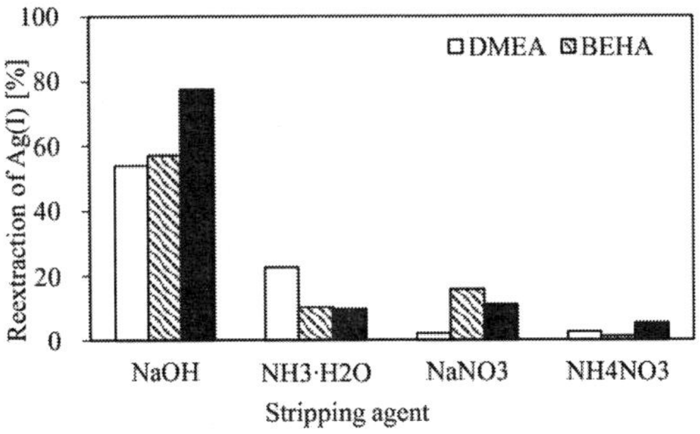

Fig. 4. Reextraction of silver(I) for amine extractants: BEHA, DMEA, and TOA ([extractant] = 10 % (v/v), [time] = 60 min).

Subsequently, the three stripping agents: NH3•H2O, NH4NO3, and NaNO3, were discarded and the effect of sodium hydroxide concentration on silver(I) reextraction was investigated (Table 4). Studies were conducted under the same conditions for the range of NaOH concentration from 0.25 to 2.0 M. In case of the organic phase containing the loaded trioctylamine (TOA), the stripping of Ag(I) from the organic phase was most efficient. The use of 1.0 M NaOH as stripping agent resulted in a 77.5% recovery of Ag(I) in one-stage process, while using a 2.0 M NaOH solution, the stripping of Ag(I) was until 78.7%. Additionally, was used a concentration of NaOH equal to 1 it is sufficient, and further increasing the concentration was not appreciably increase the recovery of silver. The maximum extraction efficiency of silver(I) (64.6 and 67.1 %) was obtained at the highest concentration of NaOH solutions for DMEA and BEHA (2M). For these extractants (DMEA and BEHA), the efficiency of the process varies considerably with the increase in concentration of NaOH solution.

Table 4. Effect of NaOH solution concentration on silver(I) reextraction using BEHA, DMEA and TOA as extractants.

Concentration of NaOH [M]	**Reextraction [%]**		
	DMEA	**BEHA**	**TOA**
0.25	16.1	28.0	36.8
0.50	44.1	54.9	71.0
1.0	53.6	57.1	77.5
1.5	56.3	59.6	77.9
2.0	64.6	67.1	78.7

3.5 Extraction of silver(I) from multicomponent solutions after chloride leaching

Silver(I) ions were also extracted from two chloride leaching solutions containing lead, copper, cobalt, nickel and zinc (composition in Table 2). Metals extraction were determined with an organic solution consisting of 10% each extractants (BEHA, DMEA, TOA) and aqueous phase at an A/O ratio of 1:1 and ambient temperature. During extraction, no third phase formation was observed. Extraction efficiency of particular metals and separation factors for silver(I) were showed in Table 5 (a,b). It can be seen that the metal extraction order was Ag>Ni>Pb>Fe>Zn>Cu>Co for DMEA, Ag>>Fe>Cu>Zn>Pb>Ni>Co for BEHA and Ag>>Cu>Zn>Fe>Co>Pb>Ni for TOA (for solutions after chloride leaching: a – without Fe, b – with Fe). The highest co-extraction of metals was observed for extractants DMEA and TOA, where separation factors (S) were significantly lower than for BEHA. The maximum extraction efficiency of silver(I) was obtained for BEHA and extraction efficiency was equal to 85.2% (a) and 86.6% (b). Additionaly, in relation to other metals the most selectivity extractant for silver(I) was also extratant BEHA and co-extraction of metals was not exceed 8.0% in case of iron-free solution. In the case of second solution (b) was show considerable co-extraction of iron(III) from solution (22.1%), which is confirmed by a lower partition coefficient in comparison to the others metals.

Table 5. Silver(I) extraction using amine extractants (BEHA, DMEA, TOA) from multicomponent solutions after atmosphere chloride leaching (*a* - containing metal such as Ag(I), Pb(II), Cu(II), Zn(II), Ni(II), Co(II), *b* - containing metal such as Ag(I), Pb(II), Cu(II), Zn(II), Ni(II), Co(II), Fe(III).

a.

Metal	**C_{intial} [mg/L]**	**DMEA**		**BEHA**		**TOA**	
		E [%]	**$S_{Ag/Me}$**	**E [%]**	**$S_{Ag/Me}$**	**E [%]**	**$S_{Ag/Me}$**
Ag(I)	69.0	58.3	-	85.2	-	70.7	-
Pb(II)	501.2	32.8	2.9	4.1	134.2	30.0	5.6
Cu(II)	4507	17.1	6.8	7.8	67.9	66.5	1.2
Zn(II)	175.4	19.6	5.7	7.1	75.6	42.0	3.3
Ni(II)	14.9	45.6	1.7	0.9	610.2	14.0	14.8
Co(II)	17.9	1.9	72.2	0.5	8.3	40.9	3.5

b.

Metal	**C_{intial} [mg/L]**	**DMEA**		**BEHA**		**TOA**	
		E [%]	**$S_{Ag/Me}$**	**E [%]**	**$S_{Ag/Me}$**	**E [%]**	**$S_{Ag/Me}$**
Ag(I)	46.2	58.4	-	86.6	-	74.0	-
Pb(II)	501.2	19.6	5.8	1.4	462.5	10.0	25.5
Cu(II)	364.2	3.3	41.4	9.3	62.8	45.3	3.4
Zn(II)	12.4	14.5	8.3	3.1	198.9	30.5	6.5
Ni(II)	15.3	49.7	1.4	0.2	3190	3.3	84.3
Co(II)	23.7	0.7	213.3	0.1	6975	21.6	10.3
Fe(III)	501.2	18.9	27.7	22.1	22.7	30.0	6.6

DOI:10.1051/e3sconf/20171801018

4 Conclusions

BEHA, DMEA and TOA were used for the solvent extraction of silver(I) ions from synthetic chloride solutions, and tested from chloride pregnant leach solutions (PLS), coming from a atmosphere leaching of the copper flotation concentrate from Lubin Concentrator (KGHM Polska Miedź S.A.) The experimental results show that Ag(I) can be efficient extracted from solutions using amine extractant. The best results were obtained using bis(2-ethylhexyl)amine as extractant using xylene as diluent. Additionally, extraction of silver(I) was efficient until the 60 minutes contact time, process significantly depends on inital pH for all extractants (BEHA, DMEA and TOA) in the aqueous phase. The most effective results were obtained for initial pH ~1.0 in the aqueous phase. Extraction process from chloride leaching solution was showed a large Ag(I) ions recovery for all using extractants, and also co-extraction of other ions (Cu(II), Pb(II), Zn(II), Ni(II), Co(II), Fe(III)) in small degree. The most selective extractant from chloride leaching solutions proved to be ekstractant BEHA, a high silver recovery rate of over 85%. The stripping of Ag(I) ions from loaded organic phase in the case of BEHA, DMEA and TOA was the most efficient using NaOH. The maximum extraction efficiency of silver(I) (64.6 and 67.1 and 78.7%) was obtained at the highest concentration of NaOH solutions equal to 2M for extractants DMEA, BEHA and TOA.

References

1. A. Piestrzynski, *Monography of KGHM Polska Miedź S.A.*, A. Piestrzynski (Ed.), (CPBM Cuprum Sp. z o.o. Publisher, Lubin, 201-237, in Polish, 2007)
2. T. Chmielewski, Sep. Sci. Technol. Separation Science and Technology **47(9)** 1264-1277 (2012)
3. S. Syed, Waste Management **50**, 234-256 (2016)
4. N. Sathaiyan, P. Adaikkalam, J.A.M.A. Kader, S. Visvanathan, *Encyclopedia of Materials: Science and Technology* (Second Edition, 8621-8623, 2001)
5. A.G. Chmielewski, T.S. Urbanski, W. Migdal, Hydrometallurgy **45**(3) 333-344 (1997)
6. M.F. Almeida, M.A. Amarante, Miner. Eng. **8**, 257-271 (1995)
7. R. Briones, G.T. Lapidus, Hydrometallurgy **50**, 243-260 (1998)
8. A.M. Shafiqul, K. Inoue, K. Yoshizuka, Y. Dong, P. Zhang, Hydrometallurgy **44**, 245-254 (1997)
9. Y. Abe, D.S. Flett,, Proc. Metall. **7**, 1127-1132 (1992)
10. A. Cieszynska, M. Wisniewski, Sep. Purif. Technol. **80**, 385-389 (2011)
11. A.P. Paiva, Sep. Sci. Technol. **28(4)** 947 (1993)
12. A.P. Paiva, Solv. Extr. Ion Exch. **18(2)** 223-271, (2000)
13 A.P. Paiva, Solvent Extr. 21st Century, Proc. ISEC '99, Society of Chemical Industry, 399-404, (2001)
14. K. Wejman-Gibas, M. Pilśniak-Rabiega, E3S Web of Conferences **8**, 01016, DOI: 10.1051/e3sconf/20160801016. (2016)
15. R.G. Pearson, J. Am. Chem. Soc. **85**, 3533 (1963)

The sorption of metal ions on nanoscale zero-valent iron

Tomasz Suponik[1,*], *Marcin* Popczyk[1], and *Piotr* Pierzyna[1]

[1]Silesian University of Technology, Faculty of Mining and Geology, 44-100 Gliwice ul. Akademicka 2, Poland

Abstract. The injection of the colloidal suspensions of nano-iron (nZVI) into an aquifer is a novel method of removing metal ions from acidic water. In the batch tests, the equilibrium study of the sorption of metal ions, Cu(II) and Zn(II), on Green Tea nanoscale Zero-Valent Ion (GT-nZVI) was carried out. The sorption of metal ions on this reactive material was described using the Langmuir, Freundlich and Sips models. This last model described in a better way the sorption equilibrium in the tested range of concentrations and temperature. The value of determination coefficient (R^2) for the Sips model, for copper and zinc, was 0.9735 to 0.9995, respectively. GT-nZVI has very good properties in removing Cu(II) and Zn(II) from acidic water. The high values of qmaxs, the maximum adsorption capacity in the Sips model, amounting to 348.0 and 267.3 mg/g for Cu(II) and Zn(II), indicate the high adsorption capacity of GT-nZVI. The analyzed metals have good or very good affinity with GT-nZVI.

1 Introduction

Waste materials resulting from the mining of non-ferrous metal ores and hard coal, as well as from the manufacturing process of non-ferrous metals, accumulated at industrial waste disposal sites contain metal minerals and, as a result, may affect the quality of ground and surface water in the southern areas of Poland. The leachate from these places may contain various toxic metals which may cause a risk to the environment.

In order to protect ground and surface water from leachates containing metal contaminants the suspension of nano-ZVI may be injected into contaminated aquifer [1]. In the study, the nanoscale zero-valent iron was obtained by reducing Fe(II) with a solution of green tea (GT). The polyphenols were responsible for the reduction properties of this tea [2]. GT is an effective, cheap and environmentally friendly alternative to other ways of obtaining nZVI. In accordance with Hoag et al. [3] the polyphenolic matrix can act as a capping agent that protects the iron nanoparticles from premature oxidation. In this way, the effect of removing metal ions by zero-valent iron can be intensified by biosorption on the coating of polyphenolic compounds. Biosorption on them is the process of selectively binding compounds (e.g. metal ions) on the surface of biological material by the presence of specific functional groups. The quality and quantity of these groups is not the same for each biosorbent. Therefore, the binding mechanism of the selected compounds is different. It may be based on physical adsorption, chelation [4], complexation, micro precipitation or ion exchange [5].

The processes of the removal of metal ions from water (directly) on Fe(0) are: reductive precipitation that lead to the metallic form, co-precipitation with iron in the form of oxides and/or hydroxides, precipitation as a result of an increase in pH and adsorption on the surface of iron corrosion products [6]. The type of processes depends on the type of metal removed, the other accompanying substances and the physicochemical conditions in the ZVI bed, mainly the pH of the water.

The aim of the study was to assess the equilibrium sorption of metal ions, Cu(II) and Zn(II), on GT-nZVI. The equilibrium of these processes were described using the Langmuir, Freundlich and Sips models.

2 Materials and methods

2.1 Synthesis of GT-nZVI

Nanoscale zero-valent iron particles was prepared in aqueous solutions by way of the reduction of ferrous iron Fe(II) using a solution of green tea (GT). The method was inspired by article written by Mystrioti et al. [7]. A method of preparing the suspension of green tea nanoscale zero-valent iron GT-nZVI was presented in the papers [1, 7, 8]. The 1:2 volume ratio of the GT solution to the Fe(II) solution were vigorously mixed in a flask reactor (see Fig. 1) at room temperature for an additional 15 minutes after the titration of GT. The solutions in the flask were mixed with a mechanical stirrer at 120 rpm. Through titration at a rate of 1.0 cm^3/s, the solution of GT was introduced to reduce ferrous ions to zero-valent iron Fe(0).

The Loyd brand of green tea was the source of polyphenolic compounds. An immersion of dry green tea in deionized water at a temperature of 80-90°C for 30 minutes, was the way of preparing the solution. It was filtered with the vacuum filtration, using a filter paper of 0.45 μm pore size. The ratio of green tea leaves to deionized water, amounted to 1g/50ml, while the ratio of

* Corresponding author: tomasz.suponik@polsl.pl

$FeCl_2{\cdot}4H_2O$ to deionized water was 1.35g/50ml. $FeCl_2x4H_2O$ was dissolved in deionised water and stirred under strong ultrasonic agitation a short time before introducing the GT solution [1].

The pH, conductivity and oxidation-reduction potential (ORP) in the GT-nZVI suspensions were 1.73, 10.7 mS/cm and 499 mV, respectively.

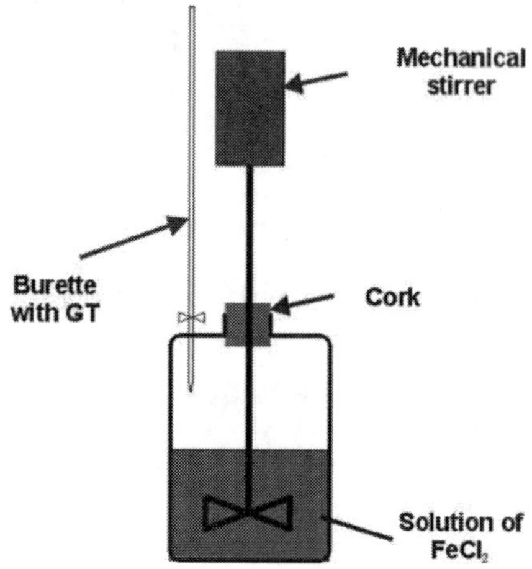

Fig. 1. Laboratory set-up for GT-nZVI synthesis [1].

2.2 Characteristics of the batch test

The several of experiments were done to study the capability of GT-nZVI in the removal of Cu(II) and Zn(II) from water in different initial concentrations of metals. For all these tests, the wastewater was prepared before. Metal ion solutions (copper, zinc) were prepared by dissolving the appropriate metal salts ($CuSO_4x5H_2O$, $ZnSO_4x7H_2O$) in distilled water. Always only one metal salts was dissolved in water. The concentrations of metal in the solutions were always higher than the allowable concentration of individual metals, within the meaning of legal regulations provided in the Journal of Laws, 2014, item 1800. The concentrations were in the ranges of 10 - 500 mg/dm^3, while the pHs for the solution of copper, zinc were 4.7, 5.3, respectively.

The tests were conducted in a programmable MULTI BIO RS-24 BIOSAN rotator equipped with plastic tubes filled with metal solution. Then the 0.5 ml of GT-nZVI was added into each of the solution. The 0.5 ml of suspension corresponds to 0.0067 g of dry matter obtained after the evaporation of water. This mass was used, as a mass of GT-nZVI, for calculations of the parameters of isotherms. However, this material is not consistent with the amount of GT-nZVI. The mass of iron is much lower. Using a lower weight, the sorption capacity of GT-nZVI would be much higher. In accordance with Mystrioti et al. [8], it was found that only ca. 27% of the total Fe was reduced to its elemental state. Namely, the produced GT-nZVI suspensions contained approximately 1.3 g/dm^3 of Fe in the form of nZVI solid particles and 3.7 g/dm^3 in the form of Fe(II) aqueous ions.

In order to eliminate the headspace (a gaseous phase above the solution) the total volume of the sample in the tube amounted to 58 cm^3. All these experiments were carried out in a laboratory temperature of 22°C±2 °C.

A two-minute orbital rotation was applied for each sample in the rotator (40 rpm), after which a six-second reciprocal motion (with a turning angle equal to 90°) with vibration followed. This sequence of shaking was repeatedly reiterated. The sequence continued for 120 minutes.

2.3 Measurement methods

After shaking the samples, the solutions were passed through filter paper of 0.45 μm pore size and assessed. The temperature, pH, the redox potential and the electrolytic conductivity were determined using electrometric methods (with the following meters: PORTAMESS 913 pH with a SenTix 41 combination electrode; PORTAMESS 913 pH with a POLYPLAST ORP electrode; PORTAMESS 913 Cound). Also, the following determinations were made with regards to water:

- concentrations of Cu_{tot}: Spectrophotometer DR5000 HachLange by the bicinchoninate method (method 8506 of Hach Co.), at wavelength 560 nm,
- concentrations of Zn_{tot}: Spectrophotometer DR5000 HachLange by the Zincon method (method 8009 of Hach Co.), at wavelength 620 nm,
- concentrations of SO_4^{2-}: Spectrophotometer DR5000 HachLange by the SulfaVer 4 method (method 8051 of Hach Co.) at wavelength 450 nm.

The measurements were carried out four times. The results were calculated as an arithmetic mean.

2.4 Mathematical description

The changes in the concentration of metal ions as a result of the use of GT-nZVI were calculated in accordance with the equation:

$$q_t = \frac{(C_0 - C_t) \cdot V}{m} \quad (1)$$

where :

q_t – the amount of metal ions adsorbed by the mass of nZVI, mg/g,

C_0 – the concentration of metal ions in water before the batch test, mg/dm^3,

C_t the concentration of metal ions in the water after 120 minutes, mg/dm^3,

V – the volume of solution, dm^3,

m – the mass of nZVI used in the test, g.

In order to determine the correlation between the amount of metals adsorbed on the surface of GT-nZVI and the amount of metals remaining in the solution, the three nonlinear mathematical models were used being the Freundlich (eq. 2) [9] and Langmuir (eq.3) [10] and Sips models.

$$q_{eq} = K_f \cdot C_{eq}^{\frac{1}{n_f}} \quad (2)$$

where :
q_{eq} – amount of metal ions adsorbed by the mass of nZVI at equilibrium, mg/g,
C_{eq} – the equilibrium concentration of metal ions in water after batch test, mg/dm^3,
K_f – Freundlich sorption equilibrium constant, (mg/g)(dm^3/mg)$^{1/n}$,
n_f – Freundlich exponent.

$$q_{eq} = \frac{q_{maxL} \cdot K_L \cdot C_{eq}}{1 + K_L \cdot C_{eq}} \quad (3)$$

where :
K_L – Langmuir equilibrium constant, dm^3/mg,
q_{maxL} – maximum sorption capacity in the Langmuir model, mg/g.
The Sips isotherm is a combination of the Langmuir and Freundlich isotherms and is given in the form [11].

$$q_{eq} = \frac{q_{maxS} \cdot K_S \cdot C_{eq}^{\frac{1}{n_S}}}{1 + K_S \cdot C_{eq}^{\frac{1}{n_S}}} \quad (4)$$

where :
K_S – the Sips equilibrium constant, (dm^3/mg)$^{1/n}$,
q_{maxS} – the Sips maximum adsorption capacity, mg/g,
n_S – the Sips model exponent.

3 Results and discussion

For the GT-nZVI, the sorption isotherms of metal ions were determined as a function of their concentration. Figures 2, 3 show isotherms for Cu(II), Zn(II), respectively. As was presented in these Figures, the sorption capacity of GT-nZVI increased for the initial concentration of metal ions in the solution in the range of 10 to 100 mg/dm^3. Then the capacity grew much slower.

The obtained results correlate well with the selected equations of adsorption equilibrium (see Table 1). This is evidenced by the high or very high values of R^2. The higher values of R2 in the Sips model, for the both of metals, indicate the best suitability fitting to the results of the experiments. These were 0.9735 and 0.9995, for copper and zinc, respectively.

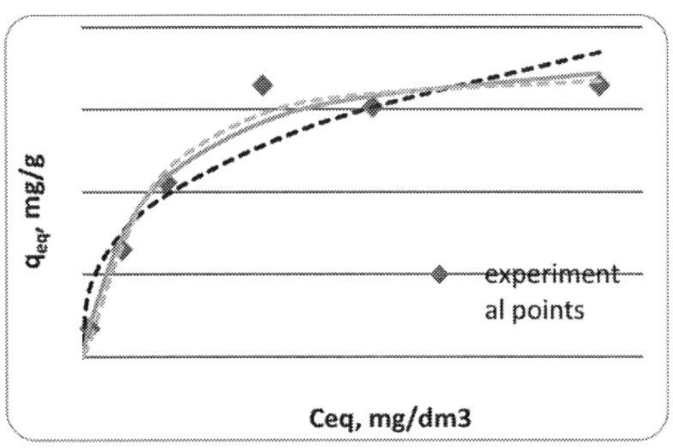

Fig. 2. Equilibrium of the sorption of Cu (II) on GT-nZVI.

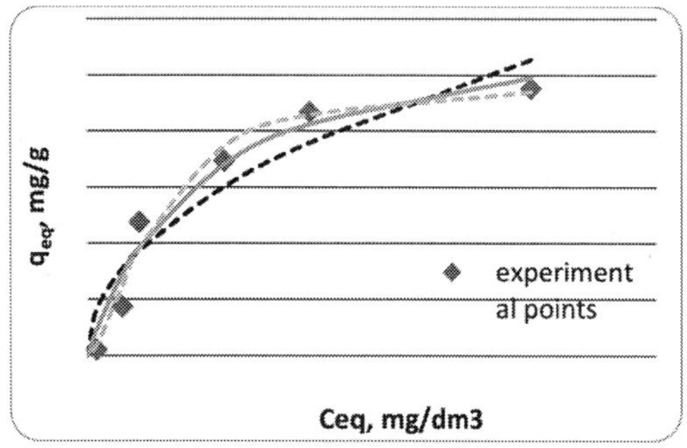

Fig. 3. Equilibrium of the sorption of Zn(II) on GT-nZVI.

Based on the data presented in the table 1, it can also be said that GT-nZVI has very good properties in removing copper and zinc from water.

The high values of $qmax_L$, in Langmuir's theory, $qmax_S$, in Sip's theory, and Kf, in Freundlich's theory, indicate the high adsorption capacity of this material for all analyzed metal ions.

These values were 387.4 mg/g, 348.0 mg/g, 43.33 (mg/g)(dm^3/mg)$^{1/n}$, for Cu(II), and 305.2 mg/g, 267.3 mg/g and 45.64 (mg/g)(dm^3/mg)$^{1/n}$, for Zn(II). It should be noted here that ionic strength of solutions was small compared to natural waters.

1/n is a function of the force with which the metal ions are sorbed by the adsorbent. The lower the 1/n (towards zero), the greater the affinity of the substance.

For both tested metals, the value of 1/n below 1 indicates a normal Sips isotherm and Freundlich isotherm, and the adsorption becomes more heterogeneous as the value of 1/n gets closer to zero.

Yang et al. [12] studied also the sorption of copper on adsorbent. This was the walnut shell supported nanoscale zero-valent iron. It was prepared from sodium borohydride, iron(II) chloride tetrahydrate, and walnut shell by liquid phase chemical reduction.

The values of q_{maxL} and K_L, in the Langmuir model, and K_f and $1/n_f$, in the Freundlich model, were, in the study of Yang et al. [12], 398.4 mg/g, 0.029dm^3/mg and 199.8 (mg/g)(dm^3/mg)$^{1/n}$, 0.084, at ca. 20°C. These values are of similar magnitude as was obtained with the use of GT-nZVI.

Another study on the sorption of copper on iron was presented by Ayob et al. [13]. They investigated the feasibility of using carboxylmethyl cellulose (CMC)-stabilized nano zero valent iron (nZVI) particles for immobilization of Cu2+ in water. The parameters of Freundlich's isotherm for this material were K_f= 9.1559 (mg/g)(dm^3/mg)$^{1/n}$ and n = 3.042.

In conclusion, it can be said that nano-iron has very good properties in the removal of metal ions from water. The process of water purification by Fe (0) was probably intensified by biosorption on the coating of polyphenolic compounds. This may be evidenced by the fact that the concentration of sulphates (see Table 2) also decreased, unlike the observations presented in the paper [14]. The parameters such as pH and ORP (see Table 2) were varied in the same way as in the case of Fe(0) in the form of fine grain steel [14] or in the form of rector with steel sheets

 DOI:10.1051/e3sconf/20171801019

positioned in parallel inside. This is proof that Fe(0) itself is reactive.

However, the changes in the values of pH, ORP and sulphate concentrations were very small, because a small amount of GT-nZVI was added to the solutions.

Table 1. The equilibrium parameters of sorption obtained from the Freundlich, Langmuir and Sips isotherms.

The type of metal removed from the solution	The parameters of the isotherm of: Langmuir									
	Freundlich			Langmuir			Sips			
	K_f (mg/g) $(dm^3/mg)^{1/n}$	n_f	R^2	K_L dm^3/mg	q_{maxL} mg/g	R^2	K_S $(dm^3/mg)^{1/n}$	q_{maxS} mg/g	n_S	R^2
Cu(II)	43.33	2.864	0.8645	0.0170	387.4	0.9637	0.0037	348.0	1.434	0.9735
Zn(II)	45.64	3.218	0.7965	0.0264	305.2	0.9567	0.0010	267.3	2.006	0.9995

Table 2. The change in the value of pH, ORP and sulphate concentrations according to initial metal ions concentration. The positive value means an increase after batch tests, while negative a decrease. The table presents the average values for the analyzed parameters/substances.

Parameter /substance	The change in the value of the parameter/substance for the initial concentration of metal ions:					
	10 mg/dm^3	50 mg/dm^3	100 mg/dm^3	200 mg/dm^3	300 mg/dm^3	500 mg/dm^3
pH	+0.11	+0.10	+0.06	+0.03	+0.01	0.01
ORP, mV	-24	-21	-18	-12	-11	-4
SO_4^{2-}, mg/dm^3	-36	-48	-61	-88	-92	-110

4 Conclusion

This study has shown that GT-nZVI is a good material for removing metal ions from acid mine water. This is determined by:

- the very good sorption capacity of GT-nZVI - In Sips model, the q_{maxS}, were 348.0 and 267.3 mg/g, for copper and zinc, respectively,
- the good affinity of copper and zinc with GT-nZVI - The low value of 1/n, in the Sips theory, amounting to 0.697 and 0.499 for copper and zinc, respectively,
- the dual action of F(0) and biosorbent - The process of water purification by Fe(0) was probably intensified by biosorption on the coating of polyphenolic compounds,
- the size of GT-nZVI and other properties – This material has a large surface area, is inexpensive to use, compared with other reactive materials, and environmentally friendly and, if kept stable, can spread without clogging the pores of the aquifer.

The obtained results of experiments correlate the best with the Sips theory. The determination coefficient (R^2) was 0.9735 and 0.9995 for copper and zinc, respectively.

References

1. T. Suponik, M. Lemanowicz, P. Wrona, E3S Web Conferences, **8,** 01048 (2016)
2. Y. Joshi, S., *Nanoscale and Microscale Phenomena*, (Springer, 2015)
3. G. E. Hoag, J. B. Collins, J. L. Holcomb, J. R. Hoag, M. N. Nadagouda, R. S. Varma, J. Mater. Chem., **19(45),** 8671-8677, (2009)
4. R. C. Hider, Z. D. Liu, H.H. Khodr, Meth. Enzymol., **335**, 190-203 (2001)
5. D. Podstawczyk, A. Witek-Krowiak, **1**, 232-237, (Creativetime, 2015)
6. T. Suponik, M. Blanko, Physicochem. Probl. Mi., **50(1)**, 359-372 (2014)
7. C. Mystrioti, N. Papassiopi, A. Xenidis, D. Dermatas. M. Chrysochoou, J. Hazard. Mater., **281**, 6 (2015)
8. C. Mystrioti, A. Xenidis, N. Papassiopi, J. Geosci. Env. Protec., **2**, 28-36 (2014)
9. H.M.F. Freundlich, Phys. Chem., **57**, 385 (1906)
10. I. Langmuir, Journal of the American Chemical Society, **38(11)**, 2221-2295 (1916)
11. O. Hamdaoui, E. Naffrechous, J. Hazard. Mater., **147(1)**, 381-394 (2007)
12. F. Yang, Y. He, S. Sun, Y. Chang, F. Zha, Z. Lei, J. Appl. Polym. Sci., **133(16)**, (2016)
13. A.F.I.Z.A.H. Ayob, N. Ismail, T. T. Teng, A.Z. Abdullah, Environ. Prot. Eng., **38(3)**, 119-131 (2012)
14. T. Suponik, Physicochem. Probl. Mi., **49(1)**, 1643-1649 (2013)

DOI:10.1051/e3sconf/20171801020

Considerations about recovery of critical metals using bio-metallurgy

Săndica Liliana Gherghe[1,*]

[1]National Research and Development Institute for Nonferrous and Rare Metals, 102 Biruintei Blvd, Pantelimon; Romania

Abstract. The increased requirement of critical metals due to green technologies needs, together with the geopolitical environment to ensure these metals, has entailed decisive measures to avoid current supply insecurities in each country. These metals are essential to the products and services made and used daily, and contribute to sustaining and growing the economy. Thus, sustainable approaches from technological, environmental, economic and social point of view are needed to recover these metals from different resources. By using resources more efficiently, innovating in the concept of circular economy, it can be assure re-using, re-manufacturing or recycling of valuable materials. In this paper are presented some considerations related to using the waste dumps as potential resources to obtain critical metals by bio-metallurgical processes.

1 Introduction

The metals are precious raw materials to the economy of any country and need to be secured for sustainable production of key components of various products such vehicle production, electronics and IT hardware, low carbon energy technologies, mechanical, electrical and process engineering.

The availability and supply of critical metals has a significant influence on the economy of any country by affecting manufacturing, export, and job creation. Most of the critical and rare metals are currently obtained through mining of primary sources, which are finite, unequally distributed in the world, and rapidly decreasing as a result of high consumption due to urbanization, increasing standards of living, and the population explosion. Therefore, it is an urgent need to source these critical and scarce elements from alternative sources using sustainable technologies. Potential alternate sources that are intensely considered include recycling from end-of-life metal wastes and recovery from mine tailings and wastewaters [1].

Continue and rapid increase of the waste materials generation and wide range of industrial applications with high-tech technologies sustain the assessment and development of innovative technologies focus on waste recycling. Biometallurgical technologies are most likely to find their niches in recovering critical metals from these waste residues. The efficacy of biometallurgical recovery of metals from waste dumps depends largely on the matrix composition and chemical species of the targeted metals in the wastes [2].

Some wastes can become valuable resources, supplying metals that are extracted today by other processes, promoting recycling, minimizing harmful waste and hazard and dissipation.

Biotechnological recovery of critical and scarce metals has received significant interest because it can be used to treat diluted wastes and required low energy input [1].

Technology is crucial to improve recycling processes and to trace and capture secondary raw materials [3].

2 Criticality assessment

Critical raw materials (CRM) are raw materials of high importance to the economy of the EU and whose supply is associated with high risk.

The development of European economy depends decisive on access to CRM. The important topic on CRM include strong and growing demand from industry as well as limited and volatile supply. The closed circuit of the suppling of raw materials potentially constrains economic growth as rising prices make key industries less profitable. In the worst cases, severe shortages of CRM may also determine even temporary production halts. This is something that could conceivably have significant negative repercussions for the competitiveness of European industry. The EU's supply of CRM is further threatened by the fact that some emerging economies, in particular China, are limiting raw materials supply by means of export restrictions [4].

The European economy is the largest in the world, with a strong industry, while the mineral production within EU does not satisfy the demand. Therefore, it lacks the mineral wealth needed to sustain the growth. The problem of addressing challenges with the minerals supply, especially concerning certain CRM, has arisen

* Corresponding author: gliliana@imnr.ro

with recent market volatility that is partly a result of dependency on certain countries.

The two main parameters, economic importance (EI) and supply risk (SR), are used to determine the criticality of the material for the EU. In the revised methodology for criticality assessment, the availability of substitutes is considered a reducing element in both the economic importance and the supply risk dimensions. The assessment only takes into account the proven substitutes that are readily available today and able to reduce the consequences of a disruption and/or influence the risk of a disruption. Commercial information and published patents are only used to identify proven substitute alternatives readily available and applicable at the market today. 'Substitutability' and 'potential future substitution' are not considered in this methodology [5].

The factors of criticality related to supply and demand of metals are presented in table 1 [6].

Table 1. The factors of criticality related to supply and demand of metals.

SUPPLY **Indicators for evaluating supply risk**	DEMAND **Indicators for evaluating vulnerability**
Country concentration	Value of products affected
Country risks	Spread of utilization
Depletion time	Strategic importance
By-product dependency	Value of the utilized material
Proved substitute	Proved substitute
Import dependence	Demand share
Price volatility	Import dependence

The increasing competition for resources, price volatility and potential interruptions in supply, caused by a combination of growing worldwide demand, concentration of supply in a few countries, trade restrictions in sometimes, lack of currently viable alternatives in key applications are the key factors in the supply response to increased demand [7].

There are several worldwide changes underway that are likely to make resource extraction from alternative sources a more and more viable option such as rapidly growing competition for resources, increasing raw material prices, the large-scale environmental problems We now face and the fact that the natural reservoirs for many valuable resources are rapidly declining [8, 9]. At the same time, several studies belonging to the research field of industrial ecology show that in many regions of the world massive amounts of strategically important materials such as metals have accumulated in landfills [10, 11, 12]. The potential for resource recovery from landfills appears significant, but depend of the performance of current technologies. In the past, the metals obtaining efficiency was not so high for some metals from some raw material and, as a consequence, important amount of such metals was discharged in tailing dams. Therefore, the mining industry has been generated waste dumps on the surface of natural terrains and caused major problem in the context of environmental protection.

Some metal-bearing and liquid wastes are considered as secondary sources of critical and scarce metals.

One measure to mitigate the criticality is recycling. Recycling is effective for the metals whose: "Criticality" is high and "Potential of recycling" is high. The "potential of recycling" of a metal depend on the "amount of metal contained in the waste-products in one country" and "the amount of the metal's demand in the same country: "Potential of recycling" of a metal =

$$\frac{\text{Amount of a metal contained in the waste} - \text{products in the country}}{\text{Amount of the metal's demand in the contry}}$$

In the assessment of the criticality, the following factors and influences are taken into account:
- technical performance (extent to which the substitute can replace the functionality of a candidate raw material in an application)
- cost performance (costs often drive decisions in business);
- substitute production (availability of substitutes in sufficient quantities need to be considered);
- substitute criticality (substituting one critical material with another is unlikely to decrease the RM supply risk in a given application). It is important to to assess the criticality of the substitute itself. If a substitute material is already critical, it might not be readily available as a substitute option.
- substitute by-/co-production (if the proposed substitute is mainly obtained as a by- or co-product, its supply is dependent on a demand for another raw material) [5].

Table 2. List of today critical raw materials at EU level.

No	Raw material
1	Antimony (Stibium)
2	Beryllium
3	Borates
4	Chromium
5	Cobalt (Cobaltum)
6	Coking coal
7	Fluorspar (Fluorite)
8	Galium
9	Germanium
10	Indium
11	Magnesite
12	Magnesium
13	Natural graphite
14	Niobium
15	Phosphate rock
16	Platinum Group Metals
17	Heavy Rare Earth Elements
18	Light Rare Earth Elements
19	Silicon metal (Silicium)
20	Tungsten (Wolframium)

Taking into account the analysis performed by experts at European level, a list of 14 critical metals has been established in 2010. Due to the fact that the criticality's degrees changed over the time, the list has been up-dated with six new materials (marked in grey in table) at 20 critical metals in 2014 [13, 14]. This new list includes thirteen of the fourteen materials identified in the previous report, with only tantalum (due to a lower supply risk) moving out of the EU critical material list. Within the framework of the EU Raw Materials Initiative, it was decided to assess and up-date the number of raw materials at least every three years.

3 Why bio-processes?

Microbes interact with metals and minerals in natural and synthetic environments, altering their physical and chemical state; on the other-hand metals and minerals are also able to affect microbial growth, activity and survival. Many minerals have biogenic genesis, and the formation of such biominerals has a geological and industrial significance, as well as providing important structural components for many organisms.

Microorganism can decisively influence the redox chemistry of minerals and therefore, could have an important effect for the extraction and processing of valuable metals.

Many microorganisms and methods are described in the literature as potential options for metals recovery/recycling, but the efficiency of these processes depends not only on the activity of the microorganisms, but also on the geochemical composition of the resources.

Naturally occurring microorganisms can sorb a variety of *PGMs, REEs* and heavy metals, including *Pt, Fe, Ni, Cu, Zn, Pb, Cu, Pd, Ag, Cd, Pt, Au, and Hg*, with binding capacities typically on the order of 10^{-5}–10^{-3} mol metal g^{-1} (dry wt) microbe [2]. The metal's chemical species influence the reduction potential and uptake efficiency.

The metals biosorption to microbes can involve a variety of processes, including absorption, ion exchange, complexation, and precipitation [15, 16]. These processes are determined by the chemical groups displayed on the extracellular surfaces of microbial cells, such as carboxyl, phosphoryl, hydroxyl and carbonyl.

The competition between ions changes in the activities of metals and/or microbial functional group, or changes to the electrical layer at the microbe-water interface is influenced by essential biological parameters (pH, ionic strength and temperature). The reduction in heavy metal uptake on Gram negative bacteria with increasing ionic strength has been attributed to changes in the activity of metal cations [17].

Most studies performed are focused on the removal of the metals from the wastewaters. Bioaccumulation is a metabolically active process in living organisms that works through adsorption, intracellular accumulation, and bioprecipitation mechanisms, individual and/or in consortia. Internal metal accumulation and/or metal biosorption by metal resistant microorganisms should be clear understand for each study case.

Biomining is often used for the processing of ore and concentrates and both bacteria and fungi are used in bioleaching [18]. A variety of microorganisms is found in leaching environments and has been isolated from leachates and acidic mine drainage. Although environmental conditions are usually described as being extreme and aggressive due to very low pH values and high metal concentrations, these systems can show high levels of microbial biodiversity [19]. Recent detailed investigations based on molecular methods revealed that microbial bioleaching communities are composed of a vast variety of microorganisms resulting in complex microbial interactions and nutrient flows. The composition of these communities is usually subjected to seasonal fluctuations and may vary between different mining locations and, in addition, organisms are not homogeneously distributed over the whole leaching environment [20]. Different biological materials, such as bacteria, algae and fungi have been received high attention for metals removal and recovery, due to their good performance, low cost and large available quantities. These microorganisms have metal-sequestering property and can be used to decrease the concentration of some metals ions in solution from ppm o ppb level. It can effectively sequester dissolved metals ions out of dilute complex solutions with high efficiency and quickly, therefore it is recommended for the treatment of high volume and low concentration complex wastewaters [21].

Better understanding of biological role of *REEs* and their acquisition by microorganisms will allow the development of bio-recovery of *REEs* from solid/liquid wastes.

Microbial reduction of soluble tellurite to insoluble elemental *Te*, represent a critical step in *Te* recovery.

Among *PGMs*, the major economic importance has Pt, Pd and Au [22]. Paladium has been obtained using diverse microorganisms and biofilms [23]. The *platinium* recovery by mixed culture has been demonstrated [24].

The *Leucobacter chromiiresistens* have been isolated and identified a strain of resistant to *chromium* [25].

The T. ferrooxidans accelerates the leaching of *gallium*, and *indium* is also solubilized. T. ferrooxidans also oxidizes chemically pure *gallium* and Ga_2S_3 associated with chalcopyrite concentrates, simultaneously by direct and indirect mechanisms.

Germanium, which replaces in isomorphic manner the Cu, Zn and Pb in the respective sulphides, can be leached by T. ferrooxidans within a wide range of pH. The *rhenium* present in molybdenite, pyrite, chalcopyrite and galena deposits is also dissolved in concentrations from 5 up to 500 mg L^{-1}. Some inorganic *selenium* compounds are oxidized/reduced by microorganisms [26].

One strain of Penicillium tricolor is capable of lowering the value of the pH from the culture medium from 10.0 to 3.0 and it can also be used to leach out *REE* and radioactive elements [27].

Metal bioaccumulation was studied in some microorganisms, especially for *U* (Rhodanobacter sp. described by Sousa et al., 201315) but also for *Th, Co, Cr, Mn, Sn and Pt.* Nanoparticles and granular biosorbents [28] were developed for water remediation, capable of agglomerating heavy metals by adsorption *Ag, Cd, Cu, Pb, Zn* and *U.* The process seems versatile and likely to have future expansion [29].

Proteus vulgaris can absorb *cobalt*, with a high efficiency under conditions of neutral pH and *Saccharomyces cerevisiae* can accumulate *cadmium and cobalt* [30, 31].

Tungsten and *molybdenum* causes changes in the bacteria *Desulfovibrio vulgaris*, because they are involved in activation of some specific enzymes [32].

The main applications of bioleaching are being used commercially for the recovery of copper, uranium and gold [33].

4 Bio-metallurgy: a way for future

The development of an EU strategy on raw materials emphasizing the concept of the "added value chain", pursuing the three-pillar strategy to: (1) ensure the fair and sustainable supply of raw materials from international markets, promoting international cooperation with developed and developing countries; (2) foster sustainable supply of raw materials from European sources, and (3) reduce consumption of primary raw materials by increasing resource efficiency and promoting recycling [34]. The above drivers serve as a backdrop rationale for the development of the innovative hybrid technologies, applicable in primary and secondary sources of raw materials.

Biotechnological recovery has received interest because it can work with dilute waste streams and requires a low energy input. In the last years, microbe–metal interactions have received renewed attention as a route to develop sustainable biotechnological metal-recovery processes for leaching from low-grade ores and solid wastes as well as via metal immobilization from leachate solutions, process streams, and wastewaters [1].

Bio-extractive methods complete the alternative approaches for extracting metals from different primary and secondary resources. The main aspects of biological processes are related to long time required to obtain economic levels of metal extraction (from days to years sometime) and concerns about unexpected behaviour of biological system involved. On the other hand, biological processes involve much lower temperature, smaller carbon footprint and could be an option for pretreatment of low-grade and/or complex primary and secondary resources for metals recovery [35].

Besides the metals recovered in the leachate, there is increasing interest in the insoluble metals left in the residues. Inferior lead sulfide concentrates can be transformed into high-value concentrates by leaching of metals (e.g., zinc, cadmium, copper) that interfere with conventional processes for the recovery of the lead. Similar procedures are being investigated for the extraction of silver and other precious metals that are finely disseminated in iron, arsenic, copper and zinc sulfides. The metal sulfides are first removed by microbial leaching and the precious metals are then recovered from the residue.

Traditionally raw materials exploitation is focused in high grade ore deposits, extracted and processed by conventional techniques, pyro and/or hydrometallurgy. The metal recovery efficiency of these techniques was variable along time and framed by a minimum efficient scale approach. As a consequence, a significant amount of metals was discarded to tailings dams, many of those in concentrations that exceed today's mines cut-off grades. Tailings from mining and metallurgical processes put some very serious environmental problems because of their metals content and have raised questions about safe disposal.

The bioleaching will become more important for zinc, nickel, cobalt and molybdenum recovery. Investment and operating costs are much lower than for conventional pyrometallurgical and hydrometallurgical processes. The processing plant can be built in the immediate vicinity of the ore deposit, saving transport costs. The procedures are not complicated and are easy to control, extensive technical knowledge is not required [36].

The studies and researches on bioleaching of high tech critical metals are at the beginning. The bioleaching processes are influenced by a wide range of parameters including physicochemical and microbiological factors, which can affect both the growth of the microorganisms and their behavior during biological processes. The optimal growth conditions for microorganisms must be maintained in order to assure the resistance of microbes to the metals and get a good efficiency of bioprocesses. [37].

5 Conclusions

There is a demand for less expensive and more environmentally friendly processes, with developments on both technical and biological issues.

A crucial aspect is related to increasing the rate of leaching and the tolerance of the microorganisms/consortia to different metals, in different types of primary/secondary materials. Biotechnological processes are already being applied at industrial scale in biomining of sulfidic ores and the treatment of wastewaters containing metals.

A little progress has been made in the development of biotechnological processes for the recovery of critical and scarce elements. Therefore, there are few or no studies on metal–microbe interactions for most critical elements. The importance of biotechnologies has to be taken into account due to the ability of microorganisms to selectively sequester diffuse elements (e.g. critical metals) from different solid/liquid wastes. This is starting point of challenges for next years, for development of new processes for the extraction and recovery of metals secondary resources such as mine tailings, stored solid wastes, end-of-use products, sewage sludge, process streams, and industrial effluents using microorganism's abilities. Present challenges faced in biological recovery include selectivity, low pH, and the presence of salts and/or other metal co-contaminants.

Progress in the development and integration of biological processes and biological recovery with hydrometallurgy methods will foster the development of novel biotechnological processes for extraction and recovery of critical metals [1].

By combining future valorization of different materials with energy production and land re-use, cost efficient resource recovery of landfills will generate economic, environmental and social issues. As primary resources become scarcer enhanced landfill mining will become more feasible. However, strategic policy decisions and tailored support systems for enhanced landfill mining is sustained by standardized life cycle assessment frameworks and will be indispensable to remove the remaining non-technical barriers [37].

The last worldwide changes and recent research findings indicate the emergence of a new perspective on landfills as reservoirs for resource extraction.

The author would like to acknowledge the financial support from the Romanian National Authority for Scientific Research and Innovation (UEFISCDI), project number 16/2016, within ERA-MIN Program.

References

1. Y. V. Nancharaiah, S. V. Mohan, P. N. L. Trends Biotechnol., **34**, No.2, (2016)
2. W-Q. Zhuang, J. Fitts, C. Ajo-Franklin, S. Maes, L. Alvarez-Cohen, T. Hennebel, Curr. Opin. Biotechnol, **33**, 327–335, (2015)
3. F. J. Dessart, L. Bontoux, Non-ferrous Metals Manufacturing: Vision for 2050 and Actions Needed, JRC Science for Policy Report, (2017)
4. European Parliament – Directorate General for Internal Policies, Substitutionability of Critical Raw Materials, (2012)
5. European Commission, Methodology for establishing the EU list of Critical Raw Materials, (2017)
6. Mitsutaka M., Criticality assessment and R&D for circulation of critical materials, The 6th Trilateral EU-US-Japan Conference on Critical Materials BREYDEL, (2016)
7. Department for Environment, Food and Rural Affairs, (2012)
8. Halada, K., Ijima, K., Shimada, M., Katagiri, N., J JPN I MET, **73**, 151–160, (2009)
9. Kapur, A., Progress of J Ind Ecol, **3**, 209–236, (2006)
10. Lifset, R.J, Gordon R.B., Graedel T.E., Spatari S., Bertram M., JOM **54,** 21-26, (2002)
11. Kapur A., Graedel T.E., Envir Sci and Tech, **40**, 3135-3141, (2006)
12. Muller D., Wang T., Duval B., Graedel T.E., PNAS **103**, 16111-16116, (2006)
13. European Commission, Critical Raw Materials for the EU, (2010)
14. European Commission, Report on critical raw materials for the EU, (2014)
15. G. M. Gadd, J. Chem. Technol Biotechnol, **84**, 13-28, (2009)
16. H. Moriwaki, H. Yamamoto, Appl. Microbiol Biotechnol, **97**, 1-8, (2013)
17. D. M. Borok, J. B. Fein, J Colloid Interface Sci, **286**, 110-126, (2005)
18. A. Schippers, S. Hedrich, J. Vasters, M. Dobre, W. Sand, S. Willscher, Adv Biochem Eng Biotechnol, **141**, 1-47, (2014)
19. J. B. Glass, H. YU, J. A. Steele, K. S. Dawson, S. Sun, K. Chourey, C. Pan, R. L. Hettich, V. J. Orphan, Environ. Microbiol. **16**, 1592–1611, (2014)
20. H. Zhiguo, G. Fengling, Z. Jiancun, H. Yuehua, Q. Guanzhou, FEMS Microbiol. Ecol, **74**, 155-164 (2010)
21. J.L. Wang, C. Chen, Biotechnol Adv, **24**, 427-51, (2006)
22. C. N. Mpinga, J. J. Eksteen, C. Aldrich, L. Dyer, Miner Eng, **78**, 93-113, (2015)
23. A. Pat-Espadas, J. A. Field, E. Razo-Flores, F. J. Cervantes, R. Sierra-Alvarez, J. Chem. Technol. Biotechnol, **91**, 4, (2016)
24. S. Maes, R. Props, J. P. Fitts, R. De Smet, R. Vilchez-Vargas, M. Vital, D. H. Pieper, F. Vanhaecke, N. Boon, T. Hennebel, Environ. Sci. Technol, **50**, 5, 2619-2626, (2016)
25. G. Sturm, J. Jacobs, C. Sproer, P. Schumann, J. Gescher, Internat. J. of System and Evol, Microb, **61**, 956-960, (2011)
26. Henry Lutz Ehrlich, *Geomicrobilogy,* Fourth edition, Chapter 20, 669-682, (2002)
27. Y. Qu, B. Lian, B. Bioresour Technol, **136**, 16-23, (2013)
28. T. Raychoudhury, T. Scheytt, Water Sci Technol, **68**, 7, 1425-39, (2013)
29. BioCriticalMetals project developed within ERAMIN Program (2016-2019).
30. M. Neyland, P. Dunkel, AL Schade, J. of. Gen. Microbiology, **7**, 409-416, (1952)
31. P. R. Norris, J. of Gen. Microbiology, **99**, 317 – 324, (1977)
32. S. M. Da Silva, J. Voordouw, C. Leitao, M. Martins, G. Voordouw, IAC Pereira, Microbiology, **159**, 1760-1769, (2013)
33. K. Bosecker, FEMS Microbiol Rev., **20**, 591-604, (1997)
34. Communication from the European Commission to the European Parliament and the Council, The Raw material initiative-meeting our critical needs for jobs and growth in Europe, COM (2008) 699 final.
35. D. Barrie Johnson, Curr. Opinion in Biotech, **30**, 24-31, (2014)
36. F. Acevedo, J. C. Gentina, S. Bustos, J. Biotechnol, **31**, 115-123 (1993)
37. A. Monballiu, N. Cardon, M. T. Nguyen, C. Cornelly, B. Meesschaert, Y. W. Chiang,
38. Bioinorg Chem Appl, (2015)
39. P. T. Jones, D. Geysen, Y. Tielemans, S. V. Passel, Y. Pontikes, B. Blanpain, M. Quaghebeur, N. Hoekstra, J Clean Prod, **55**, 92-102, (2013)

Environmentally friendly method for regeneration of copper chloride acidic solutions used in etching of printed circuits

Dorota Kopyto[1,*], *Wit* Baranek[1], *Zbigniew* Myczkowski[1], *Katarzyna* Leszczyńska-Sejda[1], *Michał* Drzazga[1], *Maciej* Matusewicz[2], *Tadeusz* Matusewicz[2], and *Wojciech* Matusewicz[2]

[1]Institute of Non Ferrous Metals (IMN), Hydrometallurgy Department, ul. Sowińskiego 5, 44-100 Gliwice, Poland
[2]MATUSEWICZ Budowa Maszyn S.A., ul. Lwowska 40, 59-620 Gryfów Śląski, Poland

Abstract. Etching of copper with acidic $CuCl_2$ solution is a widely used chemical method in printed circuits production. During the process, the solution is enriched in Cu(I) ions, resulting in reduction and then loss of etching bath capacity. In order to ensure the required etching kinetics, the solution is regenerated by oxidation of Cu(I) to Cu(II). The industrially applied Cu(I) oxidants are, e.g.: Cl_2, chlorates. Their application is expensive and associated with drawbacks related to health, safety and environmental hazards (e.g. Cl_2 emission). In the result of long-standing cooperation between the IMN and MATUSEWICZ Budowa Maszyn, an innovative, original and environmentally friendly method of acidic solutions regeneration used during printed circuits etching was developed. A new-type reactor equipped with oxidation monitoring-control systems using oxygen or oxygen-enriched air was applied. The reactor construction enables to run the process with oxygen circulation in the reactor, ensures full oxygen utilisation in the regeneration process, achievement of the required performance and process rate, and eliminates expensive, hazardous and often toxic reagents. This is an innovative method, and since there are no analogous technologies currently known and used in Europe and worldwide, the presented method is technologically, economically and ecologically unrivalled.

1 Introduction

Acidic aqueous solution of copper(II) chloride is one of the basic etching agents used in chemical processing of metallic surfaces, e.g. in production of printed electronic circuits. This solution is characterised by very high dissolution yield of metallic copper. Its additional advantage, comparing to iron(III) chloride solutions widely used in the past or ammonia solutions, is better stability and control of the set process parameters and compliance with environmental protection requirements. During the process of printed board etching part of the metallic copper present on the processed boards is dissolved. As a result, concentration of copper(I) cations in the etching solution increases. It causes reduction and finally loss of bath etching ability. In order to maintain high efficiency and stable rate of printed circuits etching it is necessary to regenerate etching solution. It is done by oxidation of monovalent copper to divalent one and removal of copper excess from the solution.

Oxidants of Cu(I) used in industry are: hydrogen peroxide (H_2O_2), gaseous chlorine (Cl_2), sodium chlorate ($NaClO_3$) [1-5]. Application of these oxidants is expensive and connected with other disadvantages like: health, safety and environmental issues (e.g. due to chlorine emissions) as well as keeping optimum composition of etching solution. Regeneration trials using air, pure oxygen or mixture of oxygen and certain ozone amount were also performed [6]. However, they did not give positive results due to too low reaction rate and process efficiency. It was caused by insufficient dispersion of gaseous oxidant in water phase.

Printed circuits industry is a huge market strictly connected to production of electronic and electric equipment and components. Currently systematic growth of global income from printed circuit board (PCB) may be observed. It is estimated that the market is worth \$78-85 billion. Despite huge and noticeable demand on PCBs neither production efficiency nor environmental protection standards improvement were observed. Every year, millions litres of solutions are used for etching of printed circuits. All the amount need to be regenerated and returned to the process. Therefore, it is so important to find economic, environmentally-friendly and efficient method of etching solution regeneration.

Currently, technology for acid etching solution regeneration using oxygen and/or oxygen-enriched air is not offered on the market. However, it is the most attractive and promising method. It has the highest application potential due to availability of oxidising agent, safety of its application, no emission of additional unwanted compounds and gases as well as low operational costs. In order to implement this method to industrial practice it is necessary to ensure sufficient rate

* Corresponding author: dorotak@imn.gliwice.pl

and efficiency of monovalent copper oxidation by e.g. application of proper reactor, which can be able to enhance oxygen dispersion in a solution.

Taking into account above-mentioned requirements company Matusewicz Budowa Maszyn S.A. (company with several years of experience in construction of equipment used in electrotyping) initiated an innovative action within the project POIG.01.04.00-02-049/11 entitled: „Development of new regeneration method for acidic copper(II) chloride solutions used in etching of printed circuits" (original Polish title :„Opracowanie nowej metody regeneracji kwaśnych roztworów chlorku miedziowego stosowanych w procesie trawienia obwodów drukowanych"). The innovation was the application of a new reactor construction equipped with regulation-control system of oxidation process using oxygen or oxygen-enriched air.

The project was implemented in close cooperation with Institute of Non Ferrous Metals (IMN) in Gliwice, partner with large experience within construction of aerators used in i.a. flue gas desulphurisation installations, gas dispersion units for process of gypsum production from wastes, or air dispersion units in flotation equipment for non-ferrous metal ores and coal [7-10].

2 Experimental

All laboratory and industrial tests of novel method for regeneration of acidic copper(I) and (II) chloride solutions used in printed circuits production technology were performed in IMN Gliwice on order and in close cooperation with company Matusewicz Budowa Maszyn S.A. Innovativeness of the method is mainly due to reactor construction, which is relatively simple. Reactor used in the experiments was composed of rotor and diffuser complex which, at sufficiently high rotor speed, enables very intense mixing and oxygenation of a solution. In order to determine kinetics of Cu(I) oxidation reaction tests using oxygen, air or oxygen-enriched air were performed. Analytical methods for determination of free HCl (titration method using NaOH), monovalent copper (potentiometric method using trivalent Fe) and overall copper (redox titration using KI and $Na_2S_2O_3$) were developed.

2.1 Laboratory scale tests

Laboratory tests were carried out using laboratory scale experimental setup (Fig. 1) composed of the following components:

- 3.0 dm^3 reactor with pneumo-mechanical rotor, made of stainless steel and additionally protected with halar coating, equipped with thermostated heating system and continuous temperature measurement inside a reactor;
- oxidising agent supply system (oxygen cylinder, supply with compressed air, reducers, flow meters).

Fig. 1. Laboratory setup for investigation on regeneration of acidic copper chloride solutions.

Table 1 presents the composition of acidic copper chloride solutions used in PCB etching tests.

Table 1. Composition of the solutions used in laboratory tests on regeneration of acidic copper chloride solutions.

Component	Concentration, g/dm³
$Cu_{overall}$	127.2
Cu(II)	126.8
Cu(I)	0.44
HCl	131.4

The solution was enriched with Cu^+ ions to the required monovalent copper content of 4.0-5.0 g/dm^3 for the experimental purposes.

Laboratory tests determining Cu(I) oxidation reaction kinetics using oxygen, air or oxygen-enriched air were performed. Monovalent copper concentration was used for determination of reoxidation rate, i.e. etching solution regeneration rate. Degree of oxygen consumption in each test was also determined. All tests were carried out at set etching solution volume v=1,3 dm^3, at 50°C. Air was supplied using IMN's compressed air system via reducer-filter system and rotameter with a valve. Air flow rate within 0-600 dm^3/h at rotor rotation rate between 720-1740 rpm was investigated. Oxygen was supplied to the reactor from a cylinder using reducing valve and rotameter. Oxygen flow rate was between 20-150 dm^3/h, while rotor speed was 720-2160 rpm. Air and oxygen streams were mixed at tee placed between rotameter outlets and the reactor. In the tests with oxygen-enriched air mixtures containing 27.3-78.9% of oxygen were investigated. Air flow rate was between 50-200 dm^3/h, oxygen flow rate was in the range 10-140 dm^3/h, while rotor speed was constant and equal 1500 rpm.

The highest regeneration rate of etching solution was equal to ca. 3.5 g Cu(I)/min·dm^3. It was obtained for oxygen used as an oxidant, at rotor speed of 1500 rpm and oxygen flow rate of 60 dm^3/h. Oxygen consumption degree for these conditions was 31.9%. Maximum oxidation rate using air was ca. 0.75 g Cu(I)/min·dm^3 and was achieved at rotor speed of 1200 rpm and air flow rate of 100 dm^3/h. Oxygen consumption degree for these conditions was only 19.2%. Regeneration rate of etching solution using oxygen-enriched air depends on oxygen partial pressure in oxidising mixture and is situated between results obtained for air and pure oxygen.

Influence of oxidant bubble size on the efficiency of investigated process was also determined. It may have an influence on reoxidation of etching solution. It was assumed that mean bubble size is smaller at large rotor speeds. In the case of oxygen slight influence of bubble size on etching solution regeneration rate was determined.

Fig. 2. Reactor of 50 dm^3 volume used for regeneration of acidic copper chloride solutions.

Relationship between concentrations of monovalent Cu, overall Cu, HCl in etching solution and its density, redox potential, electrical conductivity was investigated. Basing on obtained results it was possible to show these dependencies in the form of empirical formulas. Analysis of empirical relationships determined using factorial design 2^n type method showed that:

- specific density of chloride etching solution depends mainly on overall Cu concentration. Influence of HCl and Cu(I) concentrations on solution density are 5- and 9-fold lower, respectively;
- redox potential is de facto function of Cu(I) concentration in the solution only. Influence of other parameters like overall Cu and HCl concentrations is ca. 60- and 25-fold weaker;
- coinfluence of investigated factors (solution components) on measured physicochemical parameters is relatively insignificant;
- tests of determined empirical parameters showed, that calculated values are close to measured ones. Mean difference between calculated and measured physicochemical parameters like specific density and redox potential is ca. 1.52% and 2.8% of measured value, respectively;
- electrical conductivity of etching solution depends mainly on HCl concentration. Influence of other component concentrations – overall and monovalent Cu – is several times smaller;
- verification tests of obtained empirical formula showed that mean difference between measured and calculated value for electrical conductivity is ca. 12%, so is higher than for other investigated parameters of etching solutions.

Based on these equations regulation algorithms for optimum etching solution composition during continuous operation of the reactor were developed.

2.2 Semi-technical scale tests

Technical guidelines for the scaled-up reactor construction were based on technological parameters determined during laboratory tests. Constructed semi-technical scale reactor of 50 dm^3 volume was supplied with gaseous oxygen. Its monovalent copper oxidation rate was 10.5 kg/h (Fig. 2). The reactor was also equipped with regulation and control system.

In order to verify determined assumptions of Cu(I) oxidation process semi-industrial research was performed. It was done using the reactor composed of reaction chamber, aerator with a drive, inlet and outlet stubs, pumps and circulation chamber for measurements of selected physicochemical parameters of solution. The reactor was also equipped with measurement-control system: temperature sensor, electrode for redox potential measurement, electrode for electrical conductivity measurement, sensor of solution level inside a reactor chamber, areometer for solution density measurement and device measuring flow rate of oxygen supplied to the reactor. This measurement-control system allowed to carry out comprehensive technological tests under static and dynamic conditions. Properly situated stubs will allow eventual connection of the reactor to technological line of printed circuits etching.

Technological tests were performed using industrial etching solution, which composition is provided in Table 2. Prior to each test solution composition was modified adjusting monovalent Cu concentration to 2.0-2.5 g/dm^3. Divalent Cu concentration was maintained in the range 160-190 g/dm^3, while HCl concentration was constant and equal to ca. 6.1 g/dm^3. NaCl concentration was not adjusted assuming that it remained close to initial value during the course of the process.

 DOI:10.1051/e3sconf/20171801021

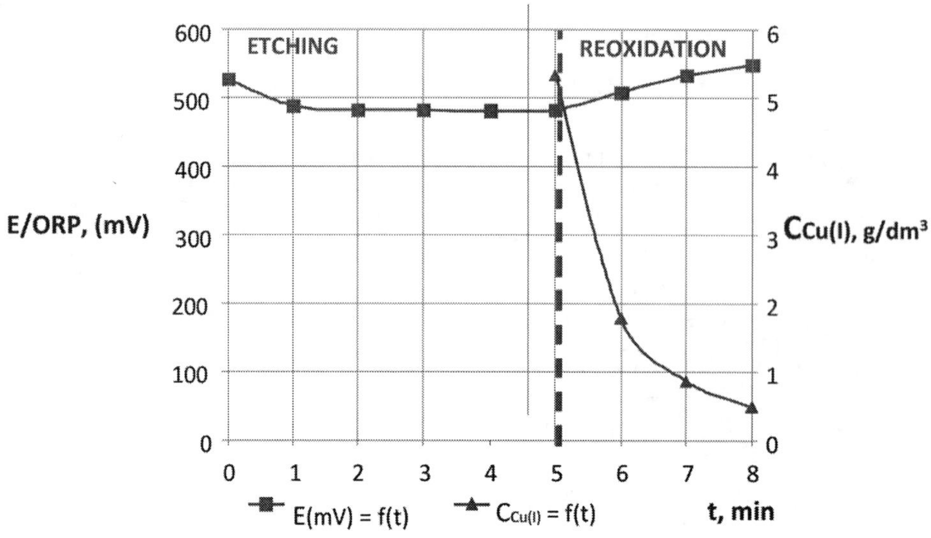

Fig. 3. Course of solution etching and regeneration process using oxygen in 50 dm^3 reactor.

Table 2. Composition of the solutions used in semi-technical test of acidic copper chlorides solution regeneration.

Component	Concentration, g/dm^3
$Cu_{overall}$	181.3
Cu(II)	126.8
Cu(I)	0.12
HCl	6.1
NaCl	123.9

Fig. 4. The technical scale reactor for acidic copper chloride solutions regeneration.

Test series for two different positions of circulation holes (closed and open) were performed. Different rotor speeds (650, 800 and 950 rpm) in the reactor were applied. Constant overpressure of 40 mbar and solution temperature of 50-55 °C inside the reactor were maintained during each test. Obtained results confirmed the process efficiency obtained during laboratory tests. Exemplary course of solution etching and regeneration (oxidation) process during performed tests is shown in Fig. 3. Optimum parameters for the periodic process using the reactor and industrial solution were selected:
- open circulation holes;
- rotor speed ca. 800 rpm;
- process temperature 50-55°C.

For these conditions, in time interval 0-2 min, the solution regeneration rate (monovalent Cu oxidation) was ca. 0.84 g Cu(I)/min·dm^3.

2.3 Technical scale tests

Technical guidelines for technical scale reactor construction were based on optimum parameters of acidic copper chloride solution regeneration process determined during semi-technical scale tests. The reactor of 0.5 m^3 volume and estimated monovalent copper oxidation rate of 20 kg/h was constructed (Fig. 4).

Basing on analysis of semi-technical scale tests results manual of reactor preparation for application tests under industrial conditions was developed. The scope of the manual covered:
- operational parameters of the reactor (rotor speed, operational volume, volume of gas phase and oxygen pressure, solution flow rate, solution temperature, density of regenerated solution);
- adaptation of the reactor to industrial conditions – incorporation of the reactor into technological line of printed circuits etching;
- pressure closure of the reactor;

* Corresponding author: dorotak@imn.gliwice.pl

 DOI:10.1051/e3sconf/20171801021

- air removal from the reactor chamber;
- technological start-up of the reactor;
- reactor shut-down;
- regulation of the reactor;

Application tests of the 0.5 m^3 reactor confirmed assumed regeneration rates of etching solutions used in PCB production. The reactor possesses 9th technological readiness level (TRL 9). It means that it is ready for direct implementation in industrial practice.

3 Summary

The result of long-standing cooperation between Institute of Non-Ferrous Metals and company Matusewicz Budowa Maszyn S.A. is development of economic, environmentally-friendly and innovative method for regeneration of acidic copper chloride solution generated during printed circuits etching process. New construction of a reactor equipped with regulation and control system of oxidation process allows effective circulation of oxidant during the process. It allows to maximize oxygen utilisation rate during monovalent copper oxidation and desired efficiency and rate of etching solution regeneration process. The advantages of the process are simplification of Cu(I) oxidation process, elimination of expensive and often toxic and harmful reagents, stable process parameters, no copper loss and wastes elimination. The additional advantage is low construction and operating costs of the reactor.

Experiments were performed within the project POIG.01.04.00-02-049/11 „Opracowanie nowej metody regeneracji kwaśnych roztworów chlorku miedziowego stosowanych w procesie trawienia obwodów drukowanych" granted by PARP in a Program Innowacyjna Gospodarka.

References

1. D. M. Allen, H.J.A. White, Processing of Advanced Materials, **2**, 19-24 (1992)
2. Robertson, Robertson's Chemicals Ltd., (1986)
3. Werner D. Bissinger, US Patent no: 46966717 (1987)
4. A. Stehlik, US Patent no: 4396475 (1983)
5. S. J. Beyer, R. M. Lukes, US Patent no: 3794571 (1974)
6. P. J. Jefferies, *PhD Thesis*, Cranfield University (2005)
7. J. Mrozowski et al. – Spr. IMN nr 6231/06 (unpublished) (2006)
8. J. Mrozowski et al. – Projekt kluczowy POIG.01.03.01–24–019/08-00 - Spr. IMN nr 7012/2012 (unpublished) (2012)
9. W. Wasilewski et al. PL Patent no: 209175 (2006)
10. J. Mrozowski, W Wasilewski, A. Chmielarz, R. Kurowski, PL Patent no: 201034 (2005)

 DOI:10.1051/e3sconf/20171801022

Romanian bentonite and fly ash characterization and their use in heavy metal "in-situ" immobilization in polluted soils

Săndica Liliana Gherghe[1,*], *Ildiko* Anger[1], *Georgiana* Moise[1], and *Roxana* Trusca[2]

[1]National Research and Development Institute for Nonferrous and Rare Metals, 102 Biruintei Blvd, Pantelimon, Romania
[2]Politehnica University Bucharest, 313 Independentei Splai, Bucharest, Romania

Abstract. This article presents the characterization of the Romanian bentonite and fly ash, using different techniques: FAAS, XRD, FT-IR, SEM and EDAX and their evaluation as sorbents for heavy metals immobilization in polluted soils coming from mining and metallurgical activities. The applicability of bentonite and fly ash for Pb (II) and Zn (II) immobilization was studied using aqueous solutions of these metals. The influence of the pH and contact time were studied. The results shown that the Romanian bentonite and fly ash could be used for Pb (II) and Zn (II) immobilization in polluted soils from brownfields.

1 Introduction

Heavy metals, such as lead, copper, zinc, nickel, cadmium, chromium are present in soil as natural components, but their presence in environment has increased due to the human activities. The heavy metals are released into the environment through various activities such as mining, energy and fuel production, electroplating, wastewater treatment [1]. The zinc is essential for life and for a healthy body and it is a common micronutrient in trace amounts [2]. In excess, zinc can be harmful and causes zinc toxicity. A long-time exposure to Zn(II) is harmful for human health. The free zinc ions in solution are highly toxic to plants, invertebrates and human. The lead is extremely toxic and it has adverse effects as it accumulates in the body. All lead compounds are cumulative poisons. In body, lead affects in special the neuronal system.

To reduce the risks associated to heavy metals presence in environment various treatment techniques are used, like: ion exchange, biodegradation, oxidation, solvent extraction and adsorption. The adsorption technique remains widely preferred due to its simplicity of design and operation, cost effectiveness, high efficiency and flexibility [3]. This method is usually used for waste water treatment, but the treatment of polluted soils with heavy metals is also based on the adsorption process. Different natural materials (clay, zeolite) or by-products (fly ash) are used for heavy metal immobilization in soil.

The fly ash is a by-product from electric power plant which utilizes lignite coal as fuel [4, 5]. The fly ash is removed from the emission control devices and it is transported and deposited in open dump where is accumulated in a large amount. For this reason, its disposal creates serious environment problems. The worldwide studies identify possibilities for fly ash use, such as: road construction, construction materials, cement industry, soil amendment and adsorbent material for the adsorption of heavy metals from waste waters [4, 6]. The major constituents of fly ash are silica, alumina, and iron oxide.

The clay minerals have a small grain size and a specific mineralogical structure which results in specific properties. The most important of these properties are the large specific surface area and the resulting ability to adsorb cations [7]. Therefore, clay minerals are used in a large variety of environmental applications. Clay minerals are composed from fine-grained aluminosilicates.

The aim of this paper is to assess the capacity of bentonite (natural clay) and fly ash to adsorb Pb(II) and Zn(II) ions from aqueous solutions. For this reason, the physical, chemical and mineralogical composition of bentonite and fly ash was determined. Batch adsorption experiments were performed to investigate the effect of pH and contact time on the adsorption of Pb(II) and Zn(II) ions onto these materials.

2 Materials and methods

The fly ash samples were obtained from power plant Govora, Romania. This power plant use lignite from Berbesti-Alunu coals mine. The fly ash was capture from the dry emission control system. In accordance with SR EN 450-1" Fly ash for concrete - Part 1, Definition, conditions and standards" these categories of fly ash, can be used as admixture type II for concrete, binder and cement slurry.

The bentonite was obtained from Orasul Nou (Maramures county) natural deposit. This bentonite is less commercialized.

* Corresponding author: gliliana@imnr.ro

2.1 Characterization

The chemical composition of the fly ash and bentonite was determined by inductive coupled plasma optical emission spectrometer (ICP-OES) SPECTROFLAME – ICP model P, Germany and Flame Adsorption Atomic Spectroscopy (FAAS).

The mineralogical composition of the bentonite and fly ash was determined by the X-ray diffraction analysis for the qualitative evaluation of the common and predominant phases within the ash, using a BRUCKER D8 ADVANCE diffractometer and software package DIFRAC^{plus} BASIC Released 2006 (Brucker) and ICDD PDF-2 database release 2006.

The optical microscopy analysis were performed using a Zeiss – Jena – Axio IMAGER Alm microscope.

The scanning electron microscopy (SEM-EDAX) was used to determine the morphological and quantitative characteristics of the materials.

The IR spectra of bentonite was obtained using a FT-IR ABB MB 3000 spectrometer.

The leaching tests of fly ash were conducted in accordance with the EN 12457-1 si EN 12457-2 standards.

2.2 Preparation of the heavy metal solutions

Stock solutions of Pb(II) and Zn(II) (containing 1000 mg metal/L) were prepared using lead nitrate ($Pb(NO_3)_2$ and zinc sulphate heptahydrate ($ZnSO_4 \cdot 7H_2O$) of analytical purity. The pH adjustments were performed using nitric acid, sulphuric acid or sodium hydroxide 0.1 M aqueous solutions.

2.3 Experimental

The batch adsorption experiments were carried out at room temperature (22^{0}C± 3^{0}C). One hundred milliliters of solution containing 100 mg/L of Pb(II) or Zn(II) were treated with the desired amount of adsorbent in 250 mL conical flasks. The mixtures were stirred for 0 – 240 min. At the end of each time period, the adsorbent was removed by filtration and the filtrate was analyzed for residual metal concentration by atomic absorption spectroscopy and FAAS.

The efficiency of the removal of Pb(II)and Zn(II) from solution was calculated using Eqn.1 [1, 8]

$$Removal\ efficiency\ \% = \frac{C_0 - C_f}{C_0} \text{ x } 100 \qquad (1)$$

where: R is the removal efficiency (%), C_0 is the initial metal concentration (mg/L), C_f is the final metal concentration (mg/L).

3 Results and Discussions

3.1 Bentonite and fly ash characteristics

The oxidic composition of the bentonite and fly ash is presented in table 1.

Table 1. The oxides composition of the bentonite and fly ash, [weight %].

	SiO_2	Al_2O_3	CaO	MgO	Fe_2O_3	Na_2O	K_2O	H_2O	LOI* (1000^{0}C)
Bentonite	68.49	10.77	1.19	1.23	1.80	4.38	0.80	11.37	-
Fly Ash	52.11	20.2	7.14	2.33	8.14	-	-	0.14	3.06

**LOI- Loss of Ignition*

The fly ash has a low content of CaO. It is siliceous type fly ash, with low loss of ignition.

The bentonite has a ratio of SiO_2/Al_2O_3 = 6.36, SiO_2/Fe_2O_3= 38.05 and MgO/Fe_2O_3= 0.69. In the fly ash, the amorphous phase is 61.5 wt% and crystalline phase is 38.5 wt%, according to results obtained from XRD analysis. The main crystalline phases are: 45 wt% quartz (SiO_2), 21 wt% anorthite (calcium aluminosilicate $Ca_{0.88}Al_{1.77}Si_{2.23}O_8$), 14 wt% anhydrite ($CaSO_4$), 9 wt% gehlenite ($Ca_2Al[AlSiO]_7$), 6 wt% hematite ($Fe_2O_3$) and 5 wt% magnetite ($Fe_3O_4$) (figure 1).

The main crystalline phases in bentonite are beidellite/montmorillonite (> 70 wt%), cristobalite, albite (potassian and calcian), quartz (figure 2).

The optical microscopy analysis for fly ash highlights very well the amorphous and crystalline phases. In these phases, the grains of hematite and magnetite appear. Figure 3 shows the photo with the amorphous phase and hematite grain. The bentonite optical analysis presents the micro-crystalline phase (clay minerals) and some albite and quartz (figure 4).

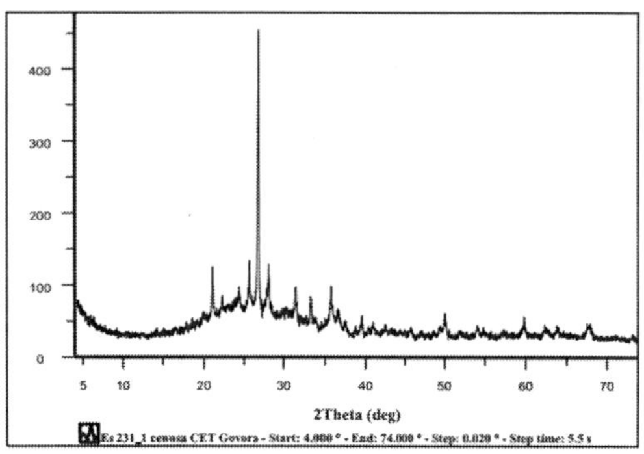

Fig. 1. Crystalline phases of fly ash determined by DRX.

The scanning electron microscopy (SEM) was used to determine the morphological characteristics and surface texture of the bentonite and fly ash.

SEM images of the fly ash Govora are presented in figure 5 and the EDAX spectrum of the fly ash is presented in figure 6.

DOI:10.1051/e3sconf/20171801022

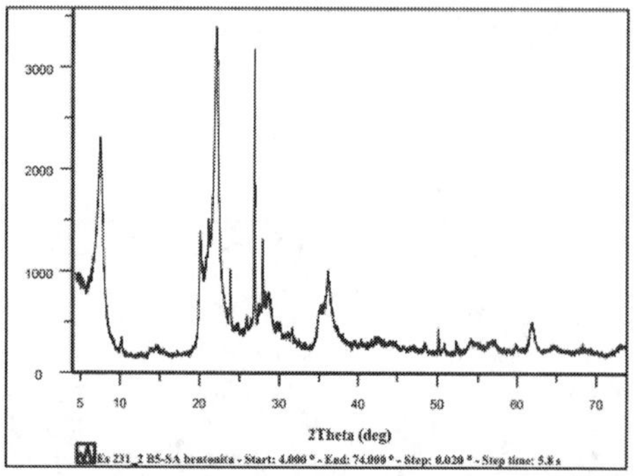

Fig. 2. Crystalline phases of the bentonite determined by DRX.

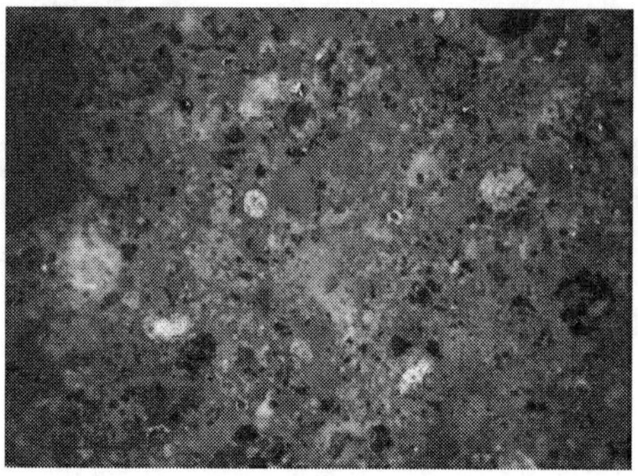

Fig. 3. The optical microscopy analysis of the fly ash.

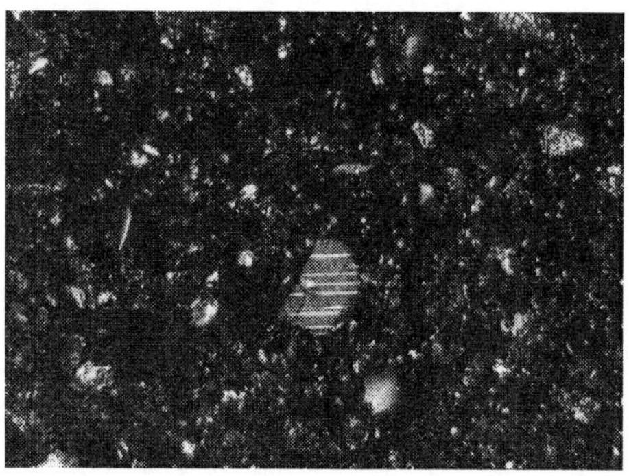

Fig. 4. The optical microscopy analysis of the bentonite.

The morphology of this ash is similar with the other ashes [9, 10, 12]. The fly ash is a fine powder, contains microspheres with diameter of 5-15 μm. The EDAX analysis shows the presence of carbon, magnesium and iron too, near the constituent elements of the minerals mentioned by DRX. The bentonite is formed from grains with submicron size (figure 7). These grains form aggregates with size of 1-10 μm.

Fig. 5. SEM micrograph of the fly ash.

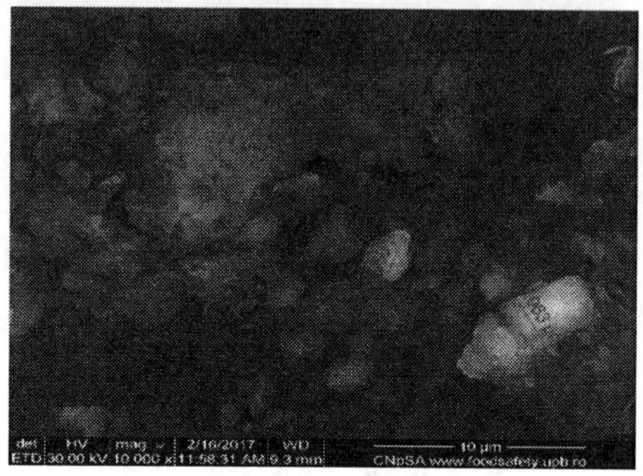

Fig. 6. SEM micrograph of the bentonite.

The EDAX analyses shown (figures 7, 8) the silicon and aluminium as main elements which form the crystalline framework together with oxygen. The other elements (Na, Mg, K, Ca and Fe) could isomorphic substitute silicon in tetrahedral layer. Fe and Mg could substitute Al in octahedral layer. The negative charges will be compensated by Na, K, Ca, Fe and Mg.

The thermal analysis (DSC) shows three peaks (endotherm effects) situated at 96,5^{0}C, 178^{0}C and 673^{0}C. At 96,5^{0}C and 178^{0}C the adsorbed water is eliminated. The peak from 673^{0}C is characteristic for bentonite (which has montmorillonite as main component) aluminium – iron – magnesium. Generally, the MgO/Fe_2O_3 ratio in bentonite is less than 0.9. The studied bentonite has the MgO/Fe_2O_3 ratio equal with 0.69.

The pH_{pzc} of bentonite was below 2. This shows that Si and Al were isomorphic substituted in the crystalline framework and the negative charges have permanent character and don't depend on pH.

The FT-IR study (figure 9) of bentonite shows the characteristic peaks at 3632 cm^{-1} assigned for stretching vibration of OH^- groups and characteristic for montmorillonite.

DOI: 10.1051/e3sconf/20171801022

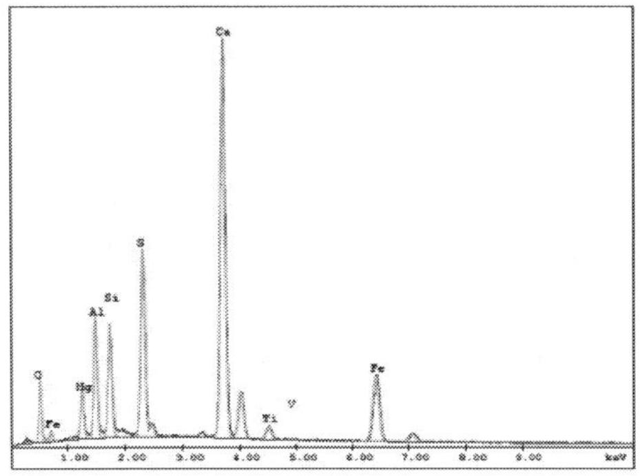

EDAX ZAF Quantification (Standardless)
Element Normalised
SEC Table : Default

Element	Wt %	At %	K-Ratio	Z	A	F
O K	13.90	27.25	0.0201	1.0711	0.1351	1.0006
MgK	3.57	4.60	0.0153	1.0270	0.4150	1.0047
AlK	8.14	9.47	0.0432	0.9966	0.5289	1.0062
SiK	7.13	7.96	0.0439	1.0254	0.5966	1.0076
S K	12.01	11.74	0.0939	1.0190	0.7578	1.0126
CaK	35.19	27.53	0.3258	0.9996	0.9190	1.0078
TiK	1.91	1.25	0.0152	0.9131	0.8611	1.0117
V K	0.08	0.05	0.0006	0.8941	0.8956	1.0181
FeK	18.08	10.15	0.1573	0.9106	0.9553	1.0000
Total	100.00	100.00				

Fig. 7. EDAX of fly ash.

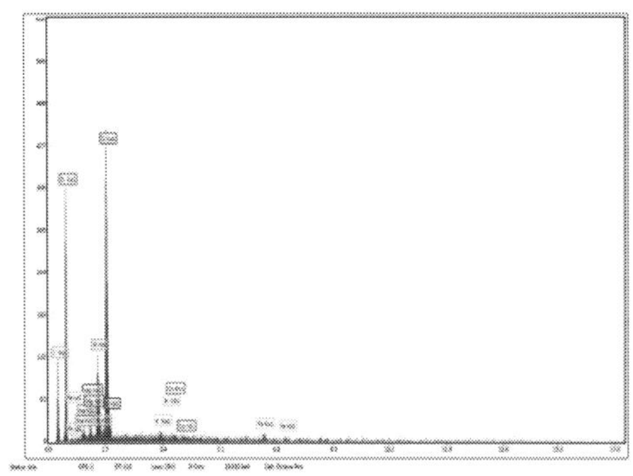

Element	Weight, %	Atomic, %	Error, %	Net Int.	K Ratio	Z	A	F
O K	59.11	71.4	9.91	57.79	0.2032	1.0336	0.3326	1
Na K	3.91	3.29	24.14	4.63	0.0085	0.947	0.2273	1.0059
Mg K	1.33	1.06	32.12	3.14	0.0043	0.9654	0.3288	1.0107
Al K	7.43	5.32	12.02	25.45	0.0325	0.9316	0.463	1.0143
Si K	26.42	18.18	7.65	106.19	0.1321	0.9539	0.5226	1.003
K K	0.6	0.29	63.65	2.69	0.0044	0.8918	0.8191	1.0196
Ca K	0.33	0.16	64.72	1.39	0.0027	0.9095	0.8745	1.025
Fe K	0.87	0.3	61.56	2.7	0.0081	0.8194	1.0105	1.1142

Fig. 8. EDAX of bentonite.

Table 2. Concentration of heavy metals in leachate [17].

Heavy metal	Concentration limits					
	Inert waste (RL)		Non - hazardous waste (RL)		Leachate	
	L/S=2 l/kg mg/kg	L/S=10 l/kg mg/kg	L/S=2 l/kg mg/kg	L/S=10 l/kg mg/kg	L/S=2 l/kg mg/kg	L/S=10 l/kg mg/kg
Cu	0.90	2.00	25	50	0.02	0.12
Ni	0.20	0.40	5.0	10	0.04	0.19
Zn	2.00	4.00	25.0	50	0.42	4.79
Pb	0.20	0.50	5.0	10	0.09	1.53
Cd	0.03	0.04	0.6	1	0.02	0.10

The peaks 3491, 3410 and 1639 cm^{-1} are assigned for stretching and deformation vibration of OH^- groups from adsorbed water in bentonite network; 918 cm^- for stretching vibration of OH^- from the Al_2OH; 795 cm^{-1} for Fe(II)Fe(III)OH group presence; peaks at 1043 and 627 cm^{-1} are attributed to the Si-O and Si-O-Al vibrations [13-16].

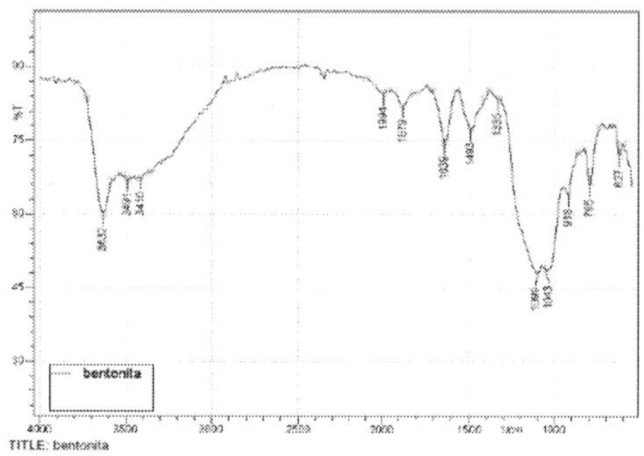

Fig. 9. The FT - IR spectra of bentonite.

The leaching test of the fly ash, presented in table 2, shows that the leachates contain heavy metals below the standards for non-hazardous waste in accordance with national legislation (RL) [17].

3.2 Adsorption tests

3.2.1 Effect of pH

The removal efficiency was studied at pH values between 2.5 and 7.0. The influence of pH on the removal efficiency of Pb(II) and Zn(II) onto bentonite (B) and fly ash (FA) is presented in figure 10. The Pb (II) removal efficiency on FA increases starting from pH = 2.5 until pH = 5.0, when the efficiency is maximum. The Zn (II) removal efficiency on FA was 95% at pH = 7.0. The Zn (II) removal efficiency is very low between pH = 2.7 - 3.6. This should be explained with the presence of silica in fly ash. The silica could uptake either positive or negative contaminants depending on the solution pH. The silicon, the central ion of silica, has electron affinity, giving the oxygen atoms bound to it low basicity. This enables the silica surface to act as a weak acid, which can react with water, forming silanol (SiOH) groups. As a result, at low pH the silica surface is positively charged and at high pH values is negatively charged [18, 19]. The other solid materials, for example iron oxide (another constituent of FA), develops positive or negative charges also. In accordance with this theory, at pH = 2.7 - 3.6, the number of sites with negative charge is very low and only a very small quantity of Zn(II) ions can be fixed with electrostatic interactions on the silica surface. Pb(II) ions can be fixed both on the silica and iron oxide surface.

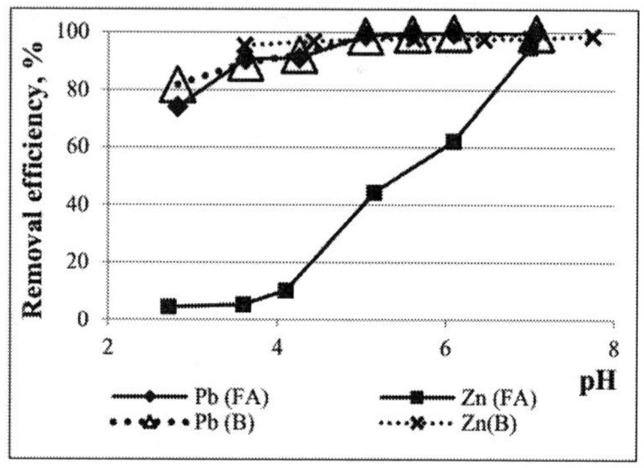

Fig. 10. Effect of pH on the Pb(II) and Zn(II) removal efficiency on bentonite (B) and fly ash(FA).

The removal efficiency of Pb(II) and Zn(II) on bentonite is high (above 80%) even at low pH. The removal of these ions from aqueous solution using B doesn't depend on pH. The pH_{pzc} of bentonite is below 2, that means the bentonite surface above pH=2 has negative charges. These negative charges result from isomorphic substitutions of Si in the tetrahedral layer and of Al in octahedral layer. The removal efficiency is maximum at pH = 5.5 for both Pb(II) and Zn(II).

3.2.2 Effect of contact time

The removal efficiency of heavy metal ions (Pb(II) and Zn(II)) from solution using bentonite and fly ash versus time is presented in figure 11.

The results indicated that the Pb(II) and Zn(II) ions uptake increase with contact time, until an equilibrium is reached. The equilibrium was reached within 60 minutes for Pb(FA) and in 30 minutes for Pb(B). For Zn(FA), the equilibrium was obtained at 180 minutes and for Zn(B) in 120 minutes. The adsorption capacity obtained for Pb(FA) was 19.9 mg/g at 60 minutes and for Zn(FA) was 5.31 mg/g at 180 minutes. The adsorption capacity of 5.31 mg/g for Zn(II) o FA is higher than the values found by other researchers for the adsorption of zinc onto fly ash [20, 21], but is very closed to q_e = 4.505 mg/g obtained by reference [22]. The adsorption capacity of 19.9 mg/g for Pb(FA) is closed to the value 18.8 mg/g obtained by authors [20].

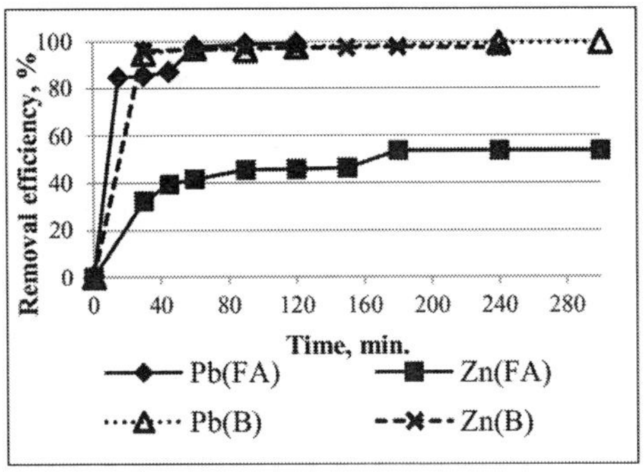

Fig. 11. Effect of contact time on the Pb(II) and Zn(II) removal efficiency on bentonite (B) and fly ash(FA).

The adsorption capacity obtained for Pb(B) was 8.96 mg/g at 30 minutes. The adsorption capacity obtained by different authors varies between 7.56 and 16.30 mg/g Pb(II) on bentonite, as function of the initial concentration of the solution and the type of bentonite used in process [23-25].

The adsorption capacity obtained for Zn(B) was 10.88 mg/g at 120 minutes. The adsorption capacity obtained by different authors varies between 3.24 and 11.84 mg/g [26-28] Zn(II) on bentonite, as function of the type of bentonite used in process.

The results showed that bentonite adsorb better Zn(II) and fly ash Pb(II).

4 Conclusions

This paper presented the chemical, structural and morphological characteristic of the bentonite and fly ash from Romania. These materials are suitable for uptake of Pb (II) and Zn(II) ions from aqueous solutions.

The removal capacity of fly ash increases with pH, when pH is between 2.5 and 5. and after that, remains constant. The removal capacity of bentonite doesn't depend on pH.

The adsorption capacity of fly ash towards Pb (II) ions is 19.9 mg/g and 5.31 mg/g for Zn(II).

The adsorption capacity of bentonite is 8.96 mg/g for Pb(II) and 10.88 mg/g for Zn(II).

The fly ash has a higher affinity for Pb(II) ions while bentonite for Zn(II).

The authors thank for financial support given by Ministry of Education and Research, through National Authority for Scientific Research and Innovation (ANCSI), within the NUCLEU research project. Special thanks for Victor FRUTH-OPRISAN, Cornel MUNTEANU from Institute of Physical Chemistry "Ilie Murgulescu" Bucharest, Romania.

References

1. U. Nadeem, M. Datta, Eur.Chem. Bull, **3**, 7 (2014)
2. AO. Dada, AP. Olalekan, AM. Olatunya, O. Dada, IOSR J. of Applied Chem., **3**, 1 (2012)
3. Ye Hengpeng and Yu Zhijuan, Nat. Res., **1** (2010)
4. T. Matsi, VZ. Jeramidas, Environmental Pollution, **104** (1998)
5. Satya Vani Yadla, Stridevil V, Chandana Lakshmi MVV, J of Chem., Biol and Physical Sci., **2**, 3 (2012)
6. M. Ulmanu, T. Matsi I. Anger, E. Gament, G. Olanescu, C. Predescu, M. Sohaciu, U.P.B. Sci. Bull., Series B, **69**, 2 (2007)
7. H. Bradl, *Encyclopedia of Surface and Colloid Science* (Ed. Marcel Dekker, 2002)
8. Y. Fernandez-Nava, M. Ulmanu, I. Anger, E. Maranon, L. Castrillon, Water Air Soil Pollut., **215**, (2011)
9. T.S. Malarvizhi, T. Santhi., J. of Water Resource and Protection, **5**, (2013)
10. Kumar Saroj, A.K. Mishra, M. Upadhyay, D. Singh, M. Mishra and Kumar Sujata, Int. Res. J. of Env, Sci, **3**, 2 (2014)
11. S.O. Bada and S. Potgieter- Vermaak, Leonardo Electronic J. of Practices and Technol., **12** (2008)
12. A. K. Agarwal, M. S Kadu, C. P. Pandhurnekar, I.L. Muthreja, The Holistic Approach to Env, **4**, 1 (2014)
13. F.G. Alabarse, R.V. Conceicao, N.M. Balzaretti, F. Schemato, A.M. Xavier, Applied Clay Science, **51** (2011)
14. C.M. Kede, M.A. Etoh, PP. Ndibewu H.M. ngomo, P.M. Ghogomu, British J. Applied Sci. Technol., **4(7)**, 1071-1088 (2014)
15. M. Barkat, S. Chegrouchel, A. Mellah, B. Bensmain, D. Nibou, M. Boufatit, J. of Surface Engineered Mat.,Adv. Technol., **4** (2014)
16. L. Aloui F. Ayan, A. Ben Othman, M. Trabelsi-Ayadi, Internat J. of Eng. and Appl. Sci. (IJEAS) **2**, 7, (2015)
17. Ministerial Order no. 95/2005 (Env. and Water Manag. Ministry)
18. T.S. Malarvizhi, T. Santhi, J. Res. Chem. Intermed, **39** (2013)
19. M. Moyo, L. Muguni, B.C. Nyamunda, Int. J. of Eng. Sci and Technology, **4**, 4 (2012)
20. P. Ricou, I. Lecuyer and P. Le Cloired, Env Technology, **19** (1998)
21. G.M. Taha, Ground Water Monitoring& Remediation, **26**, 4 (2006)
22. A. K. Agarwal, M.S. Kadu, C.P. Pandhurnekar, I.L. Muthreja, Int. J. of Environ. Sci. and Develop., **3**, 4 (2012)
23. S. Melichova, L.Hromada, Pol. J. Environ. Stud., **22**, 2(2013)

24. S.L. Bailey, T.J. Olin, K.M. Bricka, D.D. Adrian, Water Res., **33** (1999)
25. O. Abolino, M. Aceto, M. Malandrino, C. Sarzanini, E. Mentasti, Water Res., 37 (2003)
26. M. Larakeb, L. Youcef, S. Achour, Athens J of Science, **4**, 1(2017)
27. G. A. Tito, L. H. G. Chaves, R. S. De Souza, Rev. CAATINGA, **21**, 5 (2008)
28. W. Chai, Y. Huang, S. Su, G. Han, J. Liu, Y. Cao, Phys. Problems of Mineral Proc., **53**, 1 (2017)

Sulfur reduction in Sangan iron ore by flotation

Bahram Rezaee[1], *Atefe* Sarvi[*,2], *Atiyeh* Eslamian[2], *Seyed Mehdi*Jebraeeli[2], and *Abolfazl* Zabihi[3]

[1]Mineral Processing, professor, mining and metallurgical engineering department, Amirkabir university of technology, Tehran, Iran.
[2]Mineral Processing, Expert of Mineral Processing Department of KaniKavan shargh Engineering Co., Tehran, Iran.
[3] Head of technology and development of Opal Parsian Sangan industrial & Mining Co.(OPSIM), Tehran, Iran.

Abstract: in this study the flotation of pyrite as the main sulfide impurity of Sangan iron ore in Iran, was investigated. For this purpose the effect of collector type, pH, collector concentration and stage dosing on reverse flotation of iron sulfide ore from magnetite ore was investigated. Two type of thiol collectors include xanthates (sodium isopropyl xanthate (SIPX) and potassium amyl xanthate (PAX)) and dithiocarbamate (di-ethyl dithiocarbamate (DTC)) and the mixture of collectors was studied. The highest sulfur removal was obtained with potassium amyl xanthate. Stage dosing had a significant effect in sulfide flotation and the best recovery was obtained when the collector was added in 4 stages. The acidity had a positive effect on sulfide floatability and the best result was obtained at pH 3.5-4. Investigation about collector concentration showed that increasing the SIPX concentration enhanced the sulfur removal but this factor was not effective for PAX.

1 Introduction

The Sangan Iron Ore Deposit forms part of the east-west trending Kuh-e-Taleb mountain range, and is located at Sangan at latitude N 34°24', longitude E 60°16' in the Khorasan Province, North-Eastern Iran. This deposit involves several anomalies with different specification.

This study investigate anomaly C-north which is identified as high sulfur magnetite ore. The grade of sulfur in this anomaly varied from 1 to 8 percent which is higher than the norm of iron ore. Pyrite is the main source of sulfur in this ore which is accompanied with less amount pyrrhotite. So in order to reduce the sulfur content, reverse flotation was designed in concentration plant.

Thiol collectors are commonly used in the froth flotation of sulfide minerals. The xanthates, dithiophosphates and dithiocarbamates are classes of thiol collectors that each form different surface products on different sulfide minerals [1].

Xanthates have been the collector of choice for the flotation of sulfide minerals ever since the flotation process was developed and have been studied quite extensively [2]. Also it has been shown that the use of mixtures of thiol collectors can have benefits over the use of pure collectors [1, 3].

It has been observed, that an increase in xanthate concentration increased the grade of the concentrates in pyrite flotation, this is attributed to the fact that this factors increase the hydrophobicity of the pyrite and thus enhance elutriation in the froth phase [4]. Also the investigations have been shown the quality of the concentrate floated improves if collector is stage-added. Stage-adding collector involves dividing the total collector and adding it in several stages instead of one[5].The other factor which has been studied widely on sulfide flotation is the effect of pH. The previous studies say the best pyrite recovery is obtained at acidic condition but there is evidence that an acceptable pyrite recovery in alkaline condition is avoidable too [6].

In this study it has been tried to improve the quality of iron concentrate by improving the flotation of sulfur impurity. For this purpose the effect of collector type, pH, collector concentration and stage dosing on reverse flotation of iron sulfide from magnetite ore was investigated. Two type of thiol collectors include xanthates (sodium isopropyl xanthate (SIPX) and potassium amyl xanthate (PAX)) and dithiocarbamate (di-ethyl dithiocarbamate (DTC)) and the mixture of collectors were studied.

2 Material and method

2.1 Method

Considering the presence of iron sulfide (mainly pyrite) as impurity in magnetite concentrate, the reverse flotation was designed for removal of these sulfides.

Flotation experiment was carried out in a lab scale Denver flotation cell which was self air injecting for bubble generation. The sample in each test was 1 kg magnetite with d80< 38μm. Analytical grade sodium isopropyl xanthate (SIPX), potassium amyl xanthate (PAX) and di-ethyl dithiocarbamate (DTC) were used as collector in different tests and MIBC was used as frother in all tests. The pH was adjusted by addition HCl and

* Corresponding author: at.sarvi@gmail.com

NaOH and measured by pH meter. The tests were carried out by 30 % weight solid content. The floated and unfloated samples were collected separately, dried and weighted and then were sent for iron and sulfide analysis. Titration was used for iron and LECO for sulfur analysis.

2.2 Material

The samples used in this experiment were taken from Sangan concentrate plant. This plant uses low intensive magnetic separator (LIMS) for concentration of anomaly C-north from Sangan iron ore. The samples were taken from the final concentrate of LIMS, so the liberated part of iron sulfide was removed in magnetic separation. Experiment was designed to removes that part of iron sulfide which is engaged with magnetite and cannot be separated by magnetic separators. Chemical analysis of sample was given in table 1. Chemical analysis was done with Atomic Absorption Spectrometry.

According to the results the grade of sulfur exceeds the norm of sulfur in iron concentrate so it can be considered as the main impurity of the samples.

The mineralogical studies were done for identifying the type of minerals and their engagement. The result showed that magnetite is the main phase but in some places it oxidized to hematite Fig. 1.

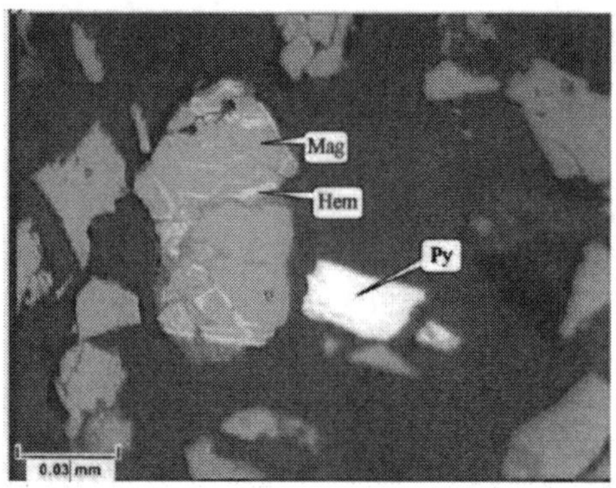

Fig. 1. Magnetite (Mag) is oxidized to hematite (Hem) with liberated pyrite (Py).

Sulfide was observed in two phase pyrite and pyrrhotite but domination is with pyrite. These minerals have different type of engagement with magnetite. In Fig. 1,2 pyrite and pyrrhotite are completely liberated while in Fig 3 pyrite has an intensive engagement with magnetite and this is the part that usually are not separated with magnetite separator.

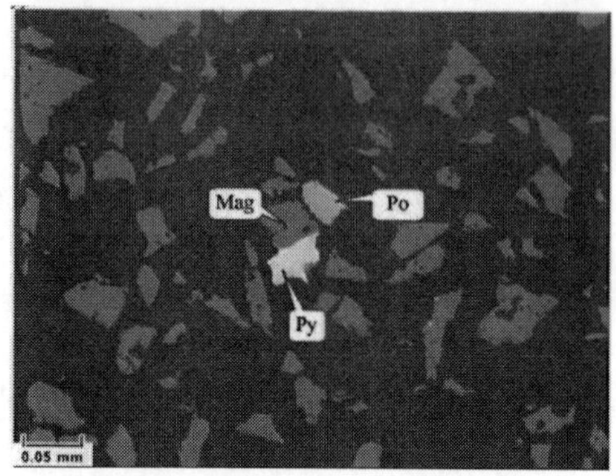

Fig. 2. Pyrite (Py) in contact with magnetite (Mag) and liberated pyrrhotite (Po).

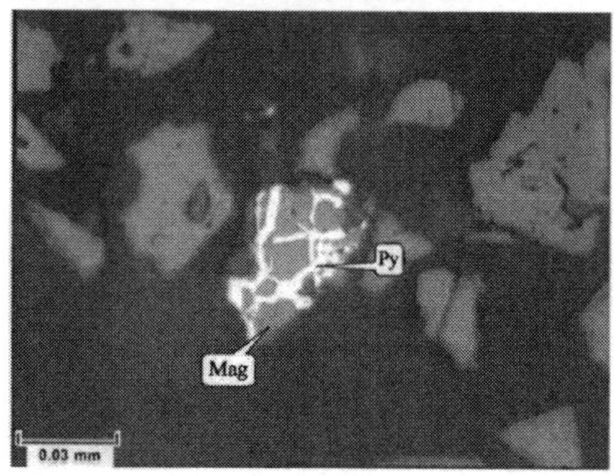

Fig. 3. Pyrite (py) and magnetite (Mag) engagement.

Table 1. Chemical analysis of sample.

Element (%) / Sample	Fe	SiO_2	Al_2O_3	CaO	MgO	P	S	Na_2O	K_2O	Mn	TiO_2	V
A	66.07	1.65	0.6	0.125	3.84	<0.01	0.85	0.01	0.014	0.205	0.07	0.004
B	66.4	1.43	0.75	0.19	4.35	<0.01	0.72	0.01	0.01	0.125	0.06	0.004
C	67.4	0.93	0.6	0.06	3.9	<0.01	0.45	0.005	0.007	0.145	0.06	0.004

3 Result

In this study the flotation of iron sulfide minerals especially pyrite as the main sulfide impurity of Sangan magnetite ore was investigated. For this purpose, Three different collectors include sodium isopropyl xanthate (SIPX), potassium amyl xanthate (PAX) and diethyl dithiocarbamate (DTC) which all of them are known as thiol collectors, were tested. The result was depicted on Fig 4.

According to the result, xanthates are better collectors for iron sulfide minerals compare to dithiocarbamate and between the two tested xanthate, PAX showed the best sulfide removal. It is supposed, because of longer carbon chain, PAX is stronger collector and more effective in iron sulfide flotation. [7, 8].

DOI:10.1051/e3sconf/20171801023

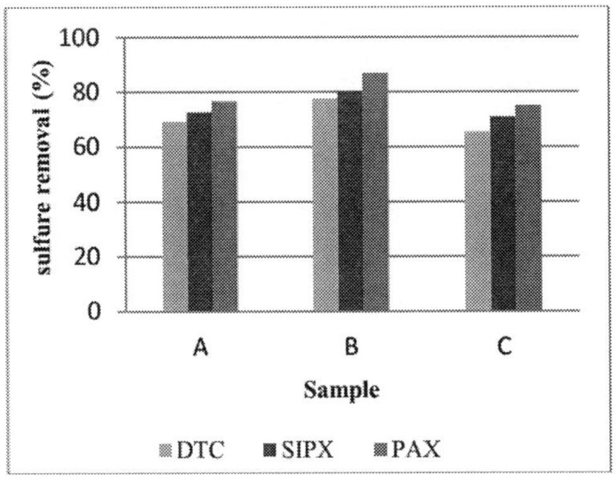

Fig. 4. The effect of collector type on removal of iron sulfide ore from magnetite ore by flotation.

It has been shown that in some cases the use of mixtures of thiol collectors has benefits over the use of pure collectors, so mixture of PAX-DTC was investigated [2]. In this order DTC and PAX (SIPX) was added respectively with some interval with this logic that when a mixture of collectors is exposed sequentially to a surface which by definition must have a heterogeneous distribution of energetically different sites, the weaker collector (DTC) will adsorb preferentially on the strong sites and the strongly adsorbing collector (PAX), added subsequently, will adsorb on the weaker sites. In this way as many sites as possible are utilized for adsorption thus enhancing the hydrophobicity [9]. The result is shown in Fig5.

According to the results, the mixture of PAX-DTC is not effective and could not enhance the floatability of iron sulfide ore and even deteriorate the effect of pure PAX. So this parameter was overruled.

The experiments showed that the sulfur removal is improved if collector is stage –added [5]. Stage adding involves adding collector in several stages. According to Fig 6, the best recovery was obtained when collector is added in four stages. It seems with this strategy the collector concentration almost remain constant during the flotation. The main part of collector is consumed at first stage of flotation and consequently the collector concentration reduced in cell, with stage dosing, collector concentration remains almost constant during the flotation.

The effect of acidity is the other parameter which was investigated. In Fig. 7, the sulfur removal percentage in different pH was shown. PAX and SIPX showed almost the same behavior. The floatability was low at high acidic pH, probably due to the decomposition of the collector and typically reaches its maximum at around pH 3.5-4. As the pH was increased, an intermediate depression in the iron sulfide recovery was observed. It seems that the formation of the hydrophilic ferric hydroxide which is the stable iron species in an alkaline medium is responsible for this phenomenon. Pyrite recovery increases gradually after pH 5 and show a second peak at pH around 9 and then it drops dramatically as pH increases. According to Göktepe (2002) at pH 5, pyrite is very soluble and iron ions released from the surface of pyrite are found to have an important effect on collector adsorption where these ions can cause rapid oxidation of xanthate to dixanthogen[10]. The decrease in recovery above pH, 9 was due to the formation of Fe $(OH)_2$ on pyrite.

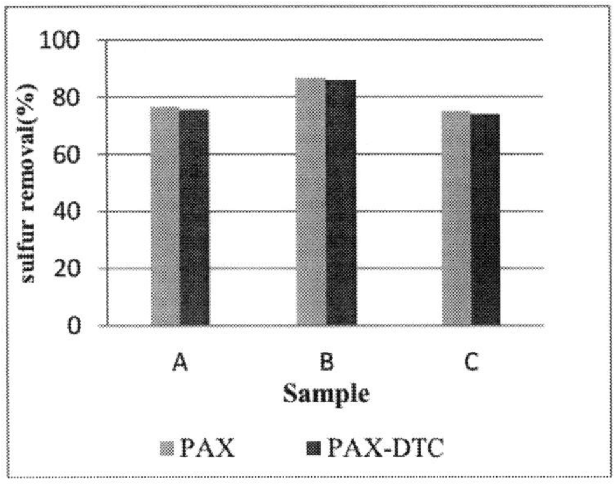

Fig. 5. The effect of collector mixture on iron sulfide flotation.

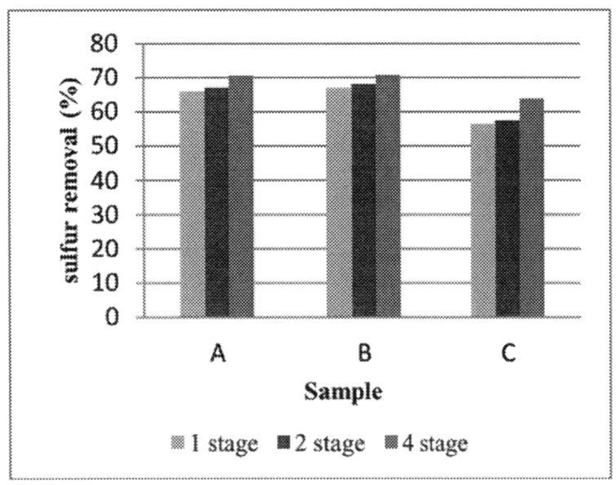

Fig. 6. The effect of stage adding of collector on flotation of iron sulfide minerals.

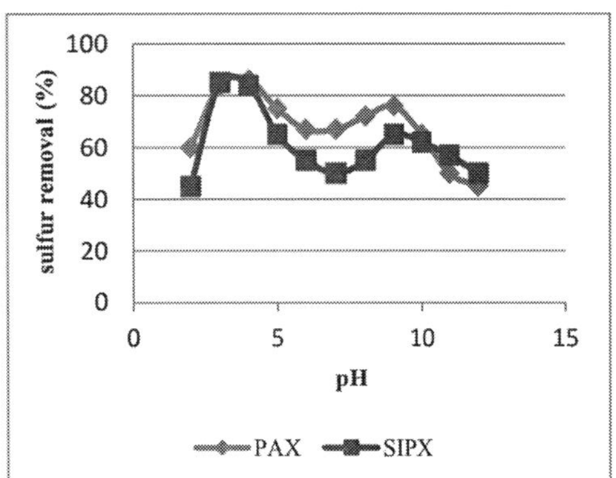

Fig. 7. The effect of pH on iron sulfide flotation.

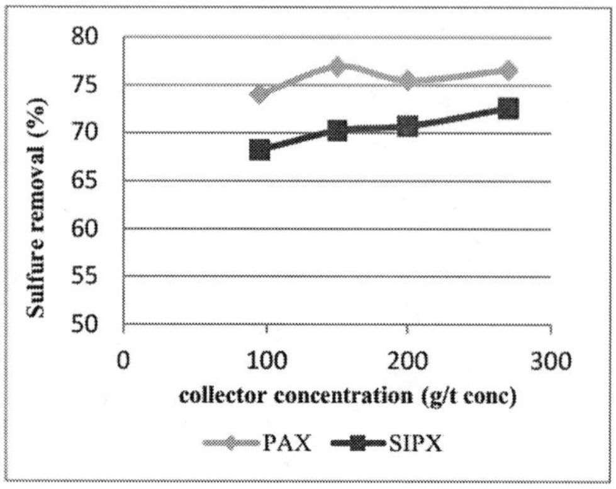

Fig. 8. The effect of collector concentration on flotation of iron sulfide minerals.

The floatability of iron sulfide minerals with variable collector concentration showed that increasing the SIPX concentration improved the sulfur removal but the PAX concentration almost did not affect the sulfide removal in the range of this investigation. The results are depicted in fig.8.

4 Conclusion

In this study the flotation of iron sulfide minerals especially pyrite as the main sulfide impurity of Sangan magnetite ore was investigated. For this purpose the effect of collector type, pH, collector concentration and stage dosing was studied. The result showed that xanthate is better collector for iron sulfide minerals compare to dithiocarbamate and between the two tested xanthates, PAX showed the best sulfide removal. The mixture of PAX as the strong and dithiocarbamate as the weak collector was ineffective in iron sulfide flotation as it was expected and even it deteriorate the effect of pure PAX.The experiments showed that the sulfur removal is improved if collector added in 4 stage during the flotation instead of one stage at the beginning.

Acidity has different effect on floatability of iron sulfide, the recovery is improved at acidic condition around pH 3.5-4, but there is a second peak in recovery at alkali condition at pH 9 albeit the second recovery is pretty weaker than the first one.

The floatability of iron sulfide minerals with variable collector concentration showed that increasing the SIPX concentration improve the sulfur removal but the PAX concentration almost did not affect the sulfide removal in the range of this investigation.

The authors would gratefully acknowledge Opal Parsian Sangan Industrial & Mining Co. (OPSIM), in providing them with funding and opportunity to complete this work. The authors would like to express their deepest gratitude to Mr. Kambiz Moazami Managing Director of Kanikavan Shargh Engineering Co. and colleagues in the company, for their support.

References

1. G. Hangone, D. Bradshaw and Z. Ekmekci, J. South. Afr. Inst. Min. Metall., **105**, 199-206 (2005)
2. R. Plessis, J.D. Miller,XXII International Mineral Processing Congress, 892-901 (2003)
3. D.J Bradshaw, C.TO'Connor, J. South. Afr. Inst. Min. Metall., 89-96 (1996)
4. A.Dimou, *thesis* for master degree (University of Cape Town, 1986)
5. https://www.911metallurgist.com/blog/stage addition of flotation-reagents
6. R Mermillod-Blondin, Centenary of Flotation Symposium, 683-692 (2005)
7. B Rezai, *Flotation*, 52-54 (1996)
8. B A. Wills, T Napier-Munn, *Mineral Processing Technology*, 270-273 (2006)
9. D.J. Bradshaw, P.J. Harris and C.T. O'Connor, J. South. Afr. Inst. Min. Metall.,189-194 (1998)
10. F Göktepe, Turkish J. Eng. Env. Sci, **26**, 309-318 (2002)

Circular Economy in Mineral Processing

Radosław Pomykała [1], and *Barbara* Tora[1,*]

[1]AGH University of Science and Technology, Faculty of Mining and Geoengineering, Kraków, Poland

Abstract. The paper aims to implementation of *Circular Economy* in mineral processing in Poland. Circular economy represents a completely new approach to product life cycle, based on the departure from the linear model of "take – make – dispose" and turning to the circular or closed-circle model of economy. Challenges and opportunities of implementation of *Circular Economy* in Mining is presented. The VERAM project, financed by The European Union (by the EUR 1.4 million) to the project is described. The examples of good practice in the area of implementation of circular economy in Poland is presented (Tauron Wydobycie - wasteless mine and ZGH Bolesław - waste management)

1 Introduction

The European Commission accepted a package of documents concerning Circular Economy on 2 December 2015. Circular economy represents a completely new approach to product life cycle, based on the departure from the linear model of "take – make – dispose" and turning to the circular or closed-circle model of economy. Increase of such activities as recycling and reuse of products brings benefits not only to the environment but also to economy, e.g. in the form of raw-material savings or reduced waste quantities.

First international conference on minerals in Circular Economy was organized by VTT Technical Research Centre of Finland, GTK Geological Survey of Finland, Aalto University, University of Oulu on 26-27 November 2014.

The modern society is based on using mineral commodities: their total consumption correlates with economic growth and urbanization. Due to population growth on the global level, the total consumption of mineral raw materials is expected to increase. At the same time, the transition towards a circular economy plays a key role in decreasing per capita consumption of minerals. For this transition to happen, top class scientific research that allows for knowledge transfer between different segments of the mineral value chain is required. Sustainable mining, enhanced material recovery, better material and product design, and sustainable resource policies are all essential parts of the science to meet the demands of the circular economy. They are also closely supported by the European Innovation Partnership (EIP) on Raw Materials and the upcoming Knowledge and Innovation Community (KIC) on raw materials, which aim to improve the mineral sector's possibilities for sustainable growth, creating more jobs in Europe and securing the European industries' raw material needs.

The Ellen MacArthur Foundation sees its mission as accelerating the transition from a linear take-make-dispose economic model to a circular model that is restorative and regenerative by design and aims to keep products, components and materials at their highest utility and value at all times. Furthermore, the European Commission has adopted a Circular Economy Package, including revised legislative proposals on waste, to stimulate Europe's transition towards a circular economy to boost global competitiveness, foster sustainable economic growth and generate new jobs.

2 Mc Kinsey report

A new McKinsey analysis on the benefits of circular economy was prepared in 2015.

The McKinsey Global Institute drafted the Report on circular economy, or the economy which emphasizes specific, super-effective treatment of raw materials and materials that does not interfere with natural balance (only biological materials are returned to nature). The so-called second-generation recycling may bring 1,8 billion euros in economic benefits [European billion], in addition to environmental and social advantages, until 2030. Europe depends on raw materials to a large extent and that also creates considerable possibilities of efficient material reuse improvement which will give impetus to better resource productivity and competitiveness.

Europe is wasteful in the area of raw-material consumption which is demonstrated by the "take – make – dispose" model. In 2012, the average European used 16 metric tons of various types of materials, with only 40 percent recycled or reused. That kind of wastefulness is also visible e.g. in the use of cars (European cars remain parked 92 percent of the time, without transporting people), or food (31 percent of food is wasted during processing), or office operation (only 35 to 50 percent of the time is used properly). And there are more surprising and original observations of that kind.

* Corresponding author: tora@agh.edu.pl

The European Commission approved a new ambitious package of documents concerning circular economy. The new approach should assist European companies and consumers in switching to circular economy in which resources are used in a more sustainable manner.

Proposed actions would contribute to "closing the circle" of product lifetime owing to increased recycling and reuse, bringing advantages to both the environment and economy. Implementation of the related plans will allow to maximise value, utmost use of all raw materials, products and waste, with the intention to save energy and reduce green-house gas emissions.

Specific proposals entail the whole product life cycle: from manufacturing to consumption to waste management to recycled materials market. That process will be financially supported by the European structural and capital-investment funds of which 5.5 billion euros will be designated for capital investments in the area of waste management. Additional support of 650 million euros will be made available by the EU under the "Horizon 2020" Programme for research and innovation, as well as capital investments in circular economy at the national level.

3 Closing the loop - An EU action plan for the Circular Economy

The European Commission adopted an ambitious Circular Economy Package, which includes revised legislative proposals on waste to stimulate Europe's transition towards a circular economy which will boost global competitiveness, foster sustainable economic growth and generate new jobs.

The Circular Economy Package [3] consists of an EU Action Plan for the Circular Economy that establishes a concrete and ambitious programme of action, with measures covering the whole cycle: from production and consumption to waste management and the market for secondary raw materials. The annex to the action plan sets out the timeline when the actions will be completed.

The proposed actions will contribute to "closing the loop" of product lifecycles through greater recycling and re-use, and bring benefits for both the environment and the economy.

The revised legislative proposals on waste set clear targets for reduction of waste and establish an ambitious and credible long-term path for waste management and recycling.

Key elements of the revised waste proposal include:

- A common EU target for recycling 65% of municipal waste by 2030;
- A common EU target for recycling 75% of packaging waste by 2030;
- A binding landfill target to reduce landfill to maximum of 10% of municipal waste by 2030;
- A ban on landfilling of separately collected waste;
- Promotion of economic instruments to discourage landfilling ;
- Simplified and improved definitions and harmonised calculation methods for recycling rates throughout the EU;
- Concrete measures to promote re-use and stimulate industrial symbiosis - turning one industry's by-product into another industry's raw material;
- Economic incentives for producers to put greener products on the market and support recovery and recycling schemes (eg for packaging, batteries, electric and electronic equipments, vehicles).

4 Environmental Protection and Efficient Resource Management

The EU cohesion policy supports sustainable development by promoting water and waste management, environmentally friendly and innovative pure technologies, as well as the activities intended to protect air, biodiversity, and nature.

The EU structural and capital-investment funds help to protect and preserve such natural resources as water, nature, biodiversity, clean air, and raw materials. Those funds are spent for capital investments in infrastructure designed for liquid waste treatment and solid waste management (e.g. recycling), as well as all necessary means for environmental monitoring systems and green infrastructure development. Owing to such activities, our environment becomes a source of economic growth and new possibilities for creating jobs.

5 Objectives

- To stimulate Europe's switching to circular economy and increase global competitiveness
- To support sustainable economic development
- To create new jobs
- To increase product lifetime through circular economy, e.g. by the following:
  - design allowing for easy repair, reuse, or processing,
  - durability improvement,
  - promotion of damaged equipment repairs instead of replacement,
  - better waste management,
  - new business and consumption models based on product sharing, borrowing, exchanging, or leasing .

6 Goal: To close the materials circulation

Through the following:

- Limitation of raw-material consumption
- Optimisation of the production processes
- Reduction of energy consumption
- Prolongation of product life cycles
- Maximisation of waste recycling and recovery
- Limitation of environmental impact (environmental trace).

7 Circular Economy: Areas of Action

1. Product design stage
2. Distribution
3. Use or consumption
4. Public procurement
5. Product labelling and product information
6. Waste management
7. Development of recovered materials market
8. Improvement of conditions in such priority sectors as:
    a) sustainable production
    b) recycled material recovery
    c) building industry
    d) critical raw materials
9. Improvement of intersectoral collaboration, e.g. by the promotion of industrial symbiosis, repairs and reuse, as well as
10. Development of new business models

7.1 Circular Economy

- Limitation of the quantity of materials required for the provision of a specific service
- Prolongation of the product availability (durability)
- Limitation of the consumption of energy and materials at the stages of manufacturing and use (effectiveness)
- Limitation of the use of hazardous or hard-to-recycle materials in products and production processes (substitution)
- Development of the recycled materials (recyclate) market, based on standards, public procurement etc.
- Designing the products that are easy to maintain, repair, modernise, remodel, or recycle (ecodesigns)
- Development of necessary services for the consumers in a given area (maintenance, repairs etc.)
- Encouraging the consumers to limit waste and implement high quality waste segregation, with offering support in those areas
- Encouraging the consumers to segregate waste and apply waste collection systems to minimise recycling and reuse;
- Facilitation of activities intended to prevent discarding side products as waste (industrial symbiosis)
- Stimulation of the conditions that facilitate wider and better consumer choices by relying on lease services
- Product renting and sharing as an alternative to product owning, with concurrent consumer interest protection (in respect of costs, safety, information, contractual conditions, insurance aspects etc.)

8 Circular Economy Implementation

During the workshop meetings conducted at this year's Green Week in Brussels (3-5 June 2017), a number of new solutions and possibilities, identified on the basis of the experiences drawn from successful initiatives, were presented. Those can be applied on a larger and broader scale involving:

- At the production stage: environmentally balanced standards of supplier selection, voluntary systems applied in industrial branches and among retail sellers, as well as industrial symbiosis for better use of raw materials and waste control, with the creation of side-product markets,
- At the distribution stage: better information on the resources contained in products and the methods of product repair or recycling, known as "product passports",
- At the consumption stage: shared consumption models, based on product borrowing, swap, replacement, and lease (e.g. lease of home appliances), and product-and-service systems designed to obtain higher value from inadequately used assets or resources (e.g. cars, tools, or premises).

9 Key Elements of Building Circular Economy in Poland

1. Innovation, increased co-operation between industrial and scientific sectors, leading to the implementation of innovative solutions in economy.
2. Creation of the European recycled materials market, with facilitated materials flows.
3. Ensuring high quality recycled materials, resulting from sustainable production and consumption,
4. Development of the services sector.

9.1 The Areas of Mining Activities in Circular Economy

- Feeding and supplementing raw-material cycles
- Dedicated products causing low emissions
- Responsibility for product quality
- Optimisation of own production cycles
- Increase of production efficiency, including:
- Reduction of raw-materials waste
- Use of methane
- Limitation of energy consumption
- Limitation of the effect of mining on the environment
- Optimisation of waste management
- Towards waste-free mining
- Intersectoral collaboration
- Development of services assisting mining operations
- Facilitation of closing cycles in other sectors, including power generation

9.2 Indicators Designed for the Waste-Free Coal Power Generation Model

To evaluate the efficiency of the implementation of circular economy, certain indicators have been proposed [2].

9.3 Waste rate indicator, Wo

The indicator determines the quantity of waste generated per base product unit. In the case of energy generation, electricity and thermal energy will be the base units. The indicator will tell us to what extent production is clean and what are the possibilities of reducing waste in a given plant.

$$Wo = O/P \quad (1)$$

O – waste; P – products

Units, e.g.: ton/MWe (metric ton per megawatt of electric power), ton/MWt (metric ton per megawatt of thermal power) etc.

The objective of waste-free coal power generation is to obtain the Wo indicator equal zero.

9.4 Productivity indicator, Wp

The indicator determines the proportion between the generated waste and the waste processed into raw materials and products.

$$Wp = S+P/O \quad (2)$$

S+P – raw materials and products generated by man and sold

O – waste

Unit: dimensionless indicator, ton/ton

The objective of waste-free coal power generation is to obtain the Wp indicator equal one.

10 Implementation of Circular Economy: Challenges and Opportunities in Mining

- Possibilities of obtaining financial resources for innovative projects
- Integration of pro-environmental activities
- Development of product-oriented approach to waste management
- Analysis of the environmental trace (including that in respect of other energy generation technologies)
- Improvement of the image of the mining sector
- Finding additional reasons in the efforts to maintain the share of coal in the energy generation mix

10.1 Implementation of the Circular Economy Principle in the Polish Mining Sector

During the International Mining Forum held in Jastrzębie Zdrój in June 2017, Mr. Tadeusz Koperski, the President of Kopex, presented a conception of organising a system of implementing circular economy in the mining sector [4].

There is a need to select the operator whose general mission will be to associate the operating system, based on the circular economy paradigm, with a broadly understood waste management business model. The necessity to integrate those two systems results from the limitation of current directions of waste management or even their eradication.

A systemic approach will allow us to optimise the structures facilitating waste management, as well as expenditures for current operations, acquire new markets, and support and develop innovations for both operator's structures and co-operating stakeholders.

Organisational innovation will allow us to create a recycled and circular-economy materials market, based on a business action model [4].

Some examples of good practice in the implementation of circular economy in Poland were presented at the International Mining Forum: "Implementation of the Waste-Free Coal Mine Policy in the Tauron Wydobycie Company", "Waste Management in the ZGH Bolesław" [5].

10.2 The Cohesion Policy

Cohesion policy investments are geared towards local needs and opportunities. They contribute to [6, 7]:

- more recycling,
- improved waste management,
- resource and energy efficiency,
- strengthening the bio-economy,
- novel solutions in product design,
- new business models,
- and the creation of green jobs.

For improved waste management, EU has €5.5 billion focusing on waste prevention, reuse and recycling. EU also invests in basic waste treatment infrastructure in less developed regions. This means 5.9 million tonnes/year of additional waste recycling capacity in Europe.For SMEs, EU invests EUR 2.3 billion in environmentally-friendly production processes and resource efficiency. This helps them save costs and creates new opportunities by turning waste into a resource.

Important research and innovation funding opportunities are available too, for instance to develop new and better products. Regions from all Member States identified priorities related to circular economy in their Smart Specialisation Strategies. The Smart Specialisation Platform helps these regions to cooperate with others along value chains, also on specific topics such as industrial modernisation, agri-food and energy.

The Circular Economy Package also promotes water reuse. Several regions in the EU already recycle their treated wastewater for park irrigation, street cleaning or to replenish groundwater reserves.

Cohesion policy can support this further, through its €15 billion investment in the water sector during the 2014-2020 period.

11 VERAM Project

The recently launched VERAM [1, 9] project will deliver a long-term 2050 Vision and Roadmap for European Raw Materials. The European Union (EU) has awarded 1.4 million to the project under Horizon 2020 programme.

Patrick Wall, project coordinator, has given more details on this coordination and support action on raw materials project, and on how a long-term vision and roadmap will be developed. [veram2050.eu]

Europe's dependence on other continents for raw materials is enough of a concern to be included among the topics that form the third pillar of Horizon 2020, Societal Challenges. One call under this pillar sought applications for a coordination and support action (CSA) to improve the coordination of raw materials research and innovation activities between the EU and member states, and to develop a long-term vision for the field.

The VERAM project is being coordinated by the European Technology Platform on Sustainable Mineral Resources. Platform is industry-led member organization and has an important role in driving innovation. The area covered energy, environment, ITC, production and processes, transport, and the bioeconomy, which is based on biotechnology research.

VERAM's Work Plan includes the identification of the current status and a gap analysis, the development of a long-term vision and a roadmap, and the dissemination of the vision and roadmap for 2030 and 2050.

The VERAM project will produce a Vision and Roadmap for European Raw Materials in 2050 based on raw materials research and innovation (R&I) coordination.

Two leading European Technology Platforms (ETPs): **ETP SMR** (Sustainable Minerals Resources) and **FTP** (Forest Technology Platform) are joining forces to develop a common vision and roadmap with the support of **ECTP** (European Construction Technology Platform), represented by UNIVPM, **SusChem**(ETP for Sustainable Chemistry), represented by Cefic, and **EuMaT** (Advanced Materials ETP), represented by VITO. This partnership provides VERAM with expertise from downstream applications and additional knowledge on secondary raw materials.

12 Conclusions - good practices in Polish mineral processing

The examples of good practice in the implementation of circular economy in Poland were presented at the International Mining Forum.

An example of good solutions in the implementation of circurlar economy are summarized below.

12.1 Wasteless mine – case TAURON Wydobycie

Tauron Mining S.A. is the producer of coal – the part of TAURON Mining are three mining plants: Sobieski and Jaworzno and Brzeszcze. These plants are seen as the moder ones. They are producting about 8 million tones/year of coal, simultaneously, apart from they are producing also about 2 mln tonnes/year of rock and additional materials [11, 12].

In the following years, the proecologic activities of TAURON Mining S.A. will focus around using the potential of partnerships of the Group TAURON to work out the model of management of produced "wastes", mainly inside of the Group. In this purpose, the project Grey2Green was initiated between areas of energetic partnerships and TAURON Mining S.A., which idea is based on using the effect of synergy in the direction of wastes management. The realization of the project's assumptions is based also on applying innovative technologies in the planned investments of constructing special installments dedicated to pro- cess of various sorts of "wastes" and transform them into products. Thanks to greener investments, realized in TAURON Mining S.A. it was possible to manage or process both own wastes as well these originating from TAURON Wytwarzanie S.A. Additionally, the planned investments as modernization of installments to direct the slimy-ash suspension in ZG Sobieski to the bottom of the mines in purpose of fire prevention as well implementation of innovative technology of complex "cleaning" of slime tanks will allow to manage totally the coal slimes. This concerns both slimes being produced directly as well these ones which are still stored in tanks what will bring back their func- tionality.

12.2 Waste management in ZGH Bolesław

ZGH „Bolesław" S.A. (Mining and Metallurgical Plant Bolesław) is a producer of Special High Grade Zinc and its products. ZGH "Bolesław" S.A. extract and process zinc and lead ores coming from his own mine, produce electrolytic zinc, zinc alloys, sulphuric acid and zinc and lead concentrates. ZGH Bolesław produce zinc using electrolysis process and imperial smelting process as well as we recycle zinc waste materials in the Waelz process. The technologies he use protect the environment [8, 9, 10].

ZGH Boleslaw SA continues the current directions of applying wastes which are significantly related to environmental protection

- Construction of levees of active pond from coarser fractions of tailings gives additional capacity for disposal being equal to volume of residues built into levees. It elongates also the time of exploitation of current ponds without the necessity of gaining additional areas for location of new ones.
- The application of tailings to construct levees and as a component of hydraulic backfill eliminates the necessity of opening sand mines causing degradation of forests and fields.
- Application of tailings within last years as reclamation material allowed to perform such action for objects hard to proceed with and to restore landscape virtues for areas degraded by mining activity.
- The application of flotation tailings is connected with necessity of their wet, periodical storage in ponds till their drying. For now, lack of technical possibilities and real new directions for total and fast application of all wastes being created excludes the possibility of liquidation of the ponds.

References

1. H. Karaś: Is the 4th Industrial Revolution in the EU Industry a chance for the RM sector. Veram Stakeholder Forum (Antverp, Belgium, 2016)
2. T. Szczygielski (red): *Mineraly antropogeniczne a gospodarka o obiegu zamknietym* (Politechnika Warszawaska, Instytut Badan Stosowanych, Warszawa, 2015)

3. Report from the commission to the european parliment on the implementation of the Circular Economy Action Plan, Annex 1, (Brussels, COM 33, 2017)
4. T. Koperski, Innowacje organizacyjne jako podstawa do systemowego wdrażania rozwiązań gospodarki opartej na zasadzie gospodarki odpadowej w obiegu zamkniętym z wykorzystaniem technologii innowacyjnych, Interantional Mining Forum (Jastrzębie Zdrój, 2017)
5. R. Pomykała, B. Tora: Rola Górnictwa w gospodarce o obiegu zamkniętym, Interantional Mining Forum (Jastrzębie Zdrój, 2017)
6. www.mos.gov.pl, Plan działań dla Zamknięcie obiegu" dokument identyfikujący główne priorytety Polski w zakresie GOZ (2015)
7. https://www.ausimmbulletin.com/feature/the-contribution-of-mining-to-the-emerging-circular-economy/
8. http://veram2050.eu
9. R. Tomanec, M. Blagojev, J. Pol. Miner. Eng. Soc. , **1(39)**, 165–172 (2017)
10. A. Jarosiński, S. Żelazny, M. Cholewa, Pol. Miner. Eng. Soc., **1(37)**, 233–240 (2016)
11. Z. Śmieszek, K. Cichy, A. Wieniewski, B.Ochab, C.Regula: Odpady flotacyjne jako potencjalne ródło zasobów surowcowych na przykładzie rud Zn- Pb, Konferencja IMN, (Gliwice, 2011)
12. A. Nowak, A. Pawluk, Pol. Miner. Eng. Soc., **2(38)**, 45–52, (2016)
13. A. Fraś, R. Przystaś, B. Tora, Ecological and economic aspects of the management of mining waste in TAURON Mining S. A., XVIII International Coal Preparation Congress, (Saint-Petersburg, Russia, Springer International Publishing, 2016.) ISBN: 978-3-319-40942-9 ; e-ISBN: 978-3-319-40943-6

E3S Web of Conferences **18**, 01025 (2017)
MEC2017
DOI:10.1051/e3sconf/20171801025

Influence of liberation of sulphide minerals on flotation of sedimentary copper ore

Alicja Bakalarz[1,*], *Magdalena* Duchnowska[1], and *Andrzej* Luszczkiewicz[1]

[1]Wroclaw University of Science and Technology, Wybrzeze Wyspianskiego 27, 50-370 Wroclaw, Poland

Abstract. Ore liberation is one of the most important parameter in mineral processing, especially in flotation. To separate the valuable components from gangue minerals, it is necessary to liberate one from others. It is achieve primarily through crushing and grinding. These stages are one of the most expensive of mineral production. It is important to determine the adequate mineral liberation which would result in huge savings in the overall cost of flotation plant. The aim of the paper was the analysis of the influence of milling time on the laboratory flotation of the copper ore from stratiform Polish deposit. Three different milling time of copper ore in laboratory ball mill was applied. The flotation results were presented as the recovery-recovery and grade-recovery upgrading curves. The liberation of sulphides and the particle size of sulphides in flotation product were analysed and compared.

1 Introduction

The success of the separation process is related to the size and liberation of feed particles [1-3]. The definition of liberation varies from author to author [4]. The feed can consist of a single mineral called 'free particles', or it may consist of two or more minerals named 'locked particles'. Locked particles can be binary, ternary, quaternary etc., as they consist of two, three, four or more minerals [5]. Liberation of particles is achieved mainly through crushing and grinding of the ore. These stages usually consume 30–50% of overall energy costs of mining operation, but it can rise even to 70% for hard and/or finely intergrown ores. Achieving 'sufficient' liberation can bring huge savings in costs of mineral production [6-8].

Flotation process strictly depends on the particle size. According to Pease et al [9], the finer particles tend to be more liberated, different than coarse particles. Nevertheless, the recovery is reduced for finer and coarser particles. Gaudin [5] as a first extensively described the liberation issue associated with comminution by two means: liberation by size reduction and liberation by detachment. According to the paper, the less abundant mineral is not fully liberated unless the particles are finer than the grain size. Moreover, the more abundant phase is always better liberated than the less abundant phase.

The liberation of the ore is dependent on the mineralogy and texture [4,5,10]. Therefore, the upgrading results of flotation stream depend on the texture of the ore and the particle size to which the processing ore is crushed or milled. In Fig. 1 is shown, how particle composition defines the theoretical grade recovery curve [11]. It is important, that this type of curve do not take account of the process kinetics and mechanical entrainment [10]. According to Cropp et al [11] the losses in tailings are caused mainly by fine and liberated valuable minerals and locked minerals. The reasons of reducing the grade of concentrate are gangue composites, entrained and activated gangue, and deleterious element distribution in various size fractions.

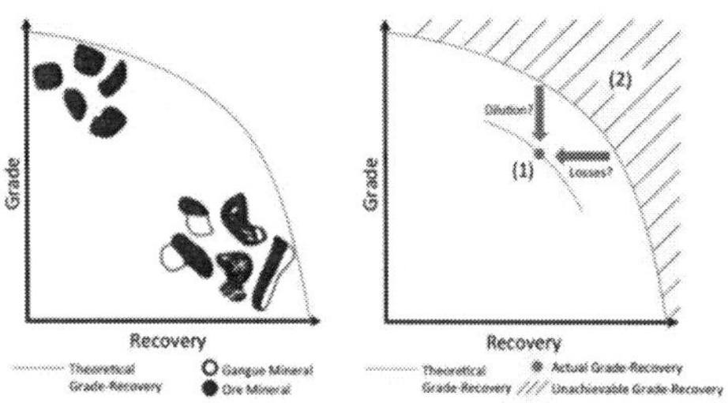

Fig. 1. A schematic of the theoretical grade recovery curve with typical particle images included (left) and the position of the actual grade/recovery relative to the theoretical potential (right); the area '1' – the upgrading can be improved by operational changes; the area '2' – grinding and classification is necessary to improve the upgrading results (based on Cropp et al [11]).

Inadequate liberation is often the main reason of the poor performance. In the literature is a lot of examples, where process mineralogy surveys identified and resolved many problems in the process. The mineralogical and liberation analysis can result in addition of a regrind mill in the circuit or an application of the metallurgical operations in the upgrading stream [10].

The modern mineral liberation measurement techniques can be separated into two main types: two and three dimensional. Two dimensional technique is used in the automated scanning electron microscopy with energy

* Corresponding author: alicja.bakalarz@pwr.edu.pl

DOI:10.1051/e3sconf/20171801025

dispersive X-ray detectors (SEM/EDS), which popularity and application have been raising since 70s of the twentieth century. Nowadays automated image analysis instruments widely available are QEMSCAN, MLA or EPMA [10,12]. Three dimensional techniques relate to the non intrusive measurements of the whole ore particle (e.g. 3D tomography).

This paper focuses on the influence of liberation of sulphide minerals in flotation of Polish copper ore from the Lubin-Glogow Basin (LGOM, SW Poland), the biggest copper deposit in Europe [13]. To prepare the flotation feed, different milling time of ore was applied. Mineralogical analysis of all upgrading products using QEMSCAN system were conducted. The influence of liberation of copper sulphide minerals in narrow particle size on the upgrading selectivity was analysed by plotting separation curves.

2 Experimental

2.1. Materials and methods

A sample of sandstone copper ore collected from Rudna mining area was investigated. The chemical and mineral composition of flotation feed is presented in Table 1. The sample contains 1.72% of copper. The flotation feed contains 59% quartz, 21% carbonates and 10% clay minerals and micas. The content of all sulphides is about 2.8% (1.3% chalcocite, 0.8% bornite, 0.2% chalcopyrite and 0.3% pyrite). Additionally, the flotation feed contained 0.1% covellite, 0.05% galena and 0.04% sphalerite.

The ore was dry-crushed in the jaw crusher to obtain particles fraction below 1 mm. A sample of 800 g in the presence of 500 cm^3 of industrial water was wet-ground in a laboratory ball mill. The grinding media (stainless balls) filled about 40% of the mill capacity. Three different milling time of copper ore was applied: 60, 80 and 100 minutes.

Table 1. Chemical and mineral composition of flotation feed.

Element/Mineral	Content, %
Cu	1.72
Chalcocite	1.35
Bornite	0.76
Chalcopyrite	0.21
Covellite	0.14
Pyrite	0.26
Galena	0.05
Sphalerite	0.04
Quartz	58.78
Ca, Mg carbonates	21.39
Clay minerals+micas	10.00
Others	5.30

The flotation methodology is presented in Fig. 2. All experiments were conducted in a Denver D12 laboratory flotation machine equipped with 2.5 and 1.5 dm^3 flotation cells. The air flow rate was regulated during each flotation test using a rotameter. Industrial mixture of sodium ethyl and isobutyl xanthate and sodium O,O-diethyl dithiophosphate in proportion 7:3 and the dose of 80 g/Mg was used as a collector. An aqueous solution of Nasfroth was utilized as a frother at a dosage of 20 g/t in all flotation tests. Both reagents were prepared directly before flotation test. After the experimental part, all the products were dried and weighted. A QEMSCAN system was used to determine mineralogy of feed samples and flotation products. The content of copper was determined by using a spectrometric method.

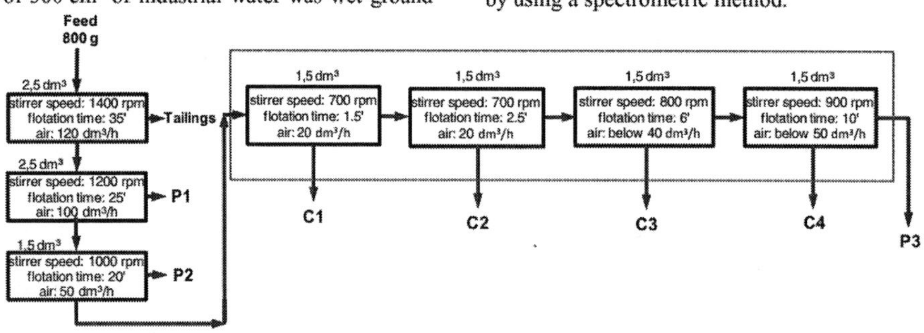

Fig. 2. The scheme of experimental methodology.

2.2 Results

2.2.1 Upgrading results

The results of flotation were evaluated by the mass balance of components and products of separation as well as by plotting recovery-recovery (the Fuerstenau upgrading curve, Fig. 3) and grade-recovery separation curves (Fig. 4).

As it is presented in Figs. 3 and 4, the best and similar results of upgrading of copper sulphides were obtained after 80 and 100 minutes of milling. In these tests the best upgrading selectivity of copper sulphides was observed. It can be assumed that the milling time equals to 80 minutes is sufficient to obtain the highest upgrading selectivity of the examined copper ore.

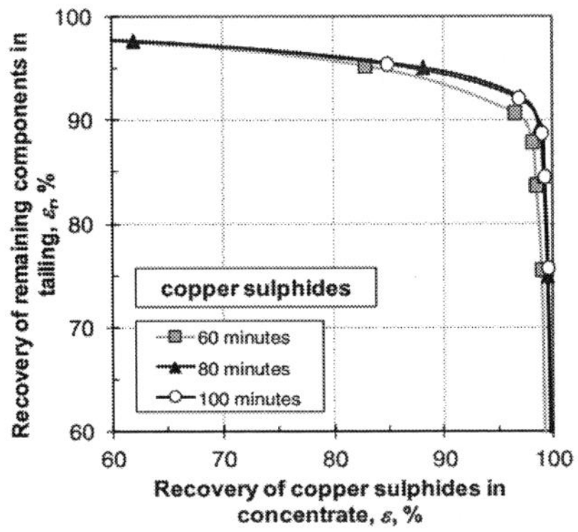

Fig. 3. A recovery of copper sulphides in concentrate vs. cumulative recovery of remaining components in tailing obtained after 60, 80 and 100 minutes of milling.

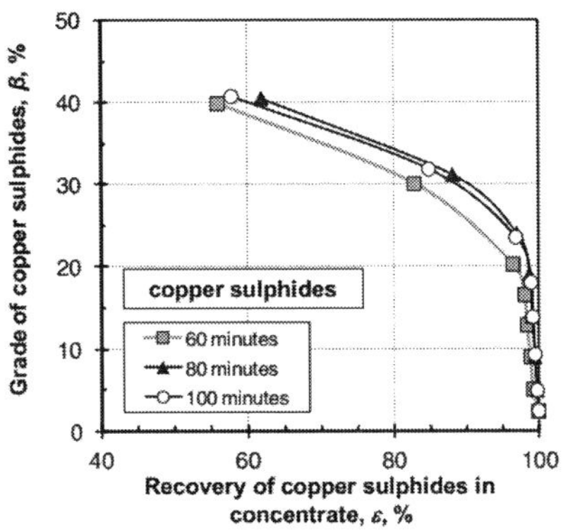

Fig. 4. Copper sulphide grade vs. recovery obtained after 60, 80 and 100 minutes of milling.

2.2.2 Analysis of liberation results

The comparison of the grain composition and liberation of sulphides in the flotation feed is presented in Fig. 5 and 6.

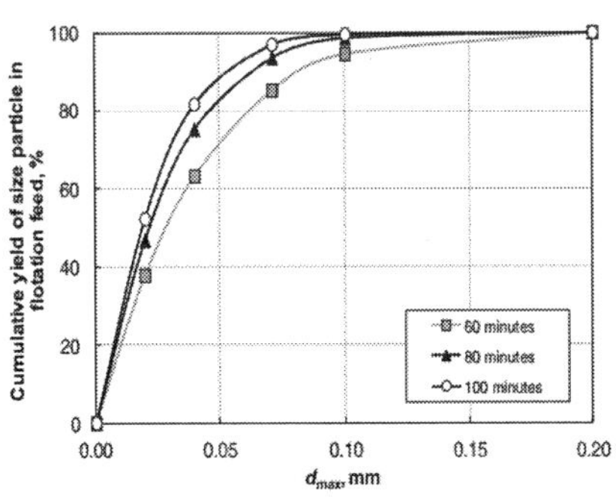

Fig. 5. Particle size distribution in flotation feeds.

As expected, the particle size decreases with increasing milling time (Fig. 5). After 60, 80 and 100 minutes of milling, P_{80} parameter equals to 52 um, 46 um and 38 um, respectively.

In table 2 the liberation of sulphides in the feed after milling is presented. The results of liberation was splited on four ranges: 0–20% (locked particles), 20–50% and 50–70% (impregnations) and 70–100% (free particles). It is important to mark that these data were calculated for sulphides of copper (chalcocite, chalcopyrite, bornite, covellite and tennantite), zinc (sphalerite), lead (galena) and iron (pyrite).

As expected, the liberation of sulphides becomes better with the milling time of the feed decreasing (Table 2). It is noticeable in each grain fraction. The highest of fully liberated particles are observed for particles size in the range of 0–20 um. It is important to note that quite small contents of impregnations are observed in each fraction in feed.

The relationship between liberation of sulphides and milling time is presented in Fig. 6. Similar results of liberation in flotation feeds after milling in 80 and 100 minutes were obtained. A significant differences are observed in liberation of sulphides in feed after 60 minutes and 80 or 100 minutes. It indicates that milling time equals to 80 minutes is sufficient to obtain the maximum of liberation of sulphide minerals in examined copper ore. However, it is important to notice, that this observation concerns in the treatment the ore in a wide size distribution. The liberation of sulphides in narrow grain fractions in flotation feed were analysed (Figs. 7-8).

Table 2. Liberation of sulphides in flotation feed.

Particle size, um	Liberation, %	Milling time of the feed, minutes		
		60	80	100
0–20	0–20	1.77	1.46	1.66
	20–50	3.95	3.38	3.86
	50–70	2.82	2.50	2.80
	70–100	35.88	41.66	45.69
20–40	0–20	1.89	1.62	1.68
	20–50	2.05	1.75	1.73
	50–70	1.61	1.54	1.43
	70–100	21.25	23.34	24.97
40–71	0–20	2.08	1.45	1.17
	20–50	1.85	1.29	1.17
	50–70	1.13	1.04	0.65
	70–100	15.17	14.09	10.88
71–100	0–20	0.92	0.33	0.13
	20–50	0.57	0.34	0.18
	50–70	0.46	0.30	0.18
	70–100	4.36	3.33	1.45
>100	0–20	0.45	0.09	0.02
	20–50	0.10	0.00	0.00
	50–70	0.12	0.00	0.00
	70–100	1.57	0.49	0.35

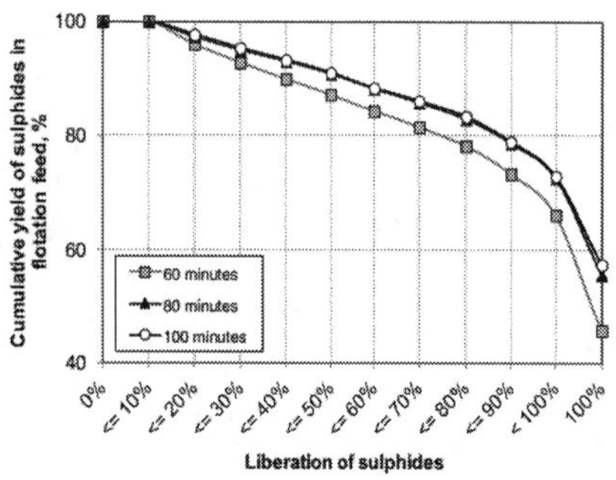

Fig. 6. Liberation of sulphide minerals in flotation feeds.

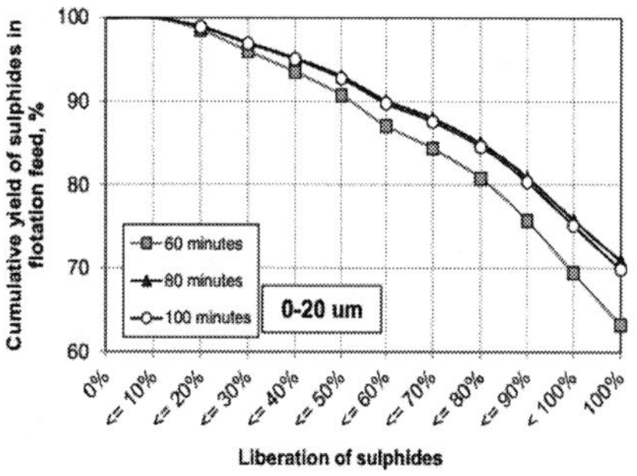

Fig. 7. Liberation of sulphide particles in the size range of 0–20 um in flotation feed.

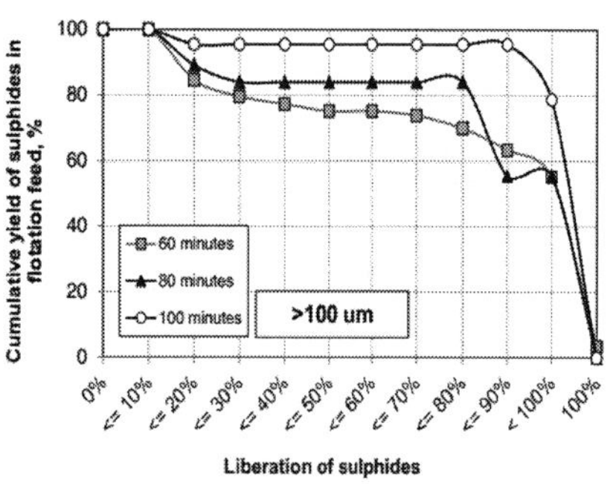

Fig. 8. Liberation of sulphide particles in the size range of >100 um in flotation feed.

As it has been shown, significantly worse liberation after 60 minutes of milling of each grain fraction was observed. It is presented in Fig. 7, that the comminution of sulphides to the size minus 20 um does not provide the good liberation of the fines. Very similar results were observed for sulphide particles in the range 20 to 40 um. Similar liberation for sulphides in the size range of 0–71 um after 80 and 100 minutes of milling was obtained. Probably, the reason of weak liberation of fines is very small dimensions of sulphides, which initially appears in the ore. It can be supposed that the grinding of narrow grain fractions separately can result in better liberation of sulphides, also for fines.

Noticeable differences between 80 and 100 minutes of milling are observed for particle size ranging from 71 to 100 um and >100 um. For the grain fraction >100 um, the liberation of sulphides increases with the milling time (Fig. 8).

The influence of liberation of sulphides was evaluated by plotting grade-recovery and recovery-recovery (α-non-sensitive) upgrading curves. The analysis was conducted for narrow particle size fractions and the liberation degree of sulphides. Only the part of figures is shown in the paper. The curves are presented in Figs. 9-21.

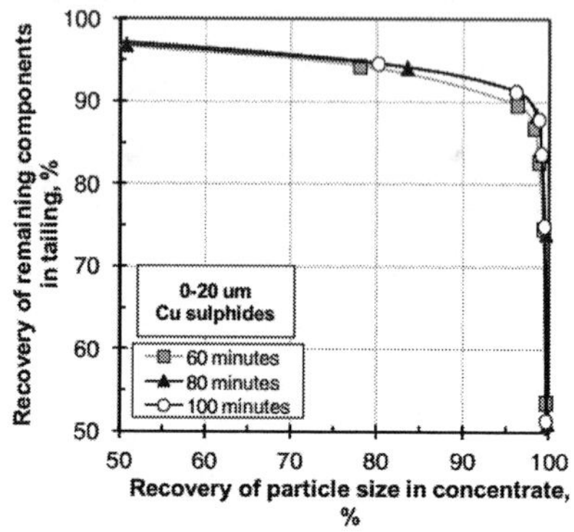

Fig. 9. A recovery of copper sulphides in concentrate vs. recovery of remaining components in tailing in size range of 0–20 um obtained after 60, 80 and 100 minutes of milling.

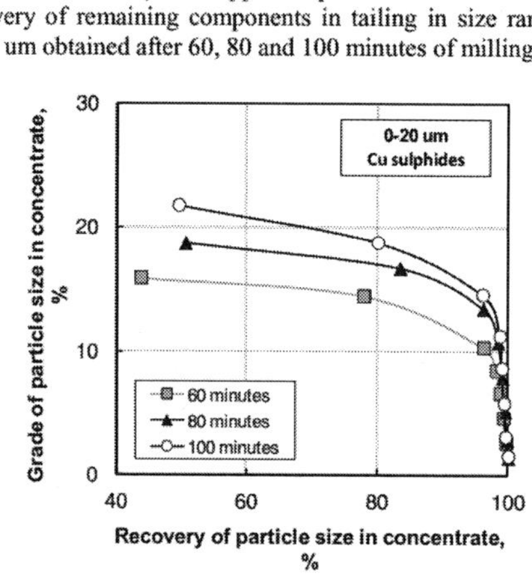

Fig. 10. Copper sulphides grade vs. recovery in size range of 0–20 um obtained after 60, 80 and 100 minutes of milling.

As it is presented in Figs. 9-10, the lowest upgrading selectivity and grade of the finest particles in concentrate is observed in flotation test in which 60 minutes of milling was applied. Similar performance was obtained for the flotation experiments after longer milling time (80 and 100 minutes). Comparable results were obtained for sulphides in the size range of 20–40 um. Since the each flotation feed contains the highest content of grain fraction between 0 and 40 um size (Fig. 3), lower upgrading selectivity of these sulphides size can result in overall poorer upgrading of copper.

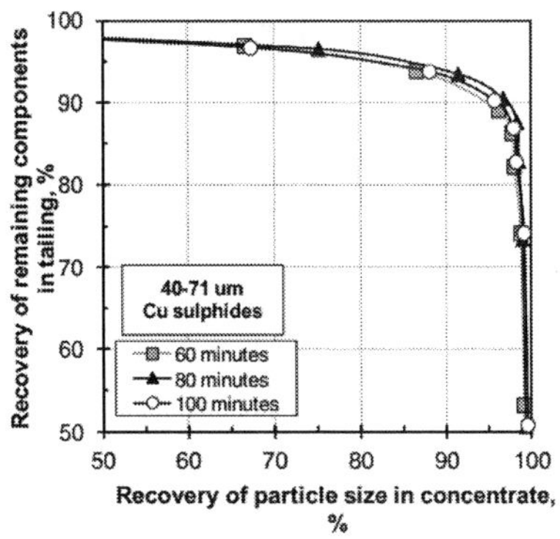

Fig. 11. A recovery of copper sulphides in concentrate vs. recovery of remaining components in tailing in size range of 40–71 um obtained after 60, 80 and 100 minutes of milling.

For the particles greater than 40 um, the differences between three curves become more significant with extending of the milling time (Fig. 11). Similar results for the particles size ranging from 71 to 100 um were obtained. The best upgrading selectivity for copper sulphides was observed after milling time equals to 80 minutes. A bit worse flotation efficiency after 100 minutes of milling was obtained. The lowest selectivity is observed for the shortest milling time. For the sulphide grains greater than 100 um the observation has changed (Fig. 12). It is quite interesting that even for coarse-grain fraction (+100 um) the best upgrading selectivity is obtained for the flotation feed after 100 minutes of milling.

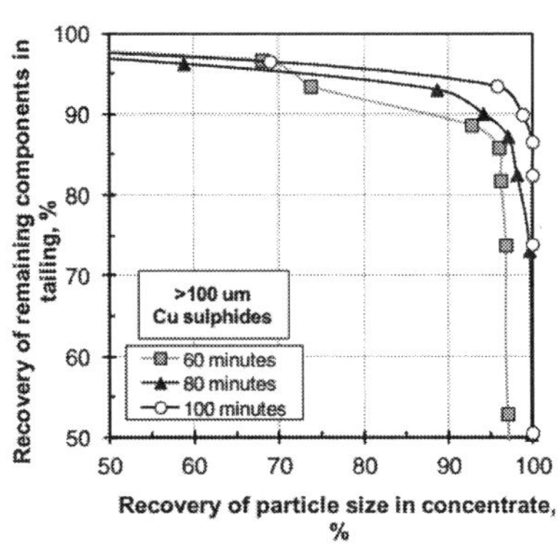

Fig. 12. A recovery of copper sulphides in concentrate vs. recovery of remaining components in tailing in grain fraction greater than 100 um obtained after 60, 80 and 100 minutes of milling.

In Figs. 13-16 the flotation results of free (liberated) sulphides particles is presented. As it can be seen in Fig. 13, the highest and similar upgrading selectivity of sulphides is observed for tests with milling time of 80 and 100 minutes. The lower selectivity was obtained after feed milling in 60 minutes. Similar results were obtained for the particle size in the range of 20–71 um. The comparable dependency is presented in Fig. 14 as grade-recovery curves. The quality of concentrate differs significantly for all tests.

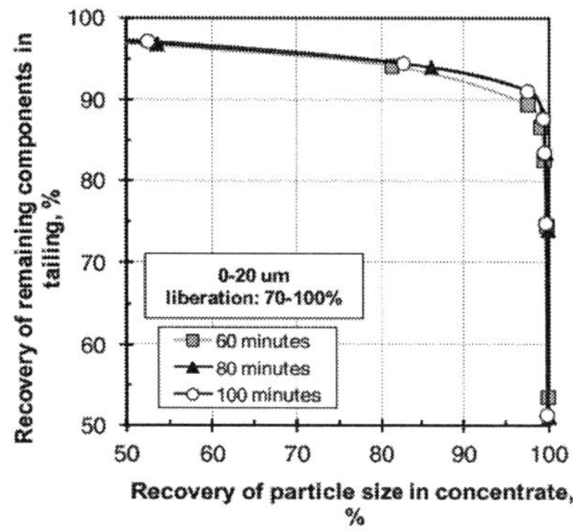

Fig. 13. A recovery of sulphides vs. recovery of remaining components in tailing in size range of 0–20 um liberated in 70–100% (free particles) obtained after 60, 80 and 100 minutes of milling.

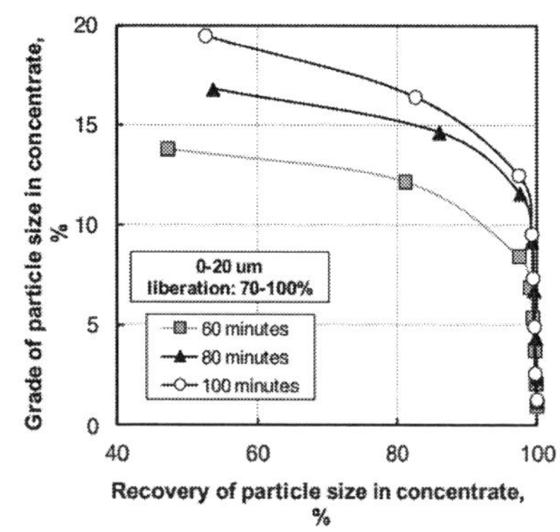

Fig. 14. Sulphides recovery vs. grade in size range of 0–20 um liberated in 70–100% (free particles) obtained after 60, 80 and 100 minutes of milling.

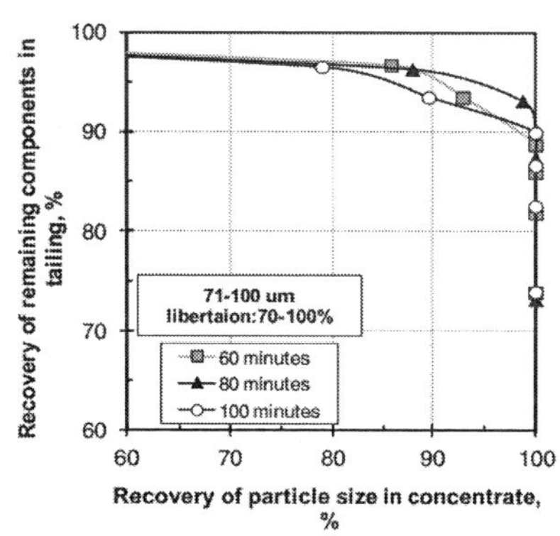

Fig. 15. A recovery of copper sulphides in concentrate vs. recovery of remaining components in tailing in size range of 71–100 um liberated in 70–100% (free particles) obtained after 60, 80 and 100 minutes of milling.

More significant differences are observed with increasing of particle size (Fig. 15). The upgrading selectivity of free particles flotation decrease in order: 80, 60 and 100 minutes of milling. The influence of the particle size on the upgrading selectivity is presented in Fig. 16. It can be seen that after milling time equals to 80 minutes, the flotation efficiency decreases while the size of the sulphides becomes smaller. The obtained results confirms data from the literature. Fines, especially in the size range of 0–20/40 um, float worse than coarse sulphides, even fully liberated. Similar dependences for the rest of flotation experiments were obtained. One of the reason of poor floatability of fully liberated fines can be too less dosage of collector. It is confirmed that fines because of a larger particle surface area demand higher doses of collector than coarse. This issue can be solved by separately upgrading fines from coarse.

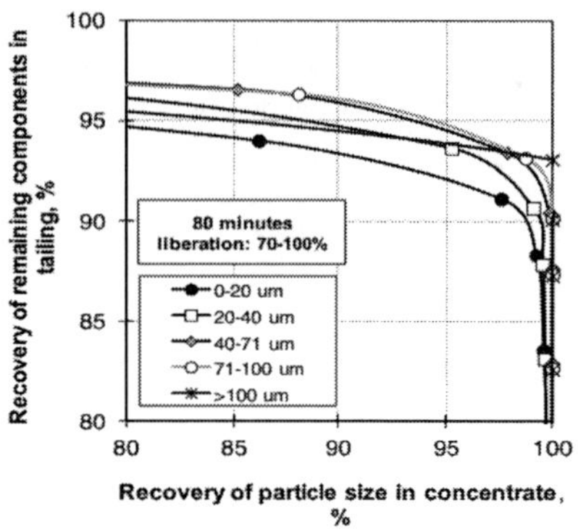

Fig. 16. A recovery of copper sulphides in concentrate vs. recovery of remaining components in tailing for sulphides liberated in 70–100% (free particles) obtained after 80 minutes of milling.

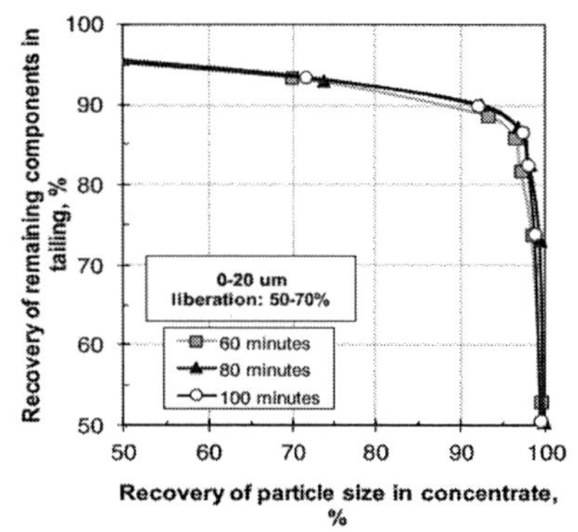

Fig. 17. A recovery of copper sulphides in concentrate vs. recovery of remaining components in tailing in particle size of 0–20 um liberated in 50–70% (impregnation) obtained after 60, 80 and 100 minutes of milling.

Next stage of the research was to study the flotation of sulphides in impregnations (Figs. 17-18). It was observed that there is no significant differences between two ranges of liberation: 20–50% and 50–70%. The results of the sulphides liberated in 50–70% flotation for particle size ranging from of 0 to 20 um and 20 to 40 um were presented. As it can be seen that for the particles size in the range of 0–20 um the best upgrading selectivity was observed after 80 and 100 minutes of milling. Insignificantly worse efficiency was obtained for the test with the feed grinded in 60 minutes.

Appreciably differences between the experiments are observed for particles in impregnations and greater than 20 um (Fig. 18). Similar results for the rest of results, not presented in the paper, were obtained. The poor upgrading selectivity was observed for the flotation test where the feed was grinded in 60 minutes.

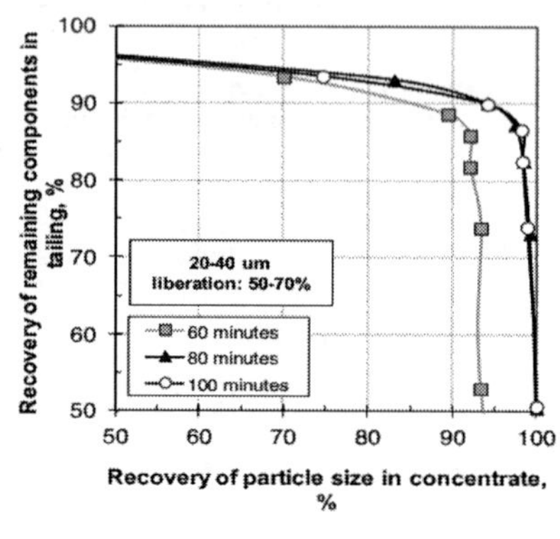

Fig. 18. A recovery of copper sulphides in concentrate vs. recovery of remaining components in tailing in size range of 20–40 um liberated in 50–70% (impregnation) obtained after 60, 80 and 100 minutes of milling.

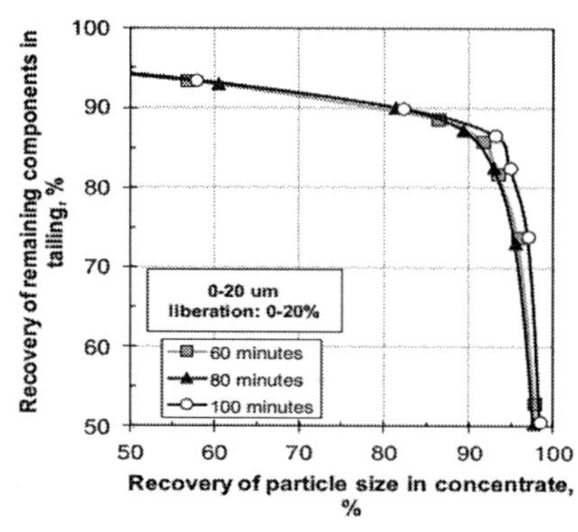

Fig. 19. A recovery of copper sulphides in concentrate vs. recovery of remaining components in tailing in particle size in the range of 0–20 um liberated in 0–20% (locked particles) obtained after 60, 80 and 100 minutes of milling.

In Figs. 19-21 the flotation results of locked sulphides particles (liberation: 0–20%) are presented. As it can be seen, the best upgrading selectivity of fines was obtained in the test with the feed grinded in 100 minutes (Fig. 19). Similar results of flotation were observed for particle size in the range of 20–40 um. For the coarser particles (+40 um) the differences between the plots courses become greater (Fig. 20). Despite the higher content of the locked

DOI:10.1051/e3sconf/20171801025

particles (+40 um) in the flotation feed after 60 minutes of milling time, better upgrading selectivity of these particles was observed.

The influence of the particle size on the upgrading selectivity for the test with milling time equals to 80 minutes is presented in Fig. 21. It can be seen that the flotation efficiency increases while the size of the sulphides decreasing. These results are contrary to date obtained for fully liberated sulphides (Fig. 16).

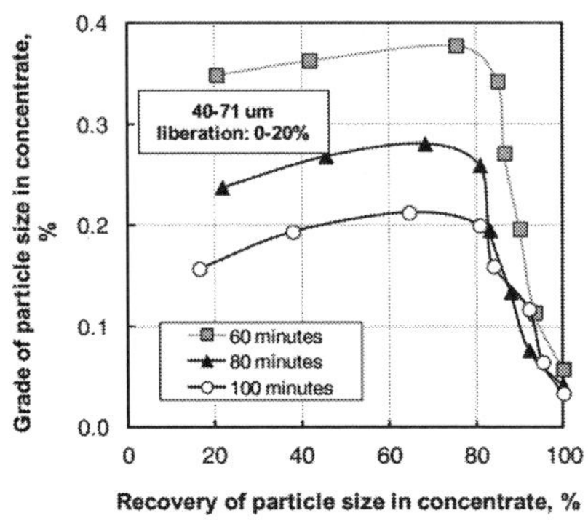

Fig. 20. Sulphides recovery vs. grade in particle size in range of 40–71 um liberated in 0–20% (locked particles) obtained after 60, 80 and 100 minutes of milling.

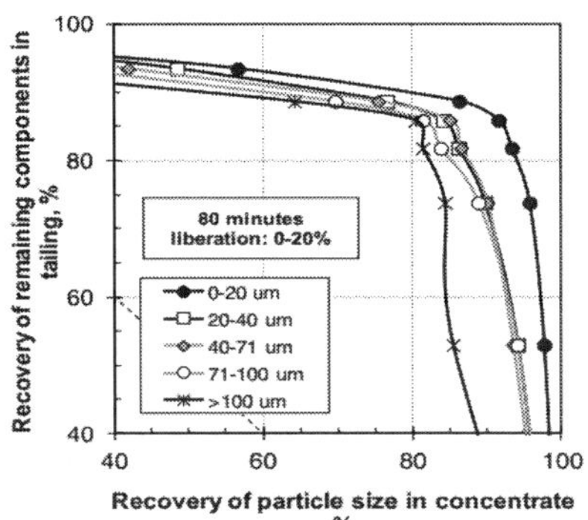

Fig. 21. A recovery of copper sulphides in concentrate vs. recovery of remaining components in tailing for sulphides liberated in 0–20% (locked particles) obtained after 80 minutes of milling.

2.3. Summary and conclusions

The influence of liberation and particle size of sulphide minerals on the laboratory flotation of sedimentary copper ore (Rudna mining area) is presented in the paper. The series of laboratory flotation experiments were performed. All flotation tests were conducted in the presence of collector and frother. Three different milling time of flotation feed were applied: 60, 80 and 100 minutes. Mineralogical analysis of all upgrading products using QEMSCAN system were conducted and analysed.

After 60, 80 and 100 minutes of milling 80% of the particles mass passing 52 um, 46 um and 38 um, respectively. In each flotation feed, the high contents of small particles (0–40 um) were observed. It was shown that the milling time of 80 minutes it is "enough" to obtained the maximum of liberation of sulphide minerals. After 100 minutes of milling the upgrading results were very similar. The milling time equals to 60 minutes is too short for the studied copper ore. Extremely similar observations were notice when it comes to liberation analysis. The comparable data of sulphides liberation for flotation feeds grinded by 80 and 100 minutes were obtained. For all flotation tests, similar results of sulphides liberation were observed for particle size in the range of 0–100 um. For the greater particles more noticeable differences between sulphides minerals in flotation feeds were obtained.

Liberation analysis confirmed upgrading selectivity of sulphides minerals in narrow fractions. The flotation efficiency was close similar for all particles in the size range of 0–100 um in each flotation test. More noticeable differences between all tests were observed for coarser particles. Moreover, even for coarse particles (+100 um) the best upgrading selectivity is obtained for the flotation feed after 100 minutes of milling.

Regardless of particle size and liberation degree, the best upgrading selectivity was obtained for experiments in which the milling time equals to 80 or 100 minutes was applied. Moreover, different dependence was observed for fully liberated and locked sulphides in analysed size fractions. With the bigger particle size, the higher upgrading selectivity of fully liberated sulphide minerals was observed. Quite different results was obtained for locked sulphide particles. The flotation efficiency increases with decrease of size of the sulphide minerals. These observations were noted for each experiment.

It is confirmed that the different flotation needs of fine and coarse particles are essential. A treatment an ore in a wide size distribution can result in poorer upgrading results [14]. One of the reason is i.e. different requirements of collector dosages for fine and coarse particles. Fines because of a larger particle surface area demand higher doses of collector than coarse. Probably, the dosage of collector which was applied in all examined experiments was insufficient for fines while enough for coarser particles. Moreover, flotation kinetics of fines is slower. Good flotation of small but locked sulphides could be the effect of hydrodynamics and low weight of these particles. It can result in mechanical entrainment. Additionally, the flotation of that kind of particles can result in poorer quality of final concentrate.

Concluding, with extending the milling time of copper sulphide ore, the liberation of valuable components becomes better, independently on particle size. In spite of poorer flotation of fully liberated and fine sulphides, deep grinding of sedimentary copper ore is essential to receive the greatest upgrading selectivity. Because of differences in flotation requirements for fine and coarser minerals, it seems to be justified to classification the feed into two or even more streams and processing in a narrow size distribution. The grinding of narrow size fraction separately and then float should result in better liberation of sulphides of each particle size, better overall upgrading efficiency and lower power consumption. This issue requires further investigations.

The research presented in the paper is the part of project titled "Development of highly effective technology of Polish copper ore beneficiation". This project is carried out by the consortium with the Wroclaw University of Science and Technology as leader, agreements No.: NCBR: *CuBR/I/7/NCBR/2015* and KGHM Polska Miedz SA: *KGHM-BZ-U-1023-2015*.

References

[1] W.J. Trahar, Int. J. Miner. Process. **8**, 289–327 (1981)

[2] T. Leißner, T. Mütze, K. Bachmann, S. Rode, J. Gutzmer, U.A. Peuker, Miner. Eng. **53**, 171–173 (2013)

[3] J. Zhang, N. Subasinghe, Miner. Eng. **49**, 68–76 (2013)

[4] M.A. Berube, J.C. Marchand, Int. J. Miner. Process. **13**, 223–237 (1984)

[5] A.M. Gaudin, *Principles of Mineral Dressing* (McGraw-Hill, New York. Henley, 1983)

[6] R. Tomanec, J. Milovanovic, Physicochem. Probl. Mi. **28**, 195–205 (1994)

[7] T.G. Vizcarra, E.M. Wightman, N.W. Johnson, E.V. Manlapig, Miner. Eng. **23**, 374–382 (2010)

[8] N. Djordjevic, Miner. Eng. 64, 131–138 (2014)

[9] J.D. Pease, D.C. Curry, M.F. Young, Miner. Eng. **19**, 831–840 (2006)

[10] D. Bradshaw, Proceedings of IMPC, Santiago, Chile, October 2014 (2014)

[11] A.F. Cropp, W.R. Goodall, D.J. Bradshaw, *AusImm GeoMet 2013 Proceedings*, 279 (2013)

[12] N.O. Lotter, Miner. Eng. **24**, 1229–1237 (2011)

[13] C. Bachowski, J. Kudelko, H. Wirth, Biul. Państw. Inst. Geol., **423**, 189–196 (2007)

[14] J.D. Pease, M.F. Young, D.Curry, N.W. Johnson, *MetPlant 2004 AusIMM Proceedings* (2004)

DOI:10.1051/e3sconf/20171801026

Solidification of hazardous waste as a part of the raw material recovery process

Daria Polek[1,*]

[1]Department of Environmental Engineering, AGH University of Science and Technology, Kraków, Mickiewicza 30, 30-059, Poland

Abstract. Paper presents the process capabilities for solidification hazardous waste. In the first part of the article the authors present general technologies and methods in a comparative model. The following section describes the results of market research for the most advanced and innovative solidification technologies. Comparative analysis of the material has shown and described the three most promising, leading-edge technologies of waste solidification avalible on the Polish market.

1 Introduction

Solidification of waste also referred as chemical immobilization, is the process of waste disposal using their physicochemical properties [1, 5].

The main idea is to bring a waste into a form, which is as neutral as it would be possible to the natural environment [3]. Successively carried out could be a process, which makes that waste:
- maximally sorption;
- increases its mechanical strength;
- increases its frost resistance.

Noticeable is the fact that the process of waste immobilization counteracts the leaching of harmful to the environment soluble metal compounds such as:
- ashes and slags from thermal processes;
- industrial dusts and muds;
- waste from the chemical industry;
- waste from galvanic processes.

Solidification is currently one of the most popular methods of waste disposal for inorganic [4] (or with a small admixture of organics) hazardous waste contained in lands and sludges. It also serves for the treatment of sewage sludge contaminated with heavy metals.
Solidification technology was developed and successively implemented in the United States for the process of hazardous waste disposal. The U.S. EPA (Environmental Protection Agency) defines this technology as the best available technology (BAT) for the disposal of about 60 types of hazardous waste.

2 Methods of hazardous waste immobilization

Modern technological processes of waste immobilization are carried out in fully automated mixers, which include hydraulic binder (eg lime, cement), water, solidifying chemicals and optional fillers (eg sand, gravel, slag) as an addition for main substratum. In an immobilization processes comprehensive chemical waste analysis is crucial because selected chemicals and the composition of the final solidification mixture are based on its results.
The minimal effect that could be obtained after the immobilization process is that the material would be no longer characterized as hazardous and becomes an inconsistent or non-hazardous waste. Very often there is a simultaneous possibility to improve the physical waste properties in order to meet the requirements for reuse in the scope of:
- road construction;
- general and industrial construction (blocks, fittings, concrete products);
- raw meal components for cement plants (substitute for alumina, iron, calcium, and silicon);
- inert material for the landfills (seals, slope profiling, reclamation, recultivation) [2].

Currently, there are four dominant technologies for solidification municipal and hazardous waste that are briefly described below.

2.1 Phoenix Ash TechnologyColour illustrations

It is a solidification process in which a mixture of fly ash, volcanic dust, and other solid materials are used to produce bricks and other concrete mixes. This is an example of the method of neutralizing pollutants using cement, which reduces the leaching of impurities from the resulting product. The process is carried out at ambient temperature and involves a significant reduction on waste volume and improving its strength. Technology could be applied to all types of waste (including hazardous and radioactive). PAT technology is one of the first pollution immobilization technologies that has received BAT certification in the USA. The average cost of waste neutralization without additional preparation is estimated at $ 60/Mg.

* Corresponding author: polek@agh.edu.pl

DOI:10.1051/e3sconf/20171801026

2.2 Solidification process with Portland cement

The second group of waste immobilization technologies is the solidification process with Portland cement [6]. Portland cement is produced on the basis of Portland clinker with a small addition of gypsum. The main Portland clinker minerals are di- and tricalcium silicate (alite and belite), which constitute together 74% by weight.

This way is the best known and refined method of solidification for all types of waste and has become the basis for many individual technological solutions such as mixtures with other mineral binding substratums, like fly ash, blast furnace slag, roasted or hydrated lime. The undisputed advantage of this technology is the ability to use it in a variety of environmental conditions, both in stationary installations and on mobile devices. The average cost of that method is about $ 40/Mg.

2.3 Encapsulation method

Another group of solidification methods is related to the use of sulfur in the polymerization processes (so-called encapsulation). This technology based on mixing material with sulfur waste at a temperature of about 120 °C to form a durable polymer bonds. The use of this method is particularly recommended for waste with high chemical (acidic and alkaline) aggressiveness. Comparing to conventional methods, the use of sulfur in the waste disposal process could reduce its cost-effectiveness by about four times (comparing to conventional cement-based technologies).

The encapsulation process could also be used as an additional barrier to the migration of toxic impurities from solidified wastes in hydraulic matrices by covering them with that particular product.

2.4 Polymerization

The last technology, which is using for permanently binding waste materials, is their polymerisation using chemical compounds. This is the latest technology group that is constantly evolving. Similarly to the sulfuric method, the high temperature of the process is important. In addition, it is necessary to ensure the consistency of the composition of the waste stream. This means carrying out costly chemical analyzes of each batch of the batch. The cost of using this technology is almost twice as expensive than the use of sulfur or cement and is therefore relatively rarely used in large-scale industries.

3 Market analysis of available technologies

The enormous diversification of the hazardous waste which could be solidified and the dynamically changing waste market have led to the rapid development and patenting of many new technological variants of the above-mentioned immobilization methods. According to a global market analysis, most solid waste companies have their own unique technologies. Each of the technologies are used under protection by appropriate patents and licensing rights and in many cases they are not a subject of sale and there is no possibility for further modifications. According to these reasons, in the article, authors decided to show three most advanced technologies offered for sale on the Polish market (with the possibility of adapting them to the individual investor's order).

3.1 GEODUR Technology

Geodur technology is based on patents EP 1 200 158 B1 and PL 192267. It is based on the appropriate stoichiometric selection of individual solidifying reagents (mainly acids and alkalis). During the waste immobilization using the Geodur system, it comes to the chemical transformation of material ("internal barriers") and physical sealing ("external barriers") and achieved physical parameters (minimal compressive strength, limited water permeability). Components of GEODUR products are readily soluble. Monomeric and polymeric substances activate the bonding properties of the added binders, such as cement, lime or fly ashes, by varying their surface tension. Components of GEODUR products create with harmful substances complex compounds or directly molecular substances. GEODUR products simultaneously act as a solvent for various components of the mix. Depending on the type of impurities contained in the immobilized waste, additional are generally used inorganic additives and binders.

One of the most important mechanisms is to change the pH value. This is possible by the addition of pozzolanic binders – the most often fly ash or lime. In addition to changing the pH of the waste to about 9-11, the accompanying element is the precipitation of heavy metals in the form of hydroxides. There is also a change in validity during the process, which is based on reducing the valence of impurities by inorganic additives such as $FeSO_4$.

Another significant mechanism is achived through insoluble mercaptans compounds in the formation insoluble complex compounds from lead and zinc.

If a waste immobilization product should be used as construction material, it is suggested, to place additional physical process that includes and change, in particular, density, surface change and compression resistance. It is also important to increase the density of the treated waste by adding cement or fly ash and adding to the production process energy mix. Accompanying factors, besides changing the density, are also the compressive strength, change of surface and permeability. Compression resistance is necessarily associated with the use of cement or pozzolanic reactive waste, which may be, for example, wastes from the combustion of sludge from a treatment plant containing calcium.

The process also has a direct impact on the consistency of waste. Depending on the moisture of the mixture and the energy supplied, solidified and stabilized material of different consistency is formed (monoliths, granulates). Pellets are produced at the highest energies transferred into a process.

The decision on which chemical additives to use in the process depends directly on the stoichiometric calculations. This is a specially adapted GEOCALC® computer program and GEODAT® data bank. There are currently over 2,300 recipes in the GEODAT® data bank for approximately 730 different types of waste.

The production process of the installation could be managed comprehensively (via internet connection) in GEOSYS® system using the above database and software. The installation control system is compatible with GEOCALC® software.

Currently, Geodur technology is applied in the processes of solidification and immobilization of ash and dust from municipal and hazardous waste incineration plants: EKOKEM (Finland), SPITTELAU (Vienna), Zurich Power Plant (Switzerland), ZUSOK (Warsaw), EKRO (Boleslaw), EKO-SERW Czajka Wastewater Treatment Plant (Warsaw).

Suggested areas of application for the products are:
- road construction (road and construction aggregates);
- general and industrial construction (blocks, fittings, concrete products);
- raw meal components for cement plants (substitute of alumina, iron, calcium and silicon);
- inert material for the construction of landfills (seals, slope profiling, reclamation).

3.2 EnviroMIX technology

EnviroMIX technology is dedicated to the disposal of soil contaminations, sludge, waste containing organic compounds and dangerous substances (heavy metals, halogen compounds, pesticides, PAH cyanides, etc.).

The EnviroMix® process is recognized as leading way of binding and stabilizing many types of hazardous waste and is a breakthrough in the field of neutralization. Stabilization of hazardous waste with using this technology is many times more effective than today's widespread stabilization methods based on the use of Portland cement. The basic distinguishing feature of this technology is several times less demand for stabilizing material with respect to the unit weight of waste. Due to the specificity of bonding of pollutants by magnesium compounds, Enviro technology is resistant to the leaching of the pollutions, which appear sometimes in cement stabilized waste.

The EnviroMix® process is classified as a chemical bonding and solidification (CFS) technology. It is a magnesium oxide based material prepared in a patented dolomite or magnesite process with a high content of magnesium carbonate. The resulting material is crushed, suitably calcined (roasted) and processed according to the individual waste stream of waste disposed. The process is based on the pollutions immobilization as thermodynamically stable mineral phases and the microencapsulation of organic pollutants into mineral structures. Machining can be carried out using commercially available mixing equipment or a suitably designed and constructed mobile installation. It is also possible to modify existing industrial applications to adapt them to stringent health and safety requirements for toxic substances. Stabilized waste could also be used as construction aggregate due to its good physical properties. Process wastes can also be used at location, eg for leveling and landfill. It is important that the waste does not have to be stored at a hazardous waste site anymore, which avoids very high costs for such storage.

The manufacturer gives many benefits of using Enviro Mix®. One of them is the possibility of using it directly at the location or storing it into lower class landfills (generally a relatively lower fee applies). Cost reduction is also achieved through favorable mixing ratios in comparition to other methods. The resulting waste could be further mixed with salted or contaminated waters. It is worth noting that the using of Enviromix® is possible for dispose of waste with high moisture content (> 70%), reducing the need for expensive and time-consuming decanting (precipitation) process. Liquid waste and sludge are solidified while clays, oils and organic waste are bound.

Enviromix® technology uses conventional mixing techniques and equipment. That is why raw materials could be delivered to *in situ* disposal sites while allowing safe transport of cleaned materials.

This method could be used for the disposal of mixed waste with a high concentration of organic and inorganic substances.

3.3 SULTECH technology

Original technical and technological solution patented for the recovery of hazardous industrial waste containing heavy metals in solid form (powders, granules, etc.) by stabilizing them with sulfur polymer in SULTECH® polymer concrete. The process of stabilizing and recovering hazardous industrial waste in polymer concrete is particularly suitable for stabilizing waste which contain heavy metals such as cadmium, chromium, barium, zinc, copper, calcium, iron, nickel, arsenic, manganese, lead, aluminum, magnesium, ashes from waste incineration plants. Solidification of waste, especially hazardous containing heavy metals, consists in forming with sulfur polymer sulphides of these metals that are insoluble in water. The received SULTECH® Polymer Concrete containing waste is an inert and excellent thermoplastic material (it can be processed many times), from which it could be formed many of very good products for excample: drainage lines, NOE® flood disks, sewerage wells, road boards, road and motorway components, railroad weights e.t.c. Polymer Concrete SULTECH® has the following properties:
- very good mechanical properties (high compressive strength and flexural strength), which exceeds similar Portland concrete;
- resistance to aggressive media including: acids, salts, seawater;
- very low absorption;
- tightness and frost resistance;
- short forming time;
- short time to achieve nominal strength;
- waste of production;

DOI:10.1051/e3sconf/20171801026

- comparable coefficient of expansion with Portland cement;
- recyclable products (thermoplastic).

4 Summary

Due to the large diversity of hazardous waste, it is difficult to distinguish one, the best method of their immobilization. It should be considered that the used technology have to be tailored to the investor needs. An important issue is the physical characteristics of waste at entry. It determines the importance of choosing the right technology.

The application areas of each of the presented methods are relatively narrow, so it is extremely important to precisely know what kind of waste we are dealing with, and what are the real possibilities of its immobilization.

Another determinant factor from the waste generator point is the processing unit cost over a ton of the material. It forces the development of new, more efficient disposal methods and continually checking the effectiveness of the currently methods, their optimization and improvement to maintain satisfactory results.

References

1. S. Caijun, A. Fernandez- Jimenez, J. Hazard. Mater. **173** (2006)
2. T. Marcinowski, K. Banaszkiewicz, Research on the mechanical durability and chemical stability of soldifier hazardous waste (Environmental engineering III, 2010)
3. V. Misra, S.D. Pandey, Environment International, **31** (2005)
4. J. Pyssa, Gospodarka Surowcami Mineralnymi, **24** (2008)
5. M.A. Tantaway, A.M. El-Roudi, A.A. Constr. Build. Mater., **30** (2012)
6. W. Wasiak, S. Różalski, Przegląd Komunalny, **6** (2003)

Silver recovery from zinc metallurgical sludge – analysis of solutions

Stanisław Pietrzyk[1], *Piotr* Palimąka[1], *Michał* Stępień[1], *Aldona* Krawczykowska[2], *Barbara* Tora[2,*], *Mirosław* Fatyga[3], *Grażyna* Pajor[3], and *Arkadiusz* Mańka[3]

[1]AGH University of Science and Technology, Faculty of Non-Ferrous Metals, Kraków, Poland
[2]AGH University of Science and Technology, Faculty of Mining and Geoengineering, Kraków, Poland
[3]Górniczo-Hutnicze „Bolesław" S.A, Kolejowa 37, 32-332 Bukowno, Poland

Abstract. During the hydrometallurgical process of zinc production, conducted in the ZGH "Bolesław" S.A. in Bukowno [Mine and Metallurgical Plant], about 40,000 tons of sludge is generated. After dehydration in the Larox filter presses, sludge contains ca. 16-18% of Zn, 20-25% of Fe, and 200-300 ppm of Ag. Next, sludge is transported to the Olkusz concentrator for flotation to obtain concentrate enriched with Ag (1,000-1,500 ppm). The concentrate is then sent to the HC "Miasteczko Śląskie" [zinc smelter], while the flotation tailings are subjected to recycling in waelz kiln in Bukowno to regain mainly Zn and Pb, in the form of oxides (also sent later to the HC "Miasteczko Śląskie").

1 Introduction

During the hydrometallurgical process of zinc production, conducted in the ZGH "Bolesław" S.A. in Bukowno [mine and metallurgical plant], about 40,000 tons of sludge is generated. After dehydration in the Larox filter presses, sludge contains ca. 16-18% of Zn, 20-25% of Fe, and 200-300ppm of Ag. Next, sludge is transported to the Olkusz concentrator for flotation to obtainconcentrate enriched with Ag (1,000-1,500 ppm). The concentrate is then sent to the HC "Miasteczko Śląskie" [zinc smelter], while the flotation tailings are subjected to recycling in waelz kiln in Bukowno to regain mainly Zn and Pb, in the form of oxides (also sent later to the HC "Miasteczko Śląskie") [1].

It seems to be reasonable to consider a possibility of sludge flotation after acid leaching, directly at the ZGH "Bolesław" plant. That will allow to reduce costs by elimination of transportation to the enrichment plant, as well as develop a flotation technology dedicated to those types of materials. Preparation of a new flotation technology is intended to increase silver output, in comparison to the presently used method, with a concurrent examination of managing the remaining solutions, in the conditions of the existing process lines.

2 Zinc production technology in the hydrometallurgical process

The hydrometallurgical zinc production process conducted in the ZGH "Bolesław" S.A. plant in Bukowno involves the following operations:

- roasting of the zinc concentrate in a fluidised condition,
- roasted concentrate leaching,
- separation of sludge and insoluble fractions of the roasted concentrate from the zinc sulphate solution (sedimentation and filtration),
- removal of impurities from the zinc sulphate solution by hydrolysis and cementation,
- electrolysis of the water solution of zinc sulphate.

Table 1. Chemical analysis of flotation zinc calcine [1].

Element	Zn	Pb	Fe	Cd	As	F	Cl
Content [%]	63.68	1.86	5.03	0.31	0.066	0.003	0.009

The basic input material for the "Bolesław" smelter consists in sulphide concentrates of zinc delivered from the "Olkusz" mine and concentrator and partly imported. Zinc concentrate is obtained by flotation. It contains ca. 50-55% of Zn, 25-30% of S, less than 2% of Pb, less than 8% of Fe, and small quantities of silver, as well as other compounds. After the flotation process, the concentrate needs to be dehydrated The concentrate s mixed, in specific proportions, with water to produce the so-called pulp. The material also contains up to 20% of added secondary zinc-containing material (Waelz zinc oxide), after previous dechlorination. The feed prepared in that way is directed into the fluidised-bed furnace. Air is supplied to the furnace by blowers and the process is conducted in the temperature of ca. 950°C. roasting of fluidised concentrate is a process which does not require any fuel supply (the sulphide oxidation reactions are strongly exothermal). The roasted concentrate produced as a result of sulphide concentrate oxidation leaves the furnace through two channels. Coarse fractions, in the

* Corresponding author: tora@agh.edu.pl

proportion of ca. 60%, are poured directly from the fluidisation zone to the cooling chamber and the calcine container. Small fractions, in the proportion of ca. 40%, in the form of dust, are removed from the fluidisation zone to the upper section of the furnace, and they leave the furnace, together with SO_2 containing off-gases. Next, gases flow through a set of cyclones for further dedusting. Clean gases are directed to the sulphuric acid plant. The composition of calcine is specified in Table 1.

The general diagram of the hydrometallurgical zinc concentrate processing in the ZGH "Bolesław" is presented in Fig. 1.

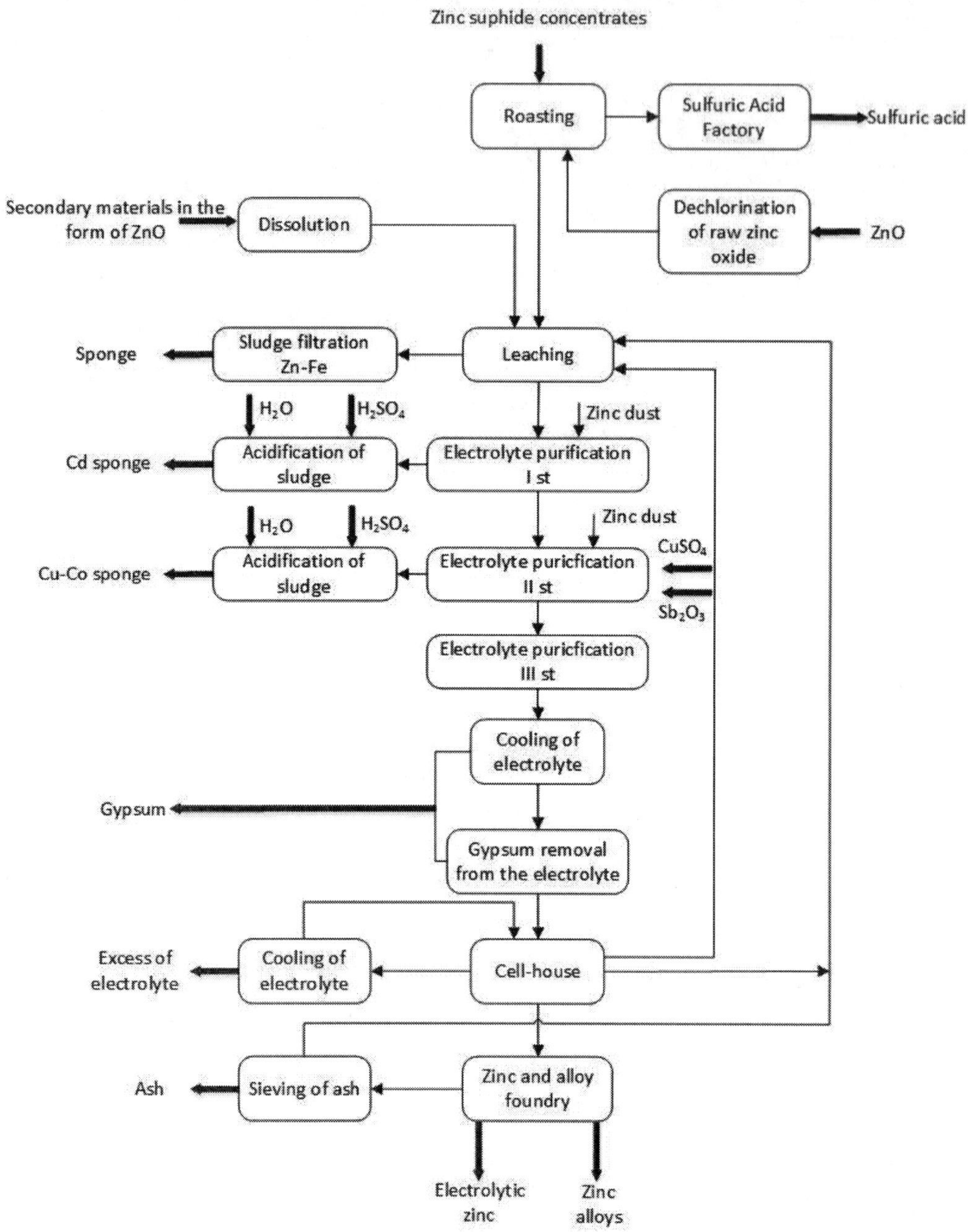

Fig. 1. Diagram of roasted zinc concentrate processing in the ZGH "Bolesław" in Bukowno (after [2]).

DOI:10.1051/e3sconf/20171801027

The roasted material originating from the fluidised-bed furnace and the dusts collected by the dedusting systems are sent to the leaching process line. The purpose of this operation is to transfer the maximum quantity of the metal from solid phase to liquid phase. Sulphuric acid solution(VI) is the leaching agent. During leaching the hydrolytical precipitation of certain impurities (Fe, As, Sb, Ge etc.) takes place. Presently, the ZGH "Bolesław" smelter applies a two-stage neutral leaching process that is conducted in the reactors with the capacity of $80m^3$ each. The process is continuous and characterised by easy control. Calcine leaching is conducted by acidic return electrolyte, containing 40-50 g/dm^3 of Zn and up to 12% of H_2SO_4. During the leaching process, pH is changing from 3.0 in the first reactor to ca. 5.0 in the last one. During that process, ca. 85% of zinc is transferred to the solution. After leaching, pulp is directed to hydrocyclones in order to separate non-leached sand. Overflow is fed to the Dorr settler, and underflowis directed to the acid leaching line. Fig. 2 presents the diagram of roasted zinc concentrate processing in the Leaching Plant of the ZGH "Bolesław" in Bukowno.

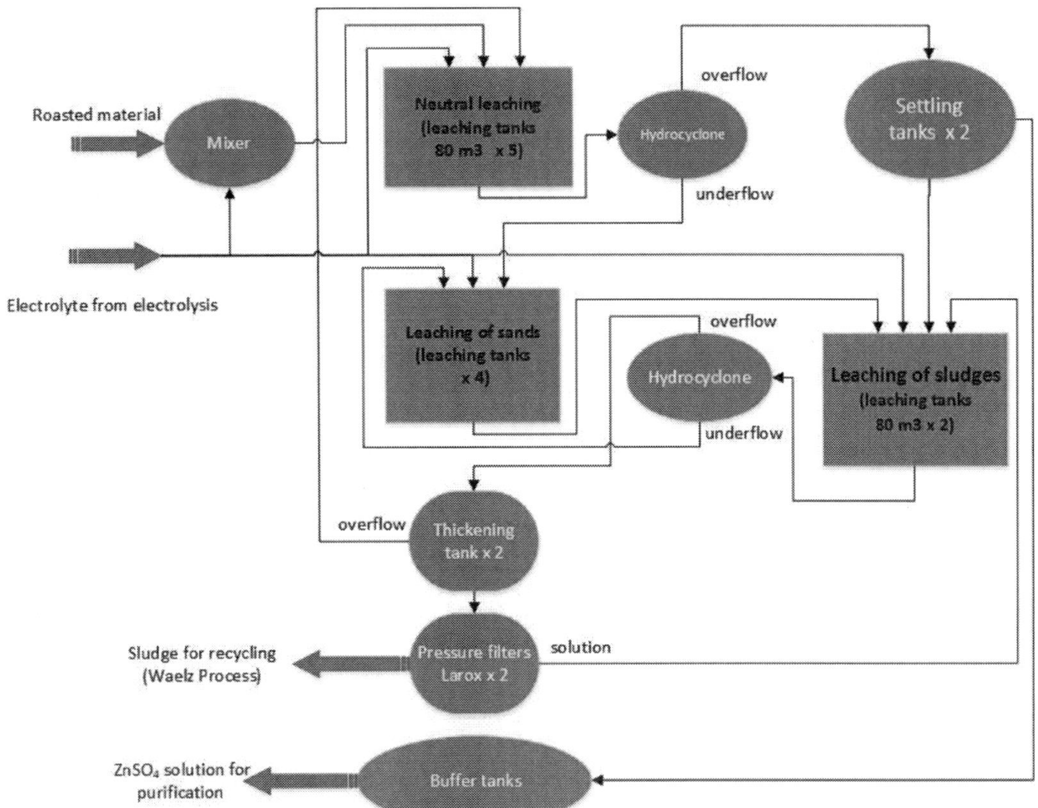

Fig. 2. Diagram of the roasted zinc concentrate processing in the Leaching Plant of the ZGH "Bolesław" in Bukowno (after [2]).

The solution collected from settling tanks is directed to the buffer tanks and later to the process of the -three- tages purification by cementation with zinc dust. After that purification process, electrolyte containing ca. 140g/cm^3 of Zn is obtained. After cooling, it is fed to electro-winning cells. Sludge originating from the Dorr settlers is subjected again to leaching in the set of acid-neutral reactors and, after concentration and dehydration at Larox press filters, the zinc content is available at the level of 15-17%. Sludge is then directed to two leaching reactors, 80 m^3 each, followed by classification in a hydrocyclone. The overflow material from the hydrocyclone is subjected to concentration and the underflow returns to acid leaching of sands. Concentrated sludge is directed to the Larox press filters. The filtration residue, containing large quantities of zinc, iron, and lead, presented in Fig. 2, is fed to waelz kiln. Finally, zinc oxide is obtained.

The correct course of the electrolytic process requires high-purity electrolyte. The most important admixtures that have to be removed from electrolyte are the following: iron (II), cobalt, nickel, copper, and cadmium and the compounds of antimony and arsenic. Other impurities are removed periodically, as they concentrate in electrolyte and after they have exceeded the allowed concentrations.

The main impurities are removed during neutral leaching, as a result of the course of hydrolytic reaction (1-7)

$$FeSO_4 + {}_2H_2O \rightarrow Fe(OH)_2\downarrow + H_2SO_4 \quad (1)$$

* Corresponding author: tora@agh.edu.pl

$$Fe_2(SO_4)_3 + 6H_2O \rightarrow 2Fe(OH)_3\downarrow + 3H_2SO_4 \quad (2)$$

$$Fe_2(SO_4)_3 + 3ZnO + 3H_2O \rightarrow$$
$$\rightarrow 2Fe(OH)_3\downarrow + 3ZnSO_4 \quad (3)$$

$$As_2(SO_4)_3 + 3ZnO + 3H_2O \rightarrow$$
$$\rightarrow As_2O3 \cdot 3H_2O\downarrow + 3ZnSO_4 \quad (4)$$

$$2FeSO_4 + MnO_2 + H_2SO_4 \rightarrow$$
$$\rightarrow Fe_2(SO_4)_3 + MnSO_4 + 2H_2O \quad (5)$$

$$10FeSO_4 + 2KMnO_4 + 8H_2SO_4 \rightarrow$$
$$\rightarrow 5Fe_2(SO_4)_3 + 2MnSO_4 + K_2SO_4 + 8H_2O \quad (6)$$

$$Sb_2(SO_4)_3 + 3ZnO + 3H_2O \rightarrow$$
$$\rightarrow Sb_2O_3 \cdot 3H_2O\downarrow + 3ZnSO_4 \quad (7)$$

The purification of zinc sulphide solution by cementation (in accordance with basic reactions 8 and 9) is conducted continually.

$$Zn + Cu^{+2} \leftrightarrow Cu\downarrow + Zn^{+2} \quad (8)$$

$$Zn + Cd^{+2} \leftrightarrow Cd\downarrow + Zn^{+2} \quad (9)$$

Neutral overflow from the sedimentation tank contains ca. 135 g/l of Zn, 470 mg/l of Cu, 170mg/l of Cd etc. The material is fed to cementation tanks situated in a cascade. All the tanks are fed with zinc dust suspension. Sludge is pumped out from the last box to filter presses. After the first stage, filtrate is directed to the second stage of hot purification that is conducted in four cementators with the capacity of 120m^3 each. Zinc dust and reaction activators, $CuSO_4$ and Sb_2O_3, are added. Before the second stage of purification, electrolyte is heated to 95°C. The generated Cu-Co sponge , containing a high proportion of zinc, is filtered at filter presses. The filtrate is sent to the zinc production line through a zinc electrolyte cooling and seasoning system, while sludge is directed to the acid treatment installation to recover zinc.

Leached and purified water solution of zinc sulphate is delivered to the cell-house where it is subjected to electrolysis. The working capacity of one cell amounts to 2.5 m^3. Cells are arranged in an individual power supply system, 12 in two rows of 6 each, and grouped in two power units. Cascades are connected in series, with the electrodes connected in parallel. Anodes used in the zinc electrolysis process are made of lead, with 1% of silver. The number of anodes is 28 in each tank. Cathodes are made of high-purity (min. 99.5%) aluminium plates. The number of cathodes in each tank is one less than that of anodes (27). Zinc deposited as a result of direct current flow is removed from cathode starter sheets every 24 hours. The last stage consists in cathode melting in low-frequency induction furnaces.. Liquid zinc flows through a system of half-pipes to cast-iron moulds where it solidifies and is extracted in the form of zinc ingot at the end of the casting line.

3 Research purpose

The technology of silver recovery from post-process sludge by the inclusion of the flotation process was implemented in 2012. Earlier, the sludge from zinc hydrometallurgy was delivered entirely to the Bolesław recycling plant and processed in waelz kiln (Fig. 3).

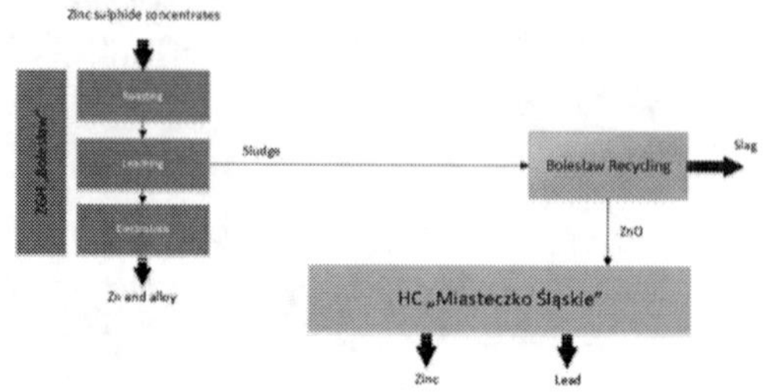

Fig. 3. Diagram of post-leaching sludge processing before 2012.

Raw zinc oxide is the main product of the waelz process, together with slag. The analysis of the above-mentioned products demonstrated that silver contained in the sludge is transferred mainly to slag (ca. 83%) of the waelz process.

4 Location of silver flotation in the technological process of zinc metallurgy - analysis of world solutions

Analysis of the possible location of flotation in the technological process was based on the data contained in the book of Brook Hunt [1].

Brook Hunt & Associates Ltd. provides research services for base and precious metal industries and markets. The company offers metals market services that track short and the long-term outlook for metal and concentrate markets; and cost services that analyze mine, smelter, and refinery production costs within base and precious metal industries. It also provides concentrates services, which analyze the flow of base metal concentrates from mines to smelters and refineries, as well as forecast concentrate market balances, and outlook for treatment and refining charges. In addition, the company offers metals consulting services. It provides its research on various metals, such as aluminum, copper gold, lead, nickel, and zinc. The company was founded in 1975 and is based in United Kingdom. On the figures 4,5 and 6 the world solution of zinc electrolytic plants are shown, fig. 5 shows the plants without silver flotation, fig. 6 with silver flotation [7].

DOI:10.1051/e3sconf/20171801027

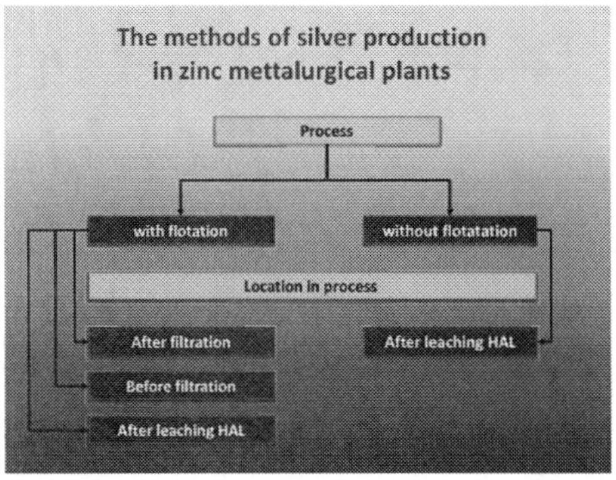

Fig. 4. Location of silver flotation in technological scheme of zinc production.

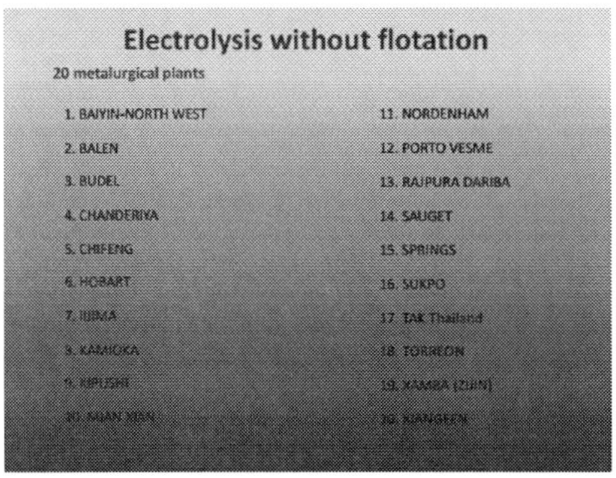

Fig. 5. List of world installations for zinc electrolysis - without silver flotation.

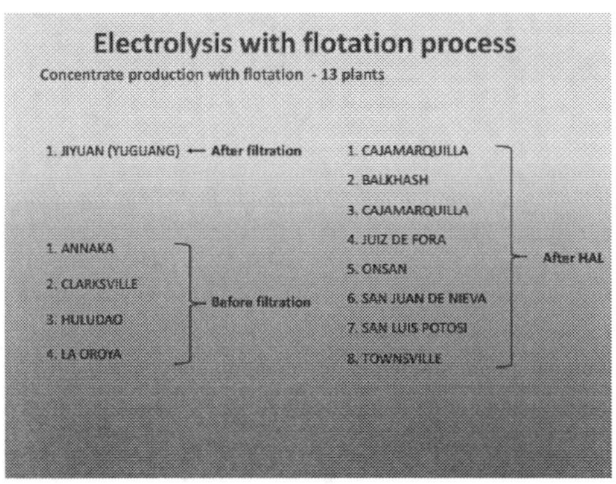

Fig. 6. Summary of world installations for zinc electrolysis - with flotation.

5 ZGH Bukowno case

Presently, sludge is transported from the leaching plant to the ZPM Olkusz Pomorzany concentrator and subjected to the flotation process. The output concentrate, enriched with zinc and silver, is delivered to the HC "Miasteczko Śląskie", while tailings to the Bolesław Recycling Sp. Z o.o., (from July 2017 is a part of ZGH Bolesław) as shown in Fig. 7.[5, 8].

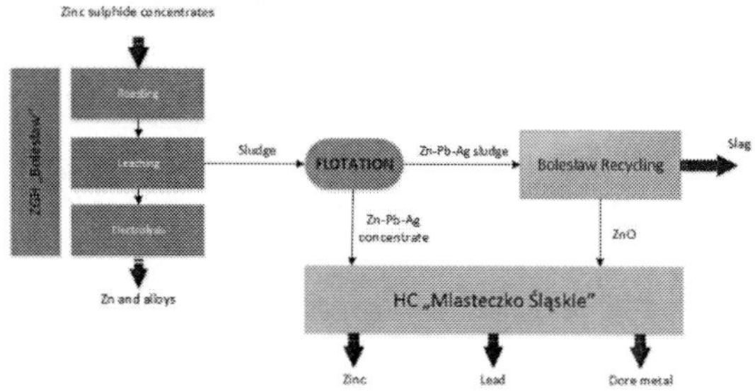

Fig. 7. Current sludge processing diagram.

The present solution has two drawbacks:

- High transportation costs to the ZPM Olkusz Pomorzany
- Low degree of post-flotation concentrate enrichment with silver (ca. 50-60%).

Upon consideration of those two items, this study was devoted to the development of a new flotation technology, allowing for higher silver recovery from the combined concentrate (Zn-Pb-Ag), in comparison to the present results. The newly prepared technology, together with the placement of the flotation process at the smelter, without the necessity to transport sludge, will allow to gain tangible economic effects.

There is only one smelter plant around the world which has the same technology as Bukowno. It is Jiyuan (YOUGUANG) Plant [1, 11, 12, 13]. Construction of the Jiyuan zinc smelter, situated 10km North of Jiyuan City on a Greenfield site, started in 2003, and the plant was commissioned in September 2005 with a nominal capacity of 100kt/a slab zinc (actual capacity possibly 120kt/a) at a cost of US$79M. The zinc smelter includes an on-site coal-fired 50MW power plant, built at a further cost of US$25M. The power plant and the Waelz kiln for residue treatment were completed several months after the main sections of the smelter were commissioned. A 10t/a indium plant costing US$2.4M was commissioned in mid-2007, a US$3.6M silver flotation unit to recover silver from leach residues came on-stream in July 2006 and cadmium production commenced in March 2006. London Metal Exchange (LME) registration for SHG zinc <YG SHG> was obtained in June 2008.

References

1. H. Brook, *Zinc Smelters and Projects Processes, Costs and Profitability* (2016)
2. P. Ostrowska-Popielska, A. Sorek, Prace Instytutu Metalurgii Żelaza, **64(4)**, 39-46 (2012)
3. Z. Blaschke, Inżynieria Mineralna **2(20)**, 26–35 (2007)
4. A. Jarosiński, L. Madejska, Gospodarka Surowcami Mineralnym, **24(4)** (2008)
5. K. Cichy, Z. Szołomicki, C. Reguła, Rudy Metale **52(10)**, 595 – 600 (2007)

6. M. Sakata, T. Imai, A. Horiuchi K. Morita, Y. Kondo, M. Kusano, Zinc Smelters Survey, International Symposium on Lead and Zinc Processing (Kyoto, Japan, 2005)
7. M. Fatyga, *Metody produkcji koncentratu zawierającego Ag* (ZGH Bolesław, 2016)
8. S. Pietrzyk et al., Ocena możliwości zwiększenia uzysku srebra w procesie flotacji szlamów po kwaśnym ługowaniu prażonki cynkowej, (ZGH Bolesław SA, Kraków 2016)
9. https://www.911metallurgist.com/blog/process-oxidized-lead-zinc-ores
10. http://minerals.usgs.gov/minerals/pubs/commodity/zinc/
11. C. Duval, D.J. Kleiner, "Lead and Zinc Production, available at: http://www.tshaonline.org/handbook/online/articles/dkl01 accessed July 23, 2017,
12. http://www.ifc.org/ifcext/enviro.nsf/Content/EnvironementalGuidelines
13. http://www.cpchem.com/bl/specchem/enus/Pages/IntroductiontoMineralProcessing
14. W. Sun, J. Su, G. Zhang, Y. Hu, J. Cent. South Univ. Technol., **19(8),** 2307–2315 (2012)

E3S Web of Conferences **18**, 01028 (2017) DOI:10.1051/e3sconf/20171801028
MEC2017

Flotation of copper-bearing shale in solutions of inorganic salts and organic reagents

Tomasz Ratajczak[1,*]

[1]Wroclaw University of Science and Technology, Faculty of Geoengineering, Mining and Geology, Wybrzeze Wyspianskiego 27, 50-370 Wroclaw

Abstract. Flotation data on copper-bearing shale in aqueous solutions of inorganic electrolytes (NaCl, Na_2SO_4, KPF_6, NH_4Cl) and organic reagents (ethylamine, propylamine) as frothers were presented and discussed . The relationships between shale flotation, surface tension of aqueous solution and foam height during bubbling with air in the flotation system were presented. It has been found that flotation of shale in the presence of inorganic salts the yield was directly proportional to the surface tension of the aqueous solution of salt and inversely proportional to the height of the foam. On the other hand, for organic reagents solutions (short chain amines), a reverse effect has been observed in relation to the inorganic compounds studied, that is the yield of copper-bearing shale flotation and the foam height were inversely proportional to the surface tension of the amine solution.

1 Introduction

Salt flotation relies on flotation in saline solutions [1, 2] without the presence of additional reagents. Salt flotation mechanism and results depend on the type, concentration and foaming properties of the saline solutions. Inorganic salts can replace organic frothers in flotation but not all inorganic salts are good frothers [1, 2, 3, 4, 5, 6, 7]. From these works it can be seen that the salts that raise or lower the surface tension can cause, respectively, an increase or decrease of flotation of naturally hydrophobic minerals. The aim of this works, which is based on flotation results of Smolska and Ratajczak [6] and Witan and Ratajczak [7], was to compare the influence of inorganic and organic regents on flotation of naturally hydrophobic copper-bearing shale and to determine how the flotation results depend on foaming and surface tension of these regents in aqueous solutions.

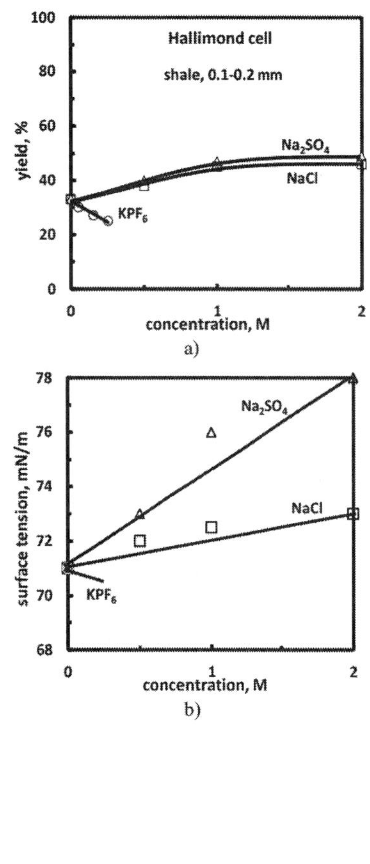

2 Flotation of copper-bearing shale in solutions of inorganic electrolytes

Flotation of copper-bearing shale, denoted by Drzymala et al. [8], as shale P, was carried out using a mono-bubble type Hallimond cell with a capacity of 200 cm^3 and height of 36 cm. The investigations were carried out in distilled water and in the presence of aqueous solutions of NaCl, Na_2SO_4 at concentrations of 0.5 to 2 mol/dm^3 and KPF_6 at concentrations of 0.05 to 0.25 mol/dm^3, that is up to its solubility limit. More precise description of methodology of investigations is given in the works of Smolska and Ratajczak [6] and Smolska [9].

* Corresponding author: tomasz.ratajczak@pwr.edu.pl

DOI:10.1051/e3sconf/20171801028

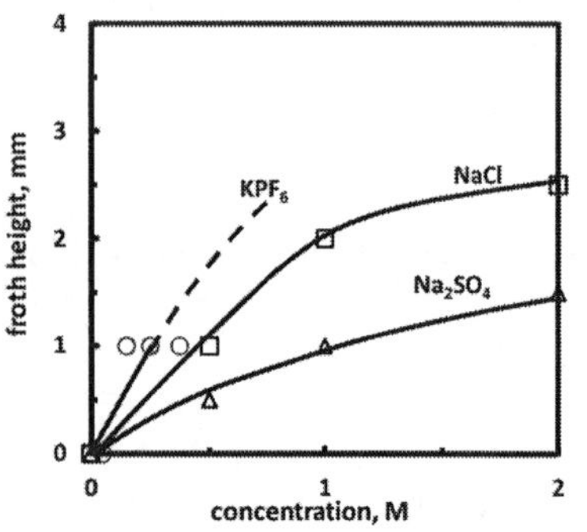

c)

Fig. 1. Dependence of a) yield of flotation of copper-bearing shale [6], b) surface tension [6]; [2]; [10]; [11], c) foam height [6] on the concentration of the investigated salt solutions.

Smolska [9] investigated flotation and the height of foam during air bubbling through water containing salts that have different surface tension of aqueous solutions (Figure 1). Figure 1a shows that flotation yield decreases in the presence of KPF6 and increases in the Na2SO4 and NaCl aqueous solutions. A comparison of yield with foam height and surface tension indicates that the yield increases in the same manner as surface tension and decreases with foam height.

3 Flotation of copper-bearing shale in solutions of organic reagents

To determine the influence of organic regents on flotation, the tests were carried out in a Mechanobr flotation machine equipped with a 0.25 dm^3 flotation cell. Reagents used for testing were ammonium chloride, as the first member of the normal short chain amines family and ethylamine as well as propylamine. For each flotation 30 g of -100 μm in sizw copper-bearing shale P was used. The time of flotation was 30 min. More details on methodology of flotation is given in the work of Witan and Ratajczak [7] and Witan [12].

Figure 2a shows the effect of concentration of ammonium chloride, ethylamine and propylamine on the yield of copper-bearing shale after 15 minutes of flotation. Figure 2b show the influence of concentration of the used frother surface tension change ($\Delta\gamma$) in relation to pure water while Fig. 2c the froth height. According to Fig. 2a, ammonium chloride makes the yield of shale slightly decreasing with a small rise in the froth height (Fig. 2c), while the surface tension increases (Fig. 2b). This agrees with the data obtained for inorganic salts show in Fig.1. On the other hand, for ethylamine and propylamine, for which the surface tension decreases with increasing concentration of frother (Figure 2b), a slight increase in yield is observed with increasing height of flotation froth (Figure 2c).

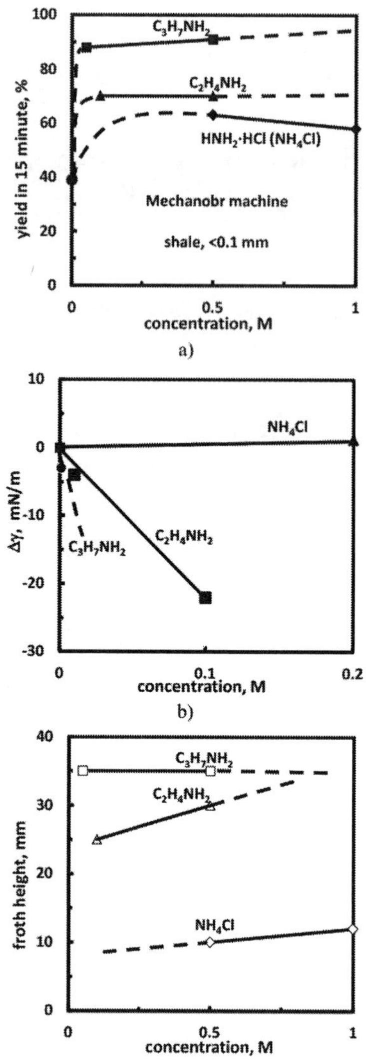

c)

Fig. 2. Effect of concentration of ammonium chloride, ethylamine and propylamine on a) yield of copper-bearing shale after 15 minutes of flotation [7], b) surface tension of their solutions [13 (for NH_4Cl, at 25 °C); 14 (for $C_2H_4NH_2$ at 30 °C, 15 (for $C_3H_7NH_2$, at 25 °C), c) froth height after 5 minutes of shale flotation) [7].

4 Summary

The results of the studies confirm influence of inorganic and organic reagents on flotation of copper-bearing shale. It was found that in the case of increasing concentrations of NaCl and Na_2SO_4, shale yield increases while for KPF_6 it decreases. Thus, it is shown that the yield of salt flotation of shale depends on the concentration and type of salt. It increases with increasing surface tension of salt and decreases as surface tension of salt solution decreases. However, the foaming of the salt solution increases with lower surface tension of the salt, that is, as in the case of organic frothers. Thus, the flotation of the shale in solutions of inorganic salts depends on the concentration, type of salt and foaming, and in turn on the surface tension of the salt.

Ammonium chloride behaves the same way as NaCl and other inorganic salts which increase surface tension. However, in the relations between yield, surface tension and froth height are different foe organic reagents. Ethylamine and propylamine, for which the surface tension decreases with increasing concentration, make flotation better while froth height increases.

In addition, it can be noticed from Fig. 2 that the maximum yield of the flotation froth product in the presence of propylamine is higher than that in salmiac and ethylamine, indicating increasing collecting properties of amine with the increasing length of the hydrocarbon radical.

This work was financed by the Polish Statutory Research Grant 0401/0124/2016.

References

1. J. Laskowski, J. Colliery Guardian, **211**, 361-366, (1965)
2. T. Ratajczak, J. Drzymała, *Flotacja solna* (Oficyna Wydawnicza Politechniki Wrocławskiej, Wrocław, 2003)
3. R. J. Pugh, P. Weissenborn, O. Paulson, Int. J. Miner. Process., **51**, 125-138, (1997)
4. B. Grabowski, J. Drzymała, Annales UMCS Sectio AA, **LXIII**, **6**, 68-72, (2008)
5. M. Lipniarski, T. Ratajczak, J. Drzymała, *Weryfikacja hipotez o roli soli we flotacji na przykładzie węgla kamiennego w wodnych roztworach NaCl i KPF$_6$*, (III Polski Kongres Górniczy, Mineralurgia i wykorzystanie surowców mineralnych, Drzymała J., Kowalczuk P.B. (red.), 14-16 września 2015, Wrocław, WGGG PWr, 35–39, 2015)
6. M. Smólska, T. Ratajczak, *Flotacja solna łupka miedzionośnego we flotowniku Hallimonda* (WGGG PWr, Wrocław, 2017)
7. J. Witan, T. Ratajczak, *Stabilność piany flotacyjnej w obecności soli* (WGGG PWr, Wrocław, 2017)
8. J. Drzymała, P. Karwowski, K. Borowski, P. Pązik, P.B. Kowalczuk, *Próba klasyfikacji łupków Legnicko-Głogowskiego Okręgu Miedziowego na podstawie zawartości węglanów, minerałów ilastych, węgla organicznego oraz miedzi*, (in: Łupek miedzionośny III, Kowalczuk P.B., Drzymała J. (red.), WGGG PWr, Wrocław, 2017, doi: 10.5277/lupek1702, 2017)
9. M. Smólska, *Flotacja solna łupka miedzionośnego we flotowniku Hallimonda.* Praca dyplomowa, opiekun T. Ratajczak (WGGG PWr, Wrocław, praca niepublikowana, 2017)
10. J. Lyklema, *Fundamentals of interface and colloid science., Vol. 2, Fundamentals*, (Academic Press, London, 1995)
11. L. Ghosh, K.P. Das, D.K. Chattoraj, J. Coll. Int. Sci., **121**, 1, 278-288, (1988)
12. J. Witan, *Stabilność piany flotacyjnej w obecności soli.* Praca dyplomowa, opiekun T. Ratajczak, (WGGG Politechniki Wrocławskiej, Wrocław, praca niepublikowana, 2017)
13. J. Lyklema, *Fundamentals of interface and colloid science. Vol. 3, Liquid-fluid interfaces* (Academic Press, London, 2000)
14. J. He, H. Liu, P. Shan, K. Zhang, Y. Qin, L. Liu, Chem. Eng. Sci., **141**, 330-341, (2016)
15. B. J. George, N. Pereira, M. A. Massum, S. D. Kolev, M. Ashokkumar, Ultrason. Sonochem., **15**, 151–156, (2008)

Effect of fly and bottom ash mixture from combustion of biomass on strength of cement mortar

Małgorzata Ulewicz[1,*], and *Jakub* Jura[1]

[1]Facutly of Civil Engineering, Czestochowa University of Technology, 42-201 Czestochowa, Poland

Abstract. The preliminary results of fly and bottom ash mixture form combustion od biomass (80% of tree waste and 20% of palm kernel shells) for the produce of ceramic mortars has been presented. Currently, bio-ash from fluidized bed are deposited in landfills. Use of this ash to production of cement mortar instead of sand will reduce the consumption of the mineral resources. The chemical composition of this waste materials was determined using X-ray fluorescence (spectrometer ARL Advant 'XP). Cement mortar were made using CEM I 42.5 R. The ash were added in an amount 20% of cement weight (in different proportions of fly and bottom ash). The results showed, that the compressive strength (after 28 days) of cement mortar containing ash is higher regardless of the type of ash mixture used. The highest compressive strength (increased by 7.0% compared to the control sample) was found for cement mortars in which the ratio of fly ash to bottom ash was 10/90. This mortars also showed the highest frost resistance (after 150 cycles freezes and unfreeze). The largest decrease the compressive strength (over 18.7%) after the frost resistance test. While cement mortars in which the ratio of fly ash to bottom ash was 90/10 showed the highest frost resistance (after 150 cycles freezes and unfreeze).

1 Introduction

The replacement of traditional raw material in building materials by fly ash and bottom ash coming from co-combustion of biomass leads to relevant environmental benefits. It avoids the extraction of large quantities of raw materials from the earth, reduces energy costs and also prevents landfill problems. Unfortunately in recent years, the energy is produced increasingly from only biomass [1, 2]. The use of biomass for energy production presents several economical and environmental advantages. Biomass is a renewable fuel meaning that it can be used to reduce the emissions associated with the greenhouse effect, while it also reduces the need to import fossil fuels such as coal, oil and natural gas. However, the fly ash generated by combustion of biomass in fluidized bed differ substantially their physicochemical and morphological properties from a traditional ash produced during combustion of coals [3-8] as well as co-combustion of biomass [9-11]. The quantity and quality of ashes produced in electrical plant during combustion of biomass are strongly influenced by the characteristics of the biomass used, such as: agriculture wastes, seed, wood or bark [12,13]. Therefore, it is important to determine the characteristics of the fly ash and bottom ash produced by the combustion of biomass, because organic and mineralogical composition of the waste and its grain size determine the possibilities of their economic use or utilization.

Several studies proposed that fly ash, from co-combustion of coal and biomass, can be used as a soil ameliorate that may improve physical, chemical and biological properties of the degraded soils and is a source of readily available plant micro-and macro-nutrients [14,15]. Several authors reported on the use of fly ash, from co-combustion of biomass, in various civil engineering applications such as construction of highways [16-18], ceramic production [19, 20], concrete and concrete like bricks, blocks and paving stones [21-23], and asphalt concrete [24].

The possibility of replacement natural fine aggregate by industrial by-products such as fly ash offers technical, economic and environmental advantages which are of great importance in the present context of sustainability in the construction sector. Unfortunately, the replacement of natural minerals by fly ashes, from co-combustion of biomass, is not easy due to the different characteristics of these materials. Agro-waste ashes, containing a large amount of silica in amorphous form, have potential for use as pozzolanic materials replacing cement. The ash from co-combustion of coal and biomass such as herbaceous, pure wood, and rice husk can be used for the production of concrete [23, 25, 26]. Unfortunately, so far there are no reports on the possibility of the use of fly ash from the combustion only of biomass to production of building materials. Typically, fly ash from biomass combustion contains more alkali metals as sodium and potassium (Na_2O, K_2O) and less alumina (Al_2O_3) than coal fly ash, traditionally used as a pozzolanic additive in cement formulations. Although many researchers have shown potential applications of fly ash from co-combustion of biomass as building and construction materials, further

* Corresponding author: ulewicz@bud.pcz.czest.pl

DOI:10.1051/e3sconf/20171801029

investigations should be carried out, especially for the ashes from combustion of biomass.

2. Experimental

To the study were used fly and bottom ash formed during the combustion of biomass in power plant from province of Swietokrzyskie. The burned biomass in a fluidized bed: 80% of tree waste and 20% of palm kernel shells. The chemical composition (Table 1) of the material was determined by using X-ray fluorescence (ARL Advant'XP spectrometer). In research, the ash were added in an amount 20% of cement weight and were used in different proportions of fly (PL) and bottom (PD) ash. The composition of the tested mortars are shown in Table 2.

Table 1. Chemical composition fly and bottom ash, %.

Component	Fly ash	Bottom ash
SiO_2	57.45	89.02
CaO	17.26	3.85
Al_2O_3	4.82	1.54
MgO	2.32	0.58
MnO	0.51	0.17
Fe_2O_3	2.94	1.03
K_2O	3.93	2.07
Na_2O	0.39	0.18
P_2O_5	2.01	0.36
TiO_2	0.30	0.11
SO_3	2.71	0.49
Cl	1.06	0.58
Other	4.30	0.02

Table 2. Composition of the studied cement mortar.

Components	M1*	M2	M3	M4	M5
Cement, g	450	450	450	450	450
Water, cm^3	225	225	225	225	225
Naturally moulding sand, g	1350	1260	1260	1260	1260
W/C	0.50	0.50	0.50	0.50	0.50
PD/PL, g	-	81/9	72/18	9/81	18/72
PD/PL, %	-	90/10	80/20	10/90	20/80

M1*-standardized cement mortar (control sample)

Bars with the following dimensions: 4x4x16 cm were made in accordance with PN-EN196-1 standards. In the research was used municipal water from the intake of the city of Czestochowa with a pH of 7.7 and the content of nitrate and chloride ions equal to 34.3 and 30.9 mg/dm^3. Electrical conductivity of water was 480 μS/m. There was used Portland cement CEM I 42.5 R (Cemex). The water–cement ratio (the ratio of the weight of water to the weight of cement used in a mix) in the cement mortars was equal 0.50. Ceramic products after 24 hours were disassembled and placed in a water bath with temperature 20 ± 1°C, where they remained for 27 days. Obtained mortars were subjected to microscopic examination (LEO Electron Microscopy Ltd.) and there was identified their compressive strength (PN-EN-196-1) and absorbability (PN-85/B-04500). For each test, three samples were made.

3 Results and discussion

In the first stage of research, the compressive strength of cement mortars was determined. As can be seen from Table 3, the compressive strength (after 28 days) of cement mortar containig fly and botttom ash mixture is higher compared to the control sample. Such a trend was observed regardless of the type of ash mixture used. The highest compressive strength (increased by 7.0% compared to the control sample) was found for cement mortars in which the percent ratio of fly ash to bottom ash was 10/90. Whereas samples containing mixture PD/PL ash equal 80/20 exhibited comparable compressive strength with the control sample. While, cement mortars with addition of ash mixture PD/PL in percent ratio equal 10/90 and 20/80 showed an increase in frost resistance about 5.7 and15.2% compared with the control sample. For this cement mortar, the absorbability increase of 10.5 and 13.8% compared with the control sample was observed. Whereas, the cement mortar containing wastes PD/PL in percent ratio equal 90/10 and 80/20 showed an increase compressive strength only about 1.6 and 4.7% compared with the control sample. This mortars showed the largest decrease the compressive strength (over 18%) after the frost resistance test (after 150 cycles freezes and unfreeze).

Table 3. Properties of the studied cement mortar (the mean from three tests).

Properties	M1*	M2	M3	M4	M5
Compressive strength, MPa	64.0	67.0	65.0	68.5	66.0
Absorbability, %	4.49	4.98	5.14	4.96	5.11
Compressive strength after testing frost resistance, MPa	52.5	54.5	52.5	60.5	55.5

Microstructural studies of cement mortars (Fig. 1) showed their heterogeneous microstructure, particularly in the contact area of ash with cement grout. The addition of ash waste and low water-cement ratio (w/c) of 0.5 has caused creation of tight structure of cement mortar and consequently not only increase in mechanical properties of the cement mortars but also fit of absorbability of mortar and increase in frost resistance.

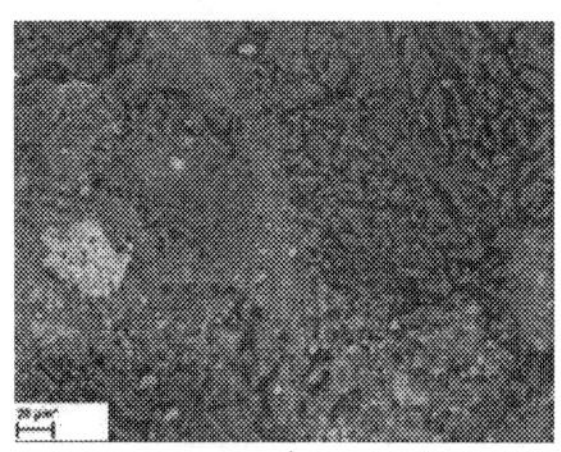

a)

DOI:10.1051/e3sconf/20171801029

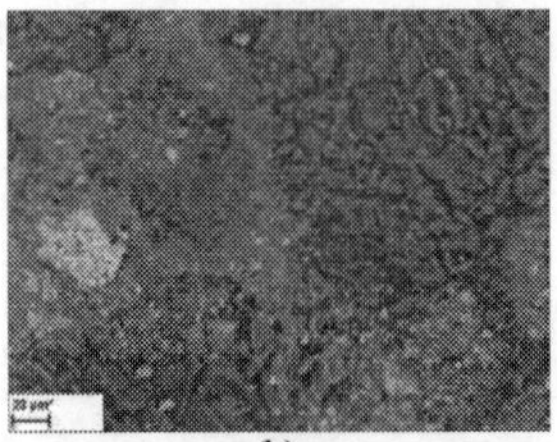

b)

c)

Fig. 1. SEM micrographs of cement mortar M1* (a), M2 (b), and M4 (c).

4 Conclusion

Cement mortar is the most extensively used material for construction of different buildings. During the cement mortar production large amounts of natural sand is used. The replacement of the natural aggregates by fly and bottom ash mixture from combustion of biomass reduce consumption of raw materials and will have a good influence on the environment.

Based on experimental studies, we can conclude that it is possible to use fly and bottom ash mixture for production of cement mortar. Studies shown that, the compressive strength of cement mortar containing bio-ash increases is higher than control simples. The highest compressive strength (increased by 7.0% compared to the control sample) showed cement mortars containing the fly ash and bottom ash mixture in the percent ratio equal 90/10. This mortar after 100 cycles freezing and defrosting showed strength loss of 11.7% compared to the control samples.

References

1. J.L Easterly, M. Durnham, Biomass Bioenergy, **10**, 79-92 (1996)
2. R. Saidura, E.A. Abdelaziza, A. Demirbasb, M.S. Hossaina, S. Mekhilef, Renew. Sust. Energ. Rev., **15**, 2262-2289 (2011)
3. D. Nordgren, H. Hedman, N. Padban, D. Boström, M. Öhman, Fuel Process. Technol., **105**, 52-58 (2013)
4. K. Trybalski, W. Kępys, A. Krawczykowska, D. Krawczykowski, D. Szponder, Pol. J. Environ. Stud., **23**, 1427-1431 (2014)
5. Y. Zheng, P.A. Jensen, A. D. Jensen, B. Sander, H. Junker, Fuel, **86**, 1008-1020 (2007)
6. Y. Shao, J. Wang, F. Preto, J. Zhu, Ch. Xu, Energies, **5**, 5171-5189 (2012)
7. R.P. Girón, B. Ruiz, E. Fuente, R.R. Gil, I. Suárez-Ruiz, Fuel, **114**, 71-77 (2013)
8. S. V. Vassilev, D. Baxter, L.K. Andersen, Ch. G. Vassileva, Fuel, **105**, 40-76 (2013)
9. Ch. J. Zygarlicke, B. C. Folkedahl, Prepr. Pap.-Am. Chem. Soc., Div. Fuel Chem., **48**, 641-642 (2003)
10. A. Gaik, A. Ściubidło, Energetyka, **11**, 834-839 (2013)
11. R. Rajamma, R.J. Ball, L.A.C. Tarelho, G. C. Allen, J.A. Labrinchad, V.M. Ferreira, J. Hazard. Mater., **172**, 1049-1060 (2009)
12. A.A.T. Masiá, B.J.P. Buhre, R.P. Gupta, T.F. Wall, Fuel Process. Technol., **88**, 1071-1081 (2007)
13. P. Kishor, A.K. Ghosh, D, Kumar, Asian J. Agr. Res., **4**,1-14 (2010)
14. S. Aggarawal, G.R. Singh, B.R. Yadav, J. Agr. Physics, **9**, 20-23 (2009)
15. C. Ferreira, A. Ribeiro, L. Ottosen, J. Hazard. Mater. B, **96**, 201-216 (2003)
16. J.R. Kleven, T.B. Edil, C.H. Benson, Trans Res Rec. Washington DC: Transportation Research Board, 40-48 (2000)
18. T. Abichou, C.H. Benson, T.B. Edil, Geotechnical Special Publication ASCE, **79**, 210-223 (1998)
19. E. S. Abdrakhimova, V. Z. Abdrakhimov, Solid Fuel Chemistry, **46**, 185-190 (2012)
20. L. Pérez-Villarejo, D. Eliche-Quesada, Fco. J. Iglesias-Godino, C. Martínez-García, Fco. A. Corpas-Iglesias, J.· Environ. Manag., **95**, 349-354 (2012)
21. T.R. Naik, R.N. Kraus, Y. M. Chun, W.B. Ramme, R. Siddique, ACI Mater. J., **101**, 99-206 (2004)
22. T.R. Naik, R.N. Kraus, Y. M. Chun, W.B. Ramme, S.S. Singh, J. Mater. Civil Eng., ASCE **15**, 400-407 (2003)
24. J. M. Khatib, D.J. Ellis, ACI Spl. Pub SP-200, ACI, 733-48 (2001)
25. S. Fiore, M. C. Zanetti, Am. J. Environ. Sci., **3**, 135-142 (2007)
26. S. Javed, C.W. Lovell, L.E. Wood, Trans. Res. Board, **1437**, 27-34 (1994)
27. H. Chao-Lung, B. L. Anh-Tuan, Ch. Chun-Tsun, Constr. Build. Mater., **25**, 3768-3772 (2011)
28. N. Van Tuan, G. Ye, K. van Breugel, A. L.A. Fraaij, B. D. Dai, Constr. Build. Mater., **25**, 2030-2035 (2011)

Evaluation of the results of coal jigging process

Agnieszka Surowiak[1,*]

[1]AGH - University of Science and Technology, Faculty of Mining and Geoengineering, Department of Environmental Engineering and Mineral Processing, Cracow, Poland

Abstract. Quality of applied hard coal fuel to combustion processes influence significantly on process efficiency and effects of its influence on surrounding environment. It is particularly important issue in time of Clean Coal Technologies (CTW). The paper presents the analysis of hard coal beneficiation in a jig for getting an optimal recovery of useful fraction in concentrate (combustible matter) and not useful fraction (ash and sulfur). On the basis of industrial sampling of coal dust jig the density analysis of collected samples of concentrate and tailings was performed in laboratory conditions. In separated fractions of separation products the yields of products were calculated and the contents of ash and total sulfur were marked in them. On the basis of the results of density and chemical analyzes, separation products balance and appropriate calculations the Fuerstenau beneficiation curves were plotted which allowed to evaluate process and compare results of beneficiation of material containing various components. This is a different approach to evaluation of coal beneficiation effects, so far being used mainly for multi-component metals ores. Furthermore, the evaluation of separation preciseness on the basis of separation curves and factors was done and the statistical analysis of mutual correlations of analyzed parameters was done.

1 Introduction

So far, the results of hard coal beneficiation processes were presented usually by means of group of Henry upgrading curves on the basis of performed qualitative and quantitative analyzes and by performing appropriate calculations. This is the simplest method of graphical presentation of beneficiation results based on analysis of contents of certain component in feed, concentrate and tailings according to obtained density fractions. The basic Henry upgrading curve presents usually relation between yield and content of certain component in beneficiation products $\gamma = f(\lambda)$. The case of ideal beneficiation is characterized on Henry curve by horizontal line presenting relation between fraction yield and content of useful component [1]. This basic information is often not sufficient because of the fact that separation preciseness determined by maximum accumulation of useful component in concentrate and possibly high amount of not useful ones in tailings becomes more and more important. Because of this there is a need of illustrate separation results in different way than the one preferred or generally accepted so far [13]. The tool making possible to evaluate efficiency of beneficiation of one product into useful component and simultaneous occurrence of not useful components of material in other product is Fuerstenau upgrading curve. Usually in technological practice this curve is being used to compare beneficiation results of metals ores as graphical relation presenting recoveries of analyzed components in concentrate and tailings. Fuerstenau curves are a good tool to compare selectivity of separation of two components occurring in material in purpose of determination of relation between recovery of one component in concentrate in function of second analyzed component in tailings [1, 3, 4]. The relation between selectivity of separation for ash content in coal was presented by Fuerstenau [1, 2]. These curves are useful to analyze quality of produced clean coals if taking into consideration quality of concentrate for volatile matter and ash or sulfur content in tailings. Fuerstenau curve allows to determine optimal quality of concentrate on the basis of the point of biggest convexity fF or the point F being the point of intersection between the plot and ideal beneficiation curve. This point is obtained by plotting diagonal on Fuerstenau graph of coordinates (0, 0) and (100, 100). In spot of crossing diagonal with Fuerstenau curve the optimal beneficiation point F is obtained being related to the equality between combustible matter recovery in concentrate and residuals recovery in tailings . In case of coal residuals is sulfur or ash recovery in tailings. Additionally, the Fuerstenau curves are beneficial because they are constructed in the way ensuring elimination of the influence of feed changeability on separation results [5, 6].

2 Industrial experiment

The experimental investigation was based on sampling of industrial jig manufactured by Allmineral of working area being equal to 17 m2. This jig is installed in mineral processing plant of one of Polish hard coal mines. The experiments were performed by keeping constant number

* Corresponding author: asur@agh.edu.pl

 DOI:10.1051/e3sconf/20171801030

of pulsations which was equal to 26 cycles per minute. The capacity of the system, which is feed flow intensity was equal to 300 Mg/h. Samples of concentrate and tailings from jig were collected by three changeable positions of additional water amounts in individual experiments. These amounts were equal to 35, 50 and 70 [m3/h], respectively. By such determined parameters, after stabilization of the process, the samples of separation products were collected from the jig in time of 3 minutes, each one at the same time. Then, the float and sink analysis in zinc chloride was performed for each collected sample of concentrate and tailings. The float and sink analysis was performed in solutions of zinc chloride of densities, respectively 1.30; 1.40; 1.50; 1.60; 1.70; 1.80; 2.00 Mg/m3. In obtained density fractions the chemical analyzes for ash and total sulfur contents were performed.

3 Elaboration of results

The mass yields of size-density particle fractions obtained from all three experiments for jigging products were presented in Tables 1-3. Data characterizing feed parameters were obtained as a result of balance calculations [10].

Table 1. Mass yields of individual particle size fractions of jigging products, additional water 35 [m^3/h].

Density fraction [Mg/m^3]	CONCENTRATE			TAILINGS		
	Mass [g]	Ash content A_t^a [%]	Sulfur content S_t^a [%]	Mass [g]	Ash content A_t^a [%]	Sulfur content S_t^a [%]
< 1.3	12960.0	3.63	0.77	4630.0	5.91	3.15
1.3 – 1.4	1375.0	9.29	2.80	5100.0	5.71	1.33
1.4 – 1.5	336.0	16.51	6.46	924.0	13.51	2.61
1.5 – 1.6	265.0	28.91	6.02	695.0	23.75	3.84
1.6 – 1.7	265.0	35.54	7.66	744.0	32.84	6.07
1.7 – 1.8	221.0	31.00	16.30	1010.0	39.55	6.68
1.8 – 2.0	367.0	42.54	14.29	3970.0	33.48	5.59
> 2.0	366.0	57.70	17.23	20034.0	43.79	7.47

Table 2. Mass yields of individual particle size fractions of jigging products, additional water 50 [m^3/h].

Density fraction [Mg/m^3]	CONCENTRATE			TAILINGS		
	Mass [g]	Ash content A_t^a [%]	Sulfur content S_t^a [%]	Mass [g]	Ash content A_t^a [%]	Sulfur content S_t^a [%]
< 1.3	12918.0	2.97	0.69	3822.0	3.62	1.04
1.3 – 1.4	2123.0	8.09	1.94	2962.0	5.47	1.12
1.4 – 1.5	284.0	18.71	4.52	1852.0	8.21	1.51
1.5 – 1.6	160.0	28.89	8.34	926.0	15.33	2.37
1.6 – 1.7	355.0	35.02	10.04	1084.0	27.08	3.69
1.7 – 1.8	565.0	42.98	15.36	1930.0	29.81	4.28
1.8 – 2.0	2812.0	56.43	22.97	8352.0	43.58	5.90
> 2.0	17091.0	68.42	5.19	31410.0	67.47	12.84

Table 3. Mass yields of individual particle size fractions of jigging products, additional water 70 [m^3/h].

Density fraction [Mg/m^3]	CONCENTRATE			TAILINGS		
	Mass [g]	Ash content A_t^a [%]	Sulfur content S_t^a [%]	Mass [g]	Ash content A_t^a [%]	Sulfur content S_t^a [%]
< 1.3	13945.0	3.49	0.73	1499.0	3.65	1.00
1.3 – 1.4	2817.0	8.83	1.38	3042.0	6.06	1.28
1.4 – 1.5	419.0	20.24	2.58	1920.0	8.32	1.49
1.5 – 1.6	108.0	36.35	4.53	1431.0	10.73	1.71
1.6 – 1.7	84.0	45.02	6.49	1448.0	14.03	2.24
1.7 – 1.8	46.0	52.98	10.26	1289.0	26.49	3.33
1.8 – 2.0	59.0	65.23	18.97	4062.0	40.91	6.91
> 2.0	13.0	72.38	35.16	36123.0	69.33	12.22

3.1 Analysis of coal dusts beneficiation

For the purposes of this work the Fuerstenau upgrading curves were used as a tool to evaluate effects of coal dusts separation process conducted in a jig. On the basis of data presented in Tables 1, 2 and 3 the balances of components were elaborated and the feed parameters were calculated in this way. According to the rules presented by Stepinski [10] the coordinates were calculated and the upgrading curves were plotted in coordinates system: combustible matter recovery in concentrate – ash recovery in tailings and combustible matter recovery in concentrate – sulfur recovery in tailings, which were positioned in Table 4 for all performed experiments. The combustible matter recovery in concentrate was calculated according to the formula:

$$\varepsilon = \gamma_k \frac{100 - \vartheta_A}{100 - \alpha_A} \quad (1)$$

where: γ_k - concentrate yield, ϑ_A - ash content in concentrate, α_A - ash content in feed.

Ash recovery in tailings was calculated as:

$$\varepsilon' = \gamma_0 \frac{\beta_A}{\alpha_A} \quad (2)$$

Sulfur recovery in tailings was calculated as:

$$\varepsilon'' = \gamma_0 \frac{\beta_S}{\alpha_S} \quad (3)$$

where: γ_o - tailings yield, β_A - ash content in tailings, β_S - sulfur content in tailings, α_A- ash content in feed, α_S - sulfur content in feed.

Table 4. Calculated coordinates to plot Fuerstenau curves.

Density fraction [Mg/m³]	Combustible matter recovery in concentrate Σε [%]			Ash recovery in tailings Σε'			Sulfur recovery in tailings Σε''		
	Amount of additional water [m³/h]			Amount of additional water [m³/h]			Amount of additional water [m³/h]		
	35	50	70	35	50	70	35	50	70
<1.3	50.74	51.09	32.05	100.00	100.00	100.00	100.00	100.00	100.00
1.3-1.4	70.76	71.23	44.63	91.38	97.45	97.56	58.76	96.03	97.39
1.4-1.5	75.58	76.06	47.64	89.16	96.91	96.79	51.84	95.21	96.70
1.5-1.6	78.71	79.20	49.61	88.94	96.85	96.68	51.15	95.12	96.61
1.6-1.7	81.94	82.44	51.63	88.77	96.81	96.62	50.68	95.07	96.56
1.7-1.8	85.71	86.21	53.97	88.52	96.75	96.53	49.95	95.00	96.50
1.8-2.0	98.34	98.43	60.93	88.14	96.60	96.16	47.62	94.80	96.21
>2.0	100.00	100.00	100.00	85.07	93.39	85.29	7.99	90.26	87.83

The Fuerstenau upgrading curves were presented on Figs 1-3 in system: combustible matter recovery in concentrate – ash recovery in tailings on curves a) and combustible matter recovery in concentrate – sulfur recovery in tailings on curves b) for various amounts of additional water directed to the jig. By means of these Figures the technologically optimal beneficiation point F was determined for each case. The best effects of separation of combustible substance in concentrate and ash in tailings were obtained in experiment with amount of additional water being equal to 50 [m3/h] in which optimal recovery of combustible matter in concentrate being equal to ash recovery in tailings = 98. Furthermore, for this case the best effect of combustible matter and sulfur separation was obtained, which recovery was equal to 95. The worst result of separation of ash and combustible matter being equal to 68 was obtained in experiment with amount of additional water being equal to 70 [m3/h]. The worst result of separation of sulfur and combustible matter in tailings being equal to only 56 was obtained for the experiment where the amount of additional water was equal to 35 [m3/h].].

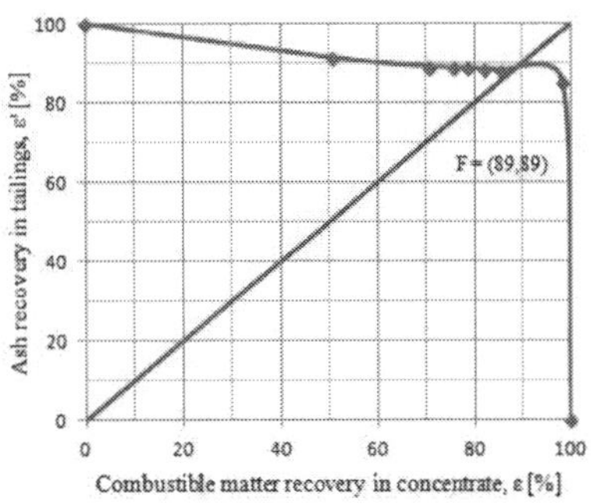

a)

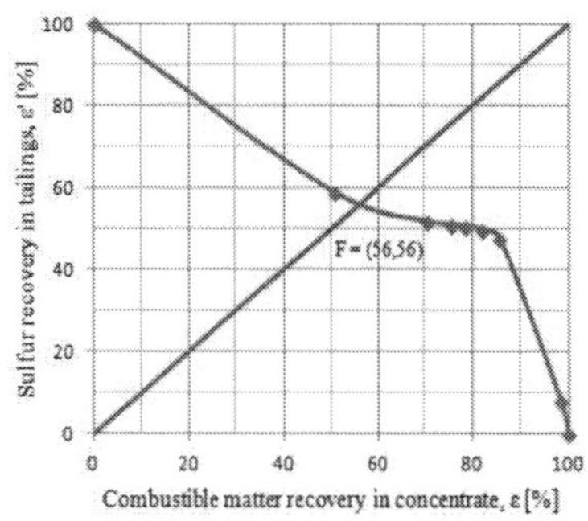

b)

Fig. 1. Fuerstenau's curves in functions of ash and sulfur recoveries. Ilość wody podsitowej 35 [m³/h], a) Fuerstenau's curve in function of ash recovery in tailings, b) Fuerstenau's curve in function of sulfur recovery in tailings.

DOI:10.1051/e3sconf/20171801030

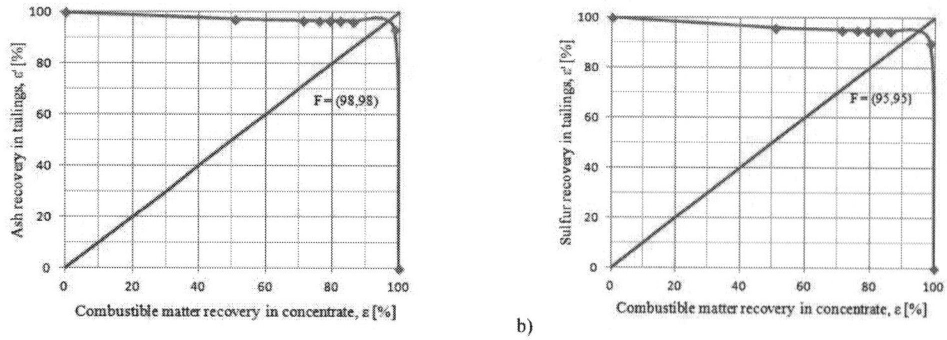

a) b)

Fig. 2. Fuerstenau's curves in functions of ash and sulfur recoveries. Ilość wody podsitowej 50 [m³/h], a) Fuerstenau's curve in function of ash recovery in tailings, b) Fuerstenau's curve in function of sulfur recovery in tailings.

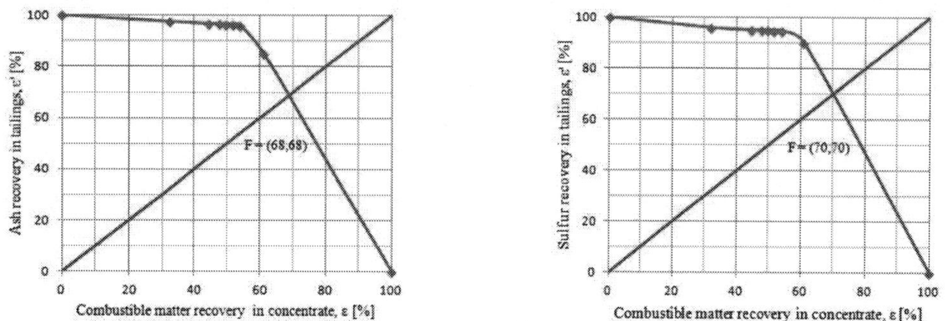

a) b)

Fig. 3. Fuerstenau's curves in functions of ash and sulfur recoveries. Ilość wody podsitowej 70 [m³/h], a) Fuerstenau's curve in function of ash recovery in tailings, b) Fuerstenau's curve in function of sulfur recovery in tailings.

3.2. Statistical analysis of obtained results

In purpose of verification of mutual relations between variables for obtained results presented in Tables 1-4 the basic statistical evaluation was done for parameters of performed experiments and obtained results of jigging. The linear correlation coefficients matrices were determined for all investigated factors. The obtained results were positioned in Tables 5-7. Analyzing the mutual correlations in individual experiments is possible to notice that the strongest correlated variables are combustible matter recoveries with density and ash contents in concentrate. It is particularly visible in first and second experiment, with amount of additional water being equal 35 and 50 [m3/h]. Furthermore, the ash recovery in tailings is highly correlated with combustible matter and sulfur recoveries, what is particularly visible in second experiment. That is why the separation of these components was the most efficient by the amount of additional water being equal to 50 [m3/h]. The statistical analysis indicated that in experiment with the amount of additional water being equal to 70 [m3/h] all recoveries are highly correlated as well there are significant relations between them and sulfur contents in concentrate and tailings. As a result worse beneficiation effects were obtained than in second experiment with amount of additional water being equal to 50 [m3/h].

Table 5. Correlation coefficients for variables from experiment with amount of additional water being equal to 35 [m³/h].

T	Determined correlation coefficients are significant with $p < 0.5000$									
	Mean	**Std deviation**	**P**	**A_C**	**A_T**	**ε**	**ε'**	**ε''**	**S_C**	**S_T**
P	1.619	0.253	1.0000	0.9686	0.9413	0.9393	-0.7668	-0.8028	0.9457	0.8874
A_C	28.140	17.872	0.9686	1.0000	0.9299	0.9279	-0.8148	-0.8638	0.8666	0.8670
A_T	24.817	14.966	0.9413	0.9299	1.0000	0.8603	-0.7517	-0.7490	0.9255	0.9607
ε	80.222	15.743	0.9393	0.9279	0.8603	1.0000	-0.9163	-0.8922	0.8932	0.7214
ε'	89.9975	4.3948	-0.7668	-0.8148	-0.7517	-0.9163	1.0000	0.9431	-0.7564	-0.5597
ε''	52.2487	24.8192	-0.8028	-0.8638	-0.7490	-0.8922	0.9431	1.0000	-0.7682	-0.6029
S_C	8.9412	6.2367	0.9457	0.8666	0.9255	0.8932	-0.7564	-0.7682	1.0000	0.8603
S_T	4.5925	2.1724	0.8874	0.8670	0.9607	0.7214	-0.5597	-0.6029	0.8603	1.0000

DOI:10.1051/e3sconf/20171801030

Table 6. Correlation coefficients for variables from experiment with amount of additional water being equal to 50 [m^3/h].

T	Determined correlation coefficients are significant with p < 0.5000									
	Mean	**Std deviation**	**P**	**Ac**	**AT**	**ε**	**ε'**	**ε''**	**Sc**	**ST**
P	1.619	0.253	1.0000	0.9978	0.9649	0.9372	-0.8139	-0.8096	0.6566	0.8879
Ac	32.688	22.834	0.9978	1.0000	0.9666	0.9424	-0.8402	-0.8361	0.6283	0.8975
AT	25.071	22.007	0.9649	0.9666	1.0000	0.8678	-0.8439	-0.8315	0.4593	0.9735
ε	80.582	15.631	0.9372	0.9424	0.8678'	1.0000	-0.8753	-0.8821	0.6681	0.7786
ε'	96.845	1.786	-0.8139	-0.8402	-0.8439	-0.8753	1.0000	0.9992	-0.2324	-0.8617
ε''	95.186	2.628	-0.8096	-0.8361	-0.8315	-0.8821	0.9992	1.0000	-0.2474	-0.8440
Sc	8.631	7.459	0.6566	0.6283	0.4593	0.6681	-0.2324	-0.2474	1.0000	0.2526
ST	4.093	3.923	0.8879	0.8975	0.9735	0.7786	-0.8617	-0.8440	0.2526	1.0000

Table 7. Correlation coefficients for variables from experiment with amount of additional water being equal to 70 [m^3/h].

T	Determined correlation coefficients are significant with p < 0.5000									
	Mean	**Std deviation**	**P**	**Ac**	**AT**	**ε**	**ε'**	**ε''**	**Sc**	**ST**
P	1.619	0.253	1.0000	0.9918	0.9195	0.8553	-0.7496	-0.7681	0.9063	0.8757
Ac	38.065	25.514	0.9918	1.0000	0.8691	0.8143	-0.7034	-0.7274	0.8559	0.8169
AT	22.440	22.629	0.9195	0.8691	1.0000	0.9523	-0.8969	-0.8940	0.9970	0.9916
ε	55.057	19.964	0.8553	0.8143	0.9523	1.0000	-0.9820	-0.9861	0.9618	0.9544
ε'	95.703	4.378	-0.7496	-0.7034	-0.8969	-0.9820	1.0000	0.9971	-0.9144	-0.9119
ε''	95.975	3.508	-0.7681	-0.7274	-0.8940	-0.9861	0.9971	1.0000	-0.9096	-0.9041
S.C.	10.012	11.778	0.9063	0.8559	0.9970	0.9618	-0.9144	-0.9096	1.0000	0.9963
ST	3.772	3.911	0.8757	0.8169	0.9916	0.9544	-0.9119	-0.9041	0.9963	1.0000

where: P - density; A_C - ash content in concentrate; A_T - ash content in tailings; ε - combustible matter recovery in concentrate; ε' - ash recovery in tailings; ε" - sulfur recovery in tailings; S_C - sulfur content in concentrate; S_T - sulfur content in tailings

3.3 Analysis of separation preciseness

To verify effects of beneficiation the analysis of separation preciseness was conducted on the basis of partition curves and factors. On the basis of data positions in Tables 1-3 the coordinates of partition curves for jigged coal dusts tailings were calculated.

Partition curves for jigs with density as an argument of separation are asymmetric curves and can be approximated well by means of Weibull distribution function what was indicated by the investigations of other authors [7, 8]. Because of it the Weibull distribution function was used to approximate empirical partition curve. The general formula of this function is given by the equation:

$$T(\rho) = 100\left\{1 - \exp\left[-\left(\frac{\rho}{\rho_o}\right)^n\right]\right\} \quad (3)$$

where: ρo and n – are distribution function parameters.

Graphs of partition curves distribution functions were presented on Figs 4-6. The continuous curve presents a model relation. The curvilinear correlation index is bigger than 0.95. Using the approximated partition curves the separation efficiency factors were calculated and the values of these parameters were positioned on Figs 4-6. These factors were partition density ρr, probable error Ep and imperfection I. The values of these parameters show that with increasing amount of additional water the values of partition preciseness factors lower, especially the values of probable error and imperfection, what means that the preciseness of jigging increases [11,12].

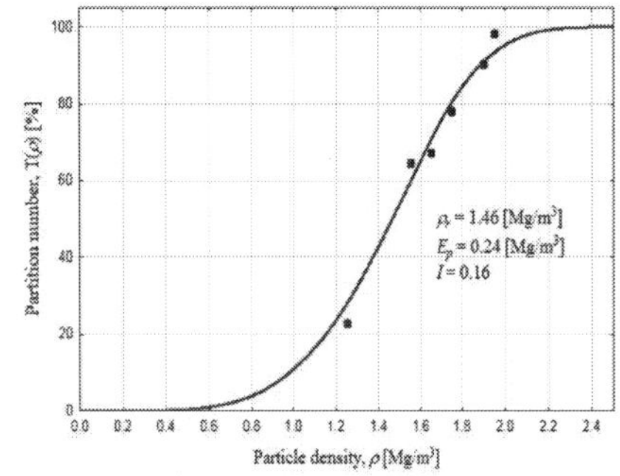

Fig. 4. Partition curve for tailings, amount of additional water 35 [m^3/h], d_o = 1.58 [Mg/m^3], n = 4.75.

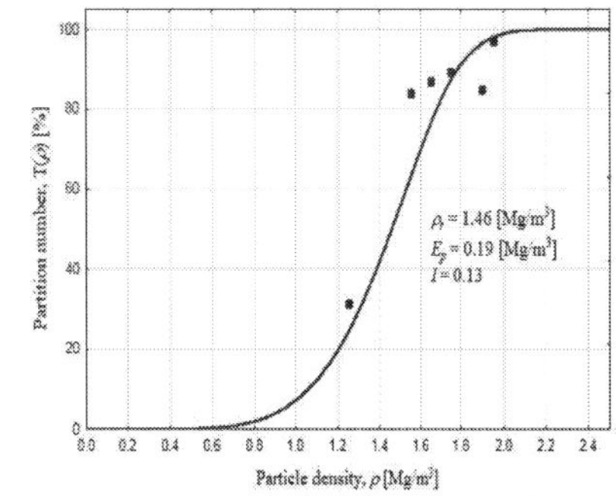

Fig. 5. Partition curve for tailings, amount of additional water 50 [m^3/h], d_o = 1.55 [Mg/m^3], n = 5.94.

 DOI: 10.1051/e3sconf/20171801030

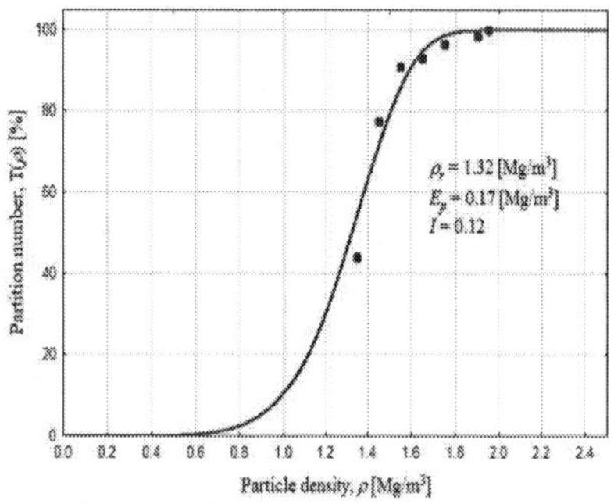

Fig. 6. Partition curve for tailings, amount of additional water 70 [m^3/h], d_o = 1.40 [Mg/m^3], n = 6.58.

4 Summary

The jigging process is realized in one of the main coal processing nodes and the evaluation of its effects because of achieving certain recoveries of both combustible matter in concentrate as well not useful components in tailings is a crucial part of process monitoring. This is true especially in situation of treating coal as a clean fuel and production of its qualitatively good assortments. The paper presents using of Fuerstenau approach to analyze separation effects on which basis is possible to state that the separation of combustible matter and ash in coal occurs much more efficiently than the separation of combustible matter and sulfur. This is caused most probably by the presence of organic sulfur which cannot be separated by means of gravitational separation processes [9]. Furthermore, it was stated that the hydrodynamic conditions of jig work are important during the process realization because dependably on amount of additional water delivered in purpose of ensuring sufficient particles liberation in working bed, various separation results can be achieved considering both qualitative and quantitative factors. The analysis of separation efficiency indicated that more efficient separation occurs by higher amounts of additional water. Implementation of Fuerstenau curves to evaluate coal dusts beneficiation process allows to perform analysis of process course because of its achievement of optimal recoveries of combustible matter in concentrate as well of ash and sulfur in tailings. This is very important in times when production of so-called clean coal is crucial.

This work was performed as a part of the University of Science and Technology Research Program No.11.11.100.276 project.

References

1. J. Drzymała, *Podstawy mineralurgii*, (Oficyna Wydawnicza Politechniki Wrocławskiej, 2001)
2. D. W. Fuerstenau, Raporty, University of California, Berkley, dane niepublikowane (1979)
3. J. Drzymała, H.A.M. Ahmed, Inter. J. Miner. Process., **76**, 55-65 (2005)
4. J. Drzymała, P.B. Kowalczuk, M. Oteng-Peprah, D. Foszcz, A. Muszer, T. Henc, Minerals Engineering, **49**, 17-23 (2013)
5. D. Foszcz, *Zasady określania optymalnych rezultatów wzbogacania wieloskładnikowych rud miedzi*, Studia, Rozprawy, Monografie, 181, (Wyd. IGSMiE PAN, Kraków, 2013)
6. D. Foszcz, T. Niedoba, T. Tumidajski, J. Pol. Miner. Eng. Soc., **16(2)** (2015)
7. B. S. Gottfried, Int. J. Miner. Process., **5**, 1-20 (1978)
8. A. D. Paul., Maronde C.P., Killmeyer R.P., Proc. XIII ICPC, Brisbane, II, 861-869 (1998)
9. T. Niedoba, A. Surowiak, Proceedings of the 24th International Mining Congress and Exhibition of Turkey, IMCET 2015, 717-721 (2015)
10. W. Stępiński, *Wzbogacanie grawitacyjne*, (Łódź-Warszawa-Kraków, Wydawnictwo PWN, 1964)
11. A. Surowiak, M. Brożek, Physicochem. Probl. Miner. Process. **52(1)**, 228–243 (2016)
12. A. Surowiak, Mineral Resources Management, **30**(1) 105–122 (2014)
13. E. Szymkowiak, J. Drzymała, Prace Naukowe Instytutu Górnictwa Politechniki Wrocławskiej, **131**, 47-56 (2010)

E3S Web of Conferences **18**, 01031 (2017) DOI:10.1051/e3sconf/20171801031
MEC2017

Physical chemical characterization of historical mining waste and ARD prediction tests

Antoneta Filcenco–Olteanu[1,*], *Lena* Alakangas[2], *Antoio* Fiuza[3], *Marius* Zlagnean[1], and *Nicolae* Tomus[1]

[1]Research and Development National Institute for Metals and Radioactive Resource, 70,Carol I, Bvd, Bucharest, Romania,
[2]Luleå University of Technology, Department of Civil, Environmental and Natural Resources Engineering, Porsön, University area, Luleå Sweden
[3]University of Porto, Engineering Faculty, Rua Dr. Roberto Frias, 4200-465 Porto, Portugal

Abstract. There are thousands of historical mine tailings and metallurgical residues present on inactive metal mining sites which have been abandoned. However, the potential release of dissolved metals, acidity, or suspended particulates from mine-waste dumps can be a serious and long-lasting problem. A variety of environmental impacts may occur at this abandoned mine sites and leading the list is acid generation, which is one of the main problems from metal mining. The objective of the present study was to characterize and evaluate the Romanian historical tailing of Sasar-Red Valley, near Baia Mare. This involved physical and chemical characterization of the materials and its acid generating potential. Sasar tailing impoundment contains 8.5 million m^3 of tailings and occupies an area of 32.5 hectares. Tailings samples were collected from different depths in three sampling points, and characterized by grain size, chemical elements concentration, elements distribution of elements in depth, paste pH, acid-base account and net acid generation tests. Acid base accounting (ABA) tests in conjunction with net acid generation (NAG) tests classified the samples into the category of 'potentially acid generating'. This paper presents a synthesis of works performed in the Improve Resource Efficiency and Minimize Environmental Footprint (REMinE) project.

1 Introduction

The key to future sustainable mining activities is resource efficiency and a holistic approach to the environmental consequences that can emerge during the mine's life, and should be based on the best-available techniques to minimize these consequences. In a long-term perspective, in combination with a more efficient extraction of minerals and metal, this could prevent the environmental impact. High content of metals and minerals in historical tailings and waste rock dumps, due to the earlier lack of efficient methods for extraction or low economical profitable at that time, could in many cases be a potential risk for environmental impact. However, certain minerals could also be resources contributing to the supply of metals and minerals to the market or could be utilized as a raw material for construction purposes.

Situated at Romania's northern border, Baia Mare, from Maramures County has been an important mining region for more than 2,500 years thanks to its abundant mineral resources. [1]

Baia Mare, Baia Sprie and Cavnic were the major ore deposits, which consist mainly of gold-bearing argentiferous galena and in some deposits gold, known in paragenesis with stibnite. [2]

There are also approximately 215 tailings dams and ponds, several ore processing facilities and copper and lead smelters as well as a sulphuric acid plant. Over the years, the industrial activity was pursued with improper waste management, treatment and disposal. [3]

The mining wastes, from Baia Mare region cause a series of major environmental damage due to their large volumes and also due to their inherent danger, such as the potential of generating acid waters, high concentrations of heavy metals and the existence of hazardous substances from ore processing. Therefore, the reprocessing, removal and safe disposal of mining, processing and metallurgical wastes were encouraged to ensure an environmental improvement area. [4]

Solid wastes (waste rocks and concentrator tailings), which often contain sulfide minerals, can generate acidic discharges known as acid mine drainage (AMD) or acid rock drainage (ARD) after exposure to atmospheric conditions (air and water). ARD is typically characterized by low pH ($pH<5$) and high levels of potentially toxic metals and sulphates [5, 6]. These contaminants remobilize under acidic conditions and migrate into the water zone and groundwater system. [7]

Static tests are geochemical analyses of sulfidic waste which are used to predict potential of waste sample to produce acid. Details on these tests are documented in the literature [8-16].

* Corresponding author: antonetafilcenco@yahoo.ro

Static tests, also called acid–base accounting (ABA) tests, are widely used to predict ARD because they are simple, inexpensive, and their results are rapidly obtained. ABA methods are also used for classifying samples according to their acid forming potential. [7,9,10]

The acid-base account involves static laboratory procedures that evaluate the balance between acid generation processes (oxidation of sulphide minerals) and acid neutralising processes (dissolution of alkaline carbonates, displacement of exchangeable bases, and weathering of silicates). Results are expressed as a net acid producing or acid consuming capacity per mass unit of rock. In addition, static tests and reporting convention vary. In Australia and the Asia Pacific region the ABA test is reported as a Net Acid Producing Potential (NAPP) value in units of kg H_2SO_4/t and the values arising from the acid-base account are referred to as the maximum potential acidity (MPA) and the acid neutralizing capacity (ANC), respectively. In North America, ABA results equivalent to NAPP are normally expressed as a Net Neutralization Potential (NNP) in units of kg $CaCO_3$/t. The components in the calculation are the same, with the maximum potential acidity (MPA) normally expressed as AP (acid potential), and ANC expressed as NP (neutralizing potential). [17]

The objective of the present study was to characterize and evaluate the Romanian historical tailing Sasar-Red Valley, located near Baia Mare. This involved physical - chemical characterization of the tailing samples collected from various depths by: grain size, chemical elements concentration, elements distribution in depth and its acid generating potential by means of paste pH acid-base accounting (ABA) tests and net acid generation (NAG) tests.

2 Materials and methods

2.1 Study area

The Maramures Country, situated at northern border of Romania with Ukraine, encloses the old "lands" of Maramures-Chioarul, Lapus and Baia Mare Basin.

Baia Mare Basin is a contact basin that interposes between the Somesana Plain and the Carpathian Mountains as a lower morphological unit, from the surrounding areas, presenting a waved surface, characterized by a convergent system of valleys and interfluves. It was formed due to the tertiary tectonic movement that took to the fragmentation and sinking of the crystalline in the North-Western part of Transylvania, as well as due to the volcanic chain of the Gutin-Oaş Mountains [4, 18].

The area belongs to the undifferentiated Quaternary characterized by deluvial deposits, andesitic blocks and alluvial deposits. Under the action of the external agents the andesites were altered and eroded and formed alluvial and deluvial deposits that make the transition from the deposits of Sasar River upper terrace. The sedimentary in Baia-Mare Basin is represented by grey-purple marls, clay marls and sands with sandstone horizons. [19]

The activities of extraction, processing and preparation of complex ore from the mining basin Baia Mare, left behind tens of dumps and mine tailings ponds, located mainly in the river valleys from this area.

The Sasar (Red Valley) tailing pond (presented in Fig. 1), located in the west part of Baia Mare, secured the storage of flotation tailings from the Processing Plant which was closed in 1982. It contains around 8.5 million m^3 of fine grained waste ores embodying the acidic water producing mineral pyrite.

Fig. 1. The Sasar tailing impoundment and sampling points.

From mineralogical point of view, the tailings disposed in the Sasar tailing pond contain silicates (from the amphiboles and pyroxenes), sulphide minerals (pyrite, FeS_2, chalcopyrite, $CuFeS_2$, arsenopyrite, FeAsS, pyrrhotite, $Fe_{(1-x)}S$, marcasite, FeS_2, jamesonite, $Pb_4FeSb_6S_{14}$, galena, PbS, sphalerite, ZnS) etc., other metallic minerals like oxides, hydroxides, as hematite, Fe_2O_3, specularite, rarely ilvaite, $CaFe_{22}+Fe_{3}+[OOH\ Si_2O_7]$, oxi-hydroxi - iron sulphides (HOF), mineral associations (muscovite), clay minerals, carbonates and some secondary minerals (goethite, α-FeO(OH), hematite Fe_2O_3). [20]

2.2 Sampling

Three sampling points were selected within tailing impoundment H1 central, H2 and H3 at the edges. (Figure 1) Drill samples were taken in September 2016 from Sasar impoundment and were preserved in polyethylene bags, transported and processed separately. The samples were taken with a soil auger from six depth intervals: 0-1m, 1-2 m, 2-3m, 3-4m, 4-5m and 5-6m.

Prior to analysis, the samples collected were split with a riffle splitter, and half of each sample was kept for potential future analysis.

2.3 Experimental methodology

A size analysis of the samples was performed to determine the size distribution of the tailings samples and the content of gold, silver, iron, calcium and sulphur, for each depth interval. Grain size distribution of tailing samples was measured by standard wet sieving for 2-0.04 mm fraction, in combination with a laser granulometric analyser type Eye Tech Anchersmid. Cumulative grain size distribution was employed to calculate D_{50} (passing diameter which allow to pass 50% of the sample weight).

The bulk chemical composition of the samples was analysed with a Perkin Elmer Optima 3100 RL ICP-AES.

The sample moisture content was determined by drying 10g soil samples at 105°C for 24h.

Static geochemical tests provide the basis for understanding potential reactivity and therefore ARD potential of a sample. The static geochemical tests were performed on pulverized samples (<75µm) collected from drill 3 from 1 m, 3 m and 6 m depth.

The paste pH and electrical conductivity (EC) measurements are good indicators of the immediate acid-base reactivity of a sample. This is demonstrated by comparison to other standard geochemical tests such as single addition net acid generation (NAG) test [22]. The paste pH/EC was determined by equilibrating the sample in deionised water for approximately 12 hours at a solid to water ratio of 1:2 (w/w) and then measuring the pH and EC. The paste pH /EC was measured by a multi-parameter system Consort C832 calibrated against standard buffers at 4.01 and 7.0 pH. This gives an indication of the inherent acidity and salinity of the waste material when initially exposed in a waste emplacement area.

The net acid production potential (NAPP) is a measure of the samples overall acid generating capacity. It represents the balance between the capacity of a sample to generate acid (MPA) and its capacity to neutralise acid (ANC). NAPP (Equation 2) is the difference between the inherent acid neutralization capacity (ANC) and the maximum potential acidity (MPA) of the sample.

$$NAPP\ (\frac{kg\ H_2SO_4}{t}) = MPA - ANC \quad (1)$$

A negative NAPP indicates that the sample has a net neutralizing capacity and a positive NAPP indicates that the sample has a net acid generating capacity [22].

MPA is based on total wt% S as determined by LECO analysis [23] multiplied by the stoichiometric conversion factor 30.6, using the following formula:

$$MPA(\frac{kg\ H_2SO_4}{t}) = (Total\%S) * 30.6 \quad (2)$$

The calculation assumes that the measured sulphur content occurs as pyrite (FeS_2) and that pyrite reacts under oxidizing conditions to generate acid according to the reaction:

$$FeS_2 + {}^{15}/_4\, O_2 + {}^{7}/_2\, H_2O \rightarrow Fe(OH)_3 + 2H_2SO_4 \quad (3)$$

The acid neutralization capacity (ANC) of the studied samples was determined using the modified Sobek et al. [14] static test. With this method, approximately 0.5 g of the sample is tested for its fizz rating by the addition of a few drops of 1:3 HCl hydrochloric acid (HCl). The resulting fizz (denoted by none, slight, moderate, or strong) determines the volume and the normality of HCl to be added to a 2g sample for ANC determination. Also, blank samples were prepared by pipetting the same volume and concentration of acid into clean bakers containing deionized water. Digestion is achieved by heating the sample at 90°C until gas bubbling stops. Afterwards, the solution and "blank" samples were back-titrated with NaOH to pH 7.0 to determine ANC of the sample. The amount of acid consumed by reaction with the sample is then calculated using the following formula: [22].

$$ANC\ (\frac{kg\ H_2SO_4}{t}) = [YxM_{HCl}/wt]xC \quad (4)$$

Where:
Y = (Vol. of HCl added) - (Vol. of NaOH titrated x B)
B = (Vol. of HCl in blank) / (Vol. of NaOH titrated in blank)
M_{HCl} = Molarity of HCl
wt = Sample weight in grams
C = Conversion factor:
C = 49.0 (to calculate kg H_2SO_4/t)
C = 5.0 (to calculate % $CaCO_3$ equivalent).

The single addition NAG test is used in association with the NAPP to classify the acid generating potential of a sample. This test involves the addition of 250 ml 15 vol % H_2O_2 to 2.5g of pulverized sample (< 75 µm), to encourage the rapid oxidation of reactive sulphides. The peroxide is allowed to react with the sample overnight and the following day the sample is gently heated to accelerate the oxidation of any remaining sulphides, then vigorously boiled for several minutes to decompose residual peroxide. When cool, the pH and acidity of the NAG liquor are measured. This pH measurement is referred to as the NAGpH. The acidity of the liquor is then used to estimate the net amount of acid produced per unit weight of sample. An indication of the form of the acidity is provided by initially titrating the NAG liquor to pH 4.5 (NAGpH4.5), then continuing the titration up to pH 7 (NAG). The titration value at a pH of 4.5 includes acidity due to free acid (i.e. H_2SO_4) as well as soluble iron. The titration value at pH 7 also includes metallic ions that precipitate as hydroxides at pH's between 4.5 and 7.

The reagents used were chemically or analytically pure (to perform the analysis) and used without additional treatment.

3 Results and discussion

3.1. Weathering profile characteristics

In Sasar impoundment three zones were identified: an upper oxidation zone (reddish-brown-yellow coloured, acid generation and metal release), a hardpan layer (acid neutralization and metal accumulation) and a reduction zone (grey coloured, limited water movement, low oxygen diffusivity). The terms oxidized / unoxidized should not be understood as a rigorous chemical oxide-reduction status but rather as a general description of sulphide tailings weathering.

The colours of the three identified zones (oxidation zone, hardpan layer and the reduction zone) for H3 drill are presented in Figure 2.

Data from H2 and H3 borehole was combined to produce single weathering characteristics. The weathering profile of Sasar tailing is shown in Figure 3.

It was noted that moisture was increasing from 18% (a dry state) on surface to 25% at 2-3 m depth for H1 and from 20% on surface to 29% at 3-4 m depth on H2 and H3, respectively. At 5-6 m depth, the moisture decreases to 12%.

Fig. 2. The colours for oxidation zone, hardpan layer and reduction zone for H3 drill.

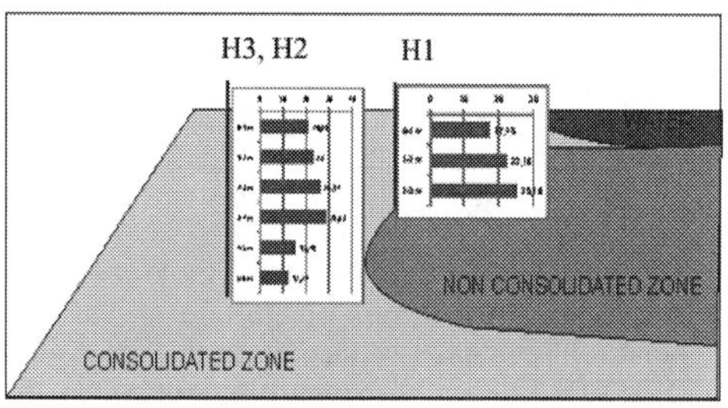

Fig. 3. Weathering profiles for samples H1, H2 and H3.

3.2. Physical characteristics of the samples

From the grain size point of view a decrease of the grain size with depth was noted. The maximum grain size on the surface was 300 microns and 150-250 microns at 6m depth.

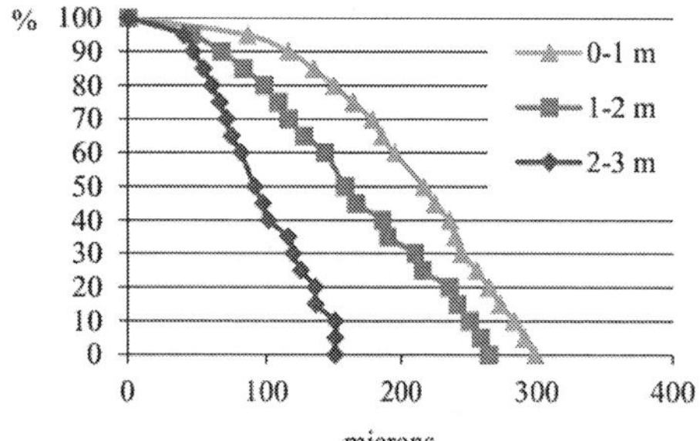

Fig. 4. Grain size distribution of H1 samples at 0-1m, 1-2 m and 2-3m depth intervals.

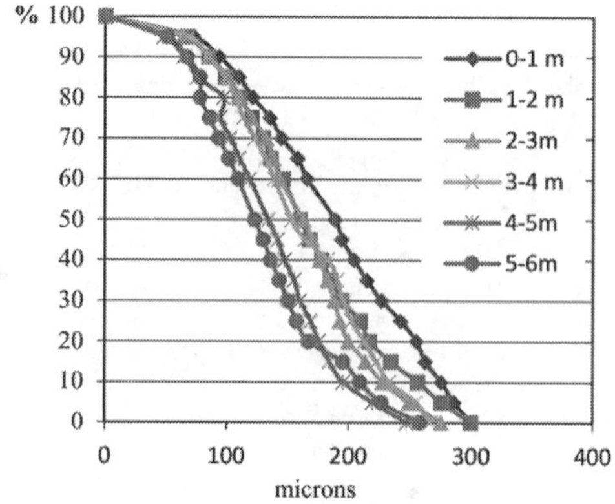

Fig. 5. Grain size distribution of H2 samples at different depth intervals (0-1m, 1-2 m, 2-3m, 3-4m, 4-5 and, 5-6m).

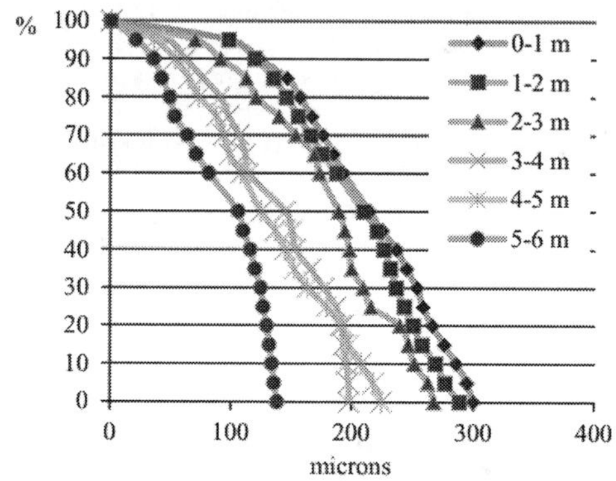

Fig. 6. Grain size distribution of H3 samples at different depth intervals (0-1m, 1-2 m, 2-3m, 3-4m, 4-5 and, 5-6m).

The results presented in Figures 4, 5 and 6 illustrate the variation of grain size distribution of samples with depth.

As a result of the oxidation and weathering of tailings on the surface, a significant decrease of the grain size with depth for all samples was noted. The reduction in D_{50} with depth is presented in Figures 7, 8 and 9.

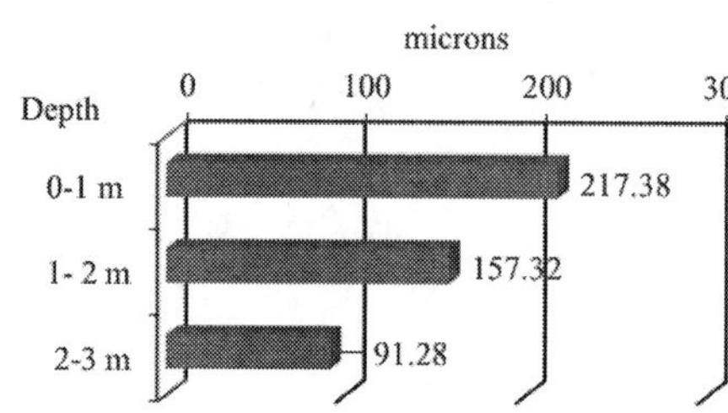

Fig. 7. D_{50} distribution of H1 sample.

 DOI:10.1051/e3sconf/20171801031

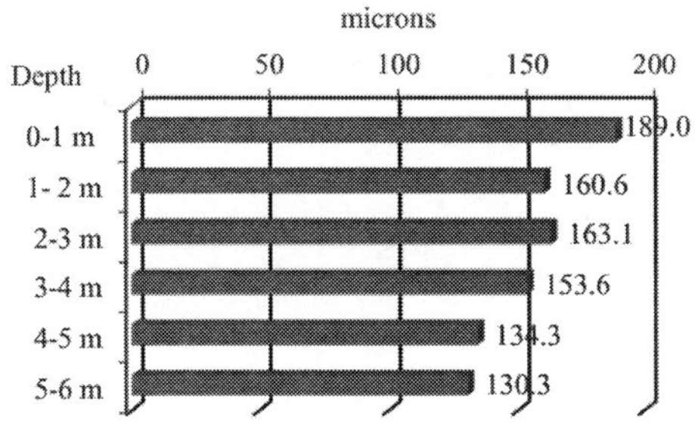

Fig. 8. D_{50} distribution of H2 sample.

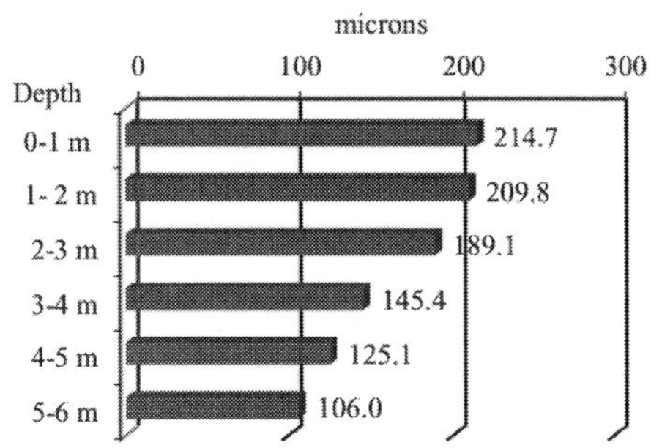

Fig. 9. D_{50} distribution of H3 sample.

This fact can be put into account of the sulphide oxidation and acid neutralization processes which entail the breakdown of minerals [26], and the formatting of fine sized weathering products.

3.3. Geochemical characterization of samples

Table 1 summarizes the chemical compositions of the tailings samples collected from drill 1 (H1), drill 2 (H2) and drill 3 (H3).

All the tailing samples show exceeding of the heavy metals concentrations, over maximum allowable limit (Annex I A of the Sewage Sludge Directive (86/278/EC)). The total concentrations of elements in the mine tailings were up to 2529 ppm manganese (Mn), 1652.89 ppm zinc (Zn), 688.44 ppm lead (Pb), 311.50 ppm copper (Cu) and 11.78 ppm cadmium (Cd). The gold and silver content of the tailing samples were 0.45 g/t and 4.59 g/t, respectively.

A size analysis of the samples was performed in order to determine the gold (Au) and silver Ag) contents of each size fraction (Table 2).

More than 62% of the gold were contained in 0.1-0.2 mm and 0–0.04 mm fractions. The 0.1-0.2 mm fraction contains around 34.67% of the gold and 19.74 % of the silver in the tailings, while the 0-0.04 mm fraction contains around 27.67% of the gold and 54.19% of the silver.

Gold and silver content distributions in depth for H3 drill samples are presented in Figure 10.

Table 1. The chemical compositions and average value of elements content in tailings profiles.

Element	H 1	H 2	H 3	Average
Au g/t	0.46	0.43	0.45	0.45
Ag g/t	6.37	3.70	3.72	4.39
Al %	1.64	1.01	1.03	1.23
As ppm	285.67	175.00	201.67	220.78
B ppm	10.00	10.00	10.00	10.00
Ba ppm	120.00	165.00	146.67	143.89
Ca %	1.29	1.02	0.78	1.03
Cd ppm	20.53	8.95	5.87	11.78
Co ppm	10.67	9.17	6.83	8.89
Cr ppm	27.08	16.83	14.67	19.53
Cu ppm	490.67	199.17	244.67	311.50
Fe %	4.82	3.57	3.94	4.11
Hg ppm	1.33	1.33	1.17	1.28
K %	0.39	0.29	0.25	0.31
Mg %	0.59	0.39	0.44	0.47
Mn ppm	3130	2940	1517	2529
Mo ppm	2.33	2.00	2.17	2.17
Na %	0.03	0.02	0.02	0.02
Ni ppm	13.00	9.00	4.33	8.78
P ppm	440.00	365.00	413.33	406.11
Pb ppm	1029.67	490.17	545.50	688.44
S %	2.81	1.79	1.64	2.08
Sc ppm	5.33	3.67	3.83	4.28
Sr ppm	23.67	18.00	15.17	18.94
V ppm	48.67	32.17	34.67	38.50
Zn ppm	2606.67	1451.33	900.67	1652.89

Table 2. Gold and silver distribution of size fractions.

Sieve (mm)	Weight (%)	Au (g/t)	Gold distribu tion (%)	Ag (g/t)	Silver distribu tion (%)
>0.2	13.33	1.125	33.31	4.2	12.75
0.1-0.2	34.67	0.225	17.32	2.5	19.74
0.075-0.1	12.67	0.26	7.22	2.50	7.21
0.04-0.075	11.67	0.26	6.76	2.30	6.11
0-0.04	27.67	0.576	35.39	8.6	54.19
	100	0.45	100.00	4.39	100.00

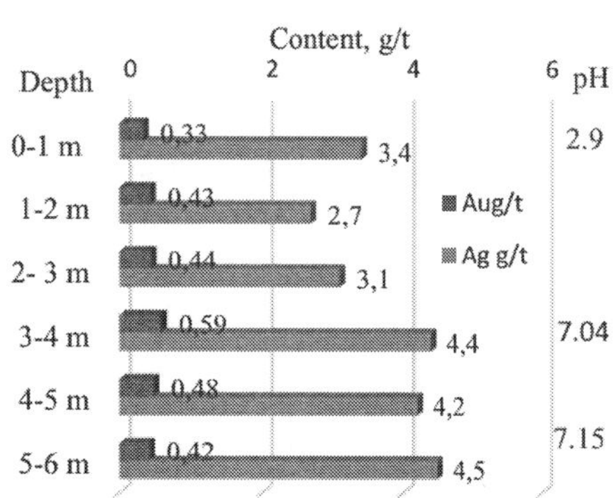

Fig. 10. Gold and silver content distributions in depth for H3 drill samples.

The obtained data showed a low content of gold and silver in oxidation zone, 0-1 m (pH=2.9). The metals from oxidation zone accumulated in hardpan layer (pH=7.04) where the highest content can be seen (3-4 m depth). The reduction zone (pH=7.15) was located in 5-6 m depth interval.

Calcium (Ca), iron (Fe) and sulphur (S) content distributions in depth for H3 drill are presented in Figure 11.

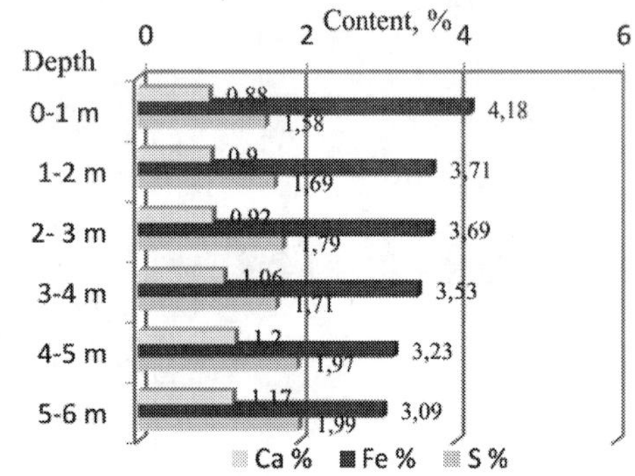

Fig. 11. Calcium, iron and sulphur content distributions in depth for H3 drill core.

In oxidation zone (pH=2,9) the lowest contents of Ca and S and the highest content of Fe can be noted. In this volume, the carbonate minerals are dissolved and the sulphides are oxidised.

The obtained data showed that the Ca content increases from 0.88% in oxidation zone (pH=2.9) to 1.17% in the reduction zone (pH=7.15) and the S content increases from 1.58% in oxidation zone (pH=2.9) to 1.99% in the reduction zone (pH=7.15). The Fe content decreases with depth from 4.18% in oxidation zone to 3.09% in reduction zone.

As Sasar tailing impoundment has high average grades of gold and silver, its reprocessing could be considered as an economically feasible possibility.

3.4 Static tests results

The static geochemical tests were performed on pulverized (<75μm) samples collected from H3 drill from 1 m, 3 m and 6 m depth.

3.4.1 Paste pH

Table 3 provides data related to paste pH and electric conductivity of samples. Samples D1, D3 and D6 were collected from H3 drill, at 1m, 3 m and 6 m depth respectively.

Table 3. Paste pH and electric conductivity.

Sample	Time, h	$pH_{1:2}$	$EC_{1:2}$(μS/cm)
D1	1	2.87	351
	12	2.93	638
D3	1	7.02	1080
	12	7.06	1466
D6	1	7.11	654
	12	7.18	1710

D1 sample showed a paste pH below 4 and an electrical conductivity > 20 μS/cm, meaning that it contains a high amount of dissolved salts. As a consequence, the D1 sample could be included in the category of samples with acid generation potential.

The values of pH over 7 obtained in the case of D3 and D6 samples indicate that those can either be potentially-acid-forming or non-acid forming, depending on their acid-base balance.

3.4.2 Acid -Base Accounting

Net acid production potential (NAPP) data, including acid neutralization capacity (ANC) and maximum potential acidity (MPA) and ANC to MPA ratio, is provided in Table 4.

Table 4. Acid - base account data for waste sample.

Sample	Fizz rate	HCl Vol (0.5M) mL	S (%)	ANC	MPA	NAPP	ANC /MPA
				(kg H_2SO_4/t)			
D1	0	4	0.46	-1.81	14.10	15.89	-0.13
D3	1	8	1.71	5.77	52.33	46.56	0.11
D6	2	20	2.05	62.3	62.73	0.43	0.99
Calculations: MPA=30.6 x %S NAPP=MPA-ANC							

A sample is classified as potentially acid forming (PAF) when NAPP has positive values. In addition, the ANC to MPA ratio of the samples were less than 1, which classified them as "possibly" or "likely acid generating" materials. [26] In our case study, the obtained net acid production potential (NAPP) values were positive, which indicated that the samples were potentially acid forming materials.

3.4.3 Single addition NAG test

The single addition net acid generating (NAG) test was used to assess the validity of NAPP predictions. Using NAG test in conjunction with ABA methods to classify the samples, can reduce the risks of misclassifying non-acid forming samples (NAF) as potentially acid forming (PAF) and PAF material as NAF. Table 5 shows ABA (NAPP) and NAG results of the samples.

Table 5. NAPP and NAG test results for D1, D3 and D6 samples.

Sample	NAGpH	NAG pH4.5	NAG	NAPP	ARD Classification
		(kgH_2SO_4/t)			
D1	3.17	3.43	6.56	15.885	PAF
D3	2.55	24.30	29.79	46.556	PAF
D6	2.70	11.95	17.05	0.430	PAF

A standard classification scheme [22] is used to present this data (Fig. 12), which compares NAG pH to NAPP and identifies non-acid forming samples from potentially-acid forming samples. Potentially acid forming (PAF), non-acid forming (NAF) and uncertain (UC) classification domains are indicated.

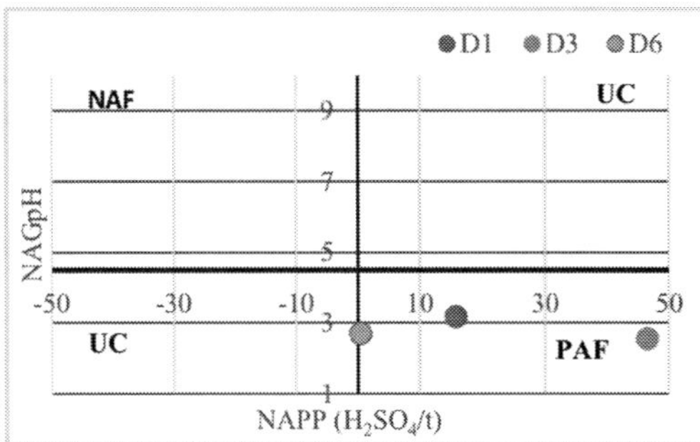

Fig. 12. ARD classification plot for a case study sample set.

A sample is classified as PAF when it has a positive NAPP and NAGpH < 4.5, and NAF when it has a negative NAPP and NAGpH ≥ 4.5. Samples are classified uncertain when there is an apparent conflict between the NAPP and NAG results, i.e. when the NAPP is positive and NAGpH ≥ 4.5, or when the NAPP is negative and NAGpH < 4.5. [22]

The NAPP results obtained compared with NAG results confirmed that samples would produce acidity upon exposure to oxygen and water.

4 Conclusions

Providing long term access to the available mineral resources therefore requires more focus on sustainable mining, both on research for environmentally sound mining and processing technology, as well as on the social and economic aspects of mining.

To counteract the steady increase in global demand for primary mineral resources and to reduce the negative social and environmental impacts associated with meeting this demand, it is necessary to increase material efficiency in manufacturing processes, to search for new substitute raw materials and to recycle materials more widely and more effectively, through technological innovations. Reprocessing of mining wastes, prioritizing those with acute environmental impacts, is a way to improve this metal recovery while protecting human being and the environment.

Tailings from mineral processing of polymetallic ore stored in Sasar (Red Valley) impoundments, located in the western part of Baia Mare (Romania), were characterized by grain size, elemental composition, elements distribution in depth and potential of ARD generation. The main conclusions related to this study can be summarized as follows:

-The results have indicated that tailings exhibit very high level of metal content, especially with Mn, Zn, Pb, Cu, exceeding the standards for soil, thus, gold mine tailings can be considered as a potential primary source of soil and groundwater contamination in mining areas

-The gold and silver content of the tailing samples were 0.45 g/t and 4.59 g/t, respectively, reprocessing of tailing could be considered as an economically feasible possibility;

-Comparison of NAG and NAPP results demonstrated the effectiveness of the NAG test in highlighting potential issues in ABA predictions;

-Both the modified ABA and the NAG test classified the samples into the category of 'potentially acid generating'.

This work was funded under the scope of the "3rd ERA-MIN Joint Call (2015) on Sustainable Supply of Raw Materials in Europe" by a grant of the Romanian National Authority for Scientific Research and Innovation, CCCDI – UEFISCDI, project: Improve Resources Efficiency and Minimize Environmental Footprint - REMinE, Contract no.13/2016. Coordinator: Luleå University of Technology (LTU), Sweden; Partners: Research and Development National Institute for Metals and Radioactive Resource, Romania, University of Porto, Portugal.

References

1. G. Morteani, J. P. Northover, *Proceedings of the NATO advanced research workshop on prehistoric gold in Europe*. Germany. 136. (Kluwer Academic Publishers 1993)
2. Gh.C. Popescu, Geol.and Seism. Inst. Bull. ASM. **N1**. (2015)
3. G. B. Lottermoser, *Mine wastes. Characterization. treatment and environmental inpacts*. 3rd Ed. Springer. 253-254. (2010)
4. C. Modoi, S.N. Vlad, I.C. Stezar, D. Manciula, A.C. Gagiu. S. Marginean. Environ Eng Man J., **10**, 43-51 (2011)
5. D.W Blowes. CJ. Ptacek. J.L. Jambor, C.G. Weisener in H.D. Holland, K.K. Turekian, *Treatise on geochemistry*. (Elsevier. Oxford.,149–204, 2003)
6. K.D. Nordstrom, C.N Alpers,. Rev Econ Geol **6A**, 133–160 (1999)
7. T. Rösner, A. van Schalkwyk. Bull Eng Geol Env.(Q Springer-Verla) **59**, 137–148 7 (2000)
8. H. Bouzahzah. M. Benzaazoua. B. Bussiere, B. Plante, Mine Water Environ. **33**, 54–65 (2014)
9. A.K. Kourtis, B. Gazea, A. Kontopoulos, Trans Inst Min Metall A **106**, A1–A8 (1997)

10 D.W. Duncan, A. Bruynesteyn, *Proceedings of the metallurgy soc AIME.* paper A79-29 (1979)

11. K. Lapakko, *Proceedings of the international land reclamation and mine drainage conference and 3rd ICAR.* Pittsburg. USA, **1**, 129-137 (1994)

12. J. Skousen, J. Simmons, L.M. McDonald, P. Ziemkiewicz, J Environ Qual, **31**, 2034–2044 (2002)

12. AA. Sobek, W.A. Schuller, J.R. Freeman, R.M Smith, EPA-**600**/2-78-054, 47-50 (1978)

13. P.A. Weber, J.E. Thomas, W.M. Skinner, R.S.C. Smart, Appl Geochem **19**, 687–694 (2004)

14. P.A. Weber, J.E. Thomas, W.M. Skinner, R.S.C. Smart. Can Mineral **43**, 1183–1192 (2005)

15. J.L. Jambor, J.E. Dutrizac, M. Raudsepp, Environ.Geol. **52**, 1019-1031 (2007)

16. A.S. Warwick, D.M. Stuart, R. Smart, *Proceedings od the 7th International Conference on Acid Rock Drainage* (ICARD) (2006)

17. I. Mac, *General Geography.* (in Romanian). Europontic Press. Cluj-Napoca, Romania (2000)

18. V. Mutihac Technical Press. Bucharest. Romania (1990)

19. F. Benciu, Did.and Ped. Publish. House. Bucharest. Romania (2007)

20. AMIRA P387A, I. W. Res.Inst. Env. Geochem. Int. Pty Ltd, **A-5** (2002)

21. J.G. Crock, B.F. Arbogast, P.J. Lamothe, In: G.S. Plumlee, M.J. Logsdon, *Reviews in Economic Geology*, **6A,** 264-287(Chapter 13) (1999)

22. R.W. Lawrence, P.B. Marchant. MEND Project 1.16.1b. CAMNET. Ottawa (1989)

23. J.L. Jambor, J.E. Dutrizac, M. Raudsepp. L.A. Groat, J.Environ Qual **32**, 2373-2378, (2003).

24. B. Dold, L. Fontboté, J. of Geochem. Expl.**74**, 3-55, (2001)

25. M.J. Brodie, L.M. Broughton. A. MacG. Robertson, *Proceedings of the second international conference on the abatement of acidic drainage,* **1 - 4**, Montreal, Canada. (1991)

Possibilities of use of waste sulfur for the production of technical concrete

Łukasz Grabowski[1,*], *Maciej* Gliniak[2], and *Daria* Polek[1]

[1]AGH University of Science and Technology, Faculty of Mining and Geoengineering, Department of Environmental Engineering and Mineral Processing, Mickiewicza 30 Av., 30-059 Krakow, Poland
[2]University of Agriculture in Krakow, Department of Technical Infrastructure and Ecoenergetics, Institute of Agriculture Engineering and Informatics, Faculty of Production and Power Engineering, Mickiewicza 21 Av., 30-120 Krakow, Poland

Abstract. The article presents the results of research that focused on the selected physical properties of sulphur waste stabilized in concrete. The influence of replacing part of the cement on the physical properties of concrete was studied in the article. The concrete mixtures was modified by pure waste sulphur in amounts of 10, 20 and 30wt% of the cement. The results show that using sulphur admixture can reduce concrete absorptivity and make concrete more watertight.

1 Introduction

The growing demand of the construction industry for specialized technical concretes contributes to the increasing emissions of carbon dioxide to the atmosphere. One way to reduce the scale of this phenomenon is to use in the concrete production process selected groups of industrial waste. The most commonly used additives are dust from coal burning and biomass burning [1] and used car tires [2-3].

Because of the increasing requirements for chemical resistance of concrete, many years of work on the addition of polymers improving their resistance to corrosion and mechanical strength have been underway. Waste sulfur is one example of a substance that easily polymerizes and improves the properties of technical concretes. The melting point of the sulfur is about 110 °C, flowing 150-400 °C and boiling above 400 °C. Thanks to these properties, it is easy to carry out the polymerization of waste by changing its alotropic variation and molecular weight [4-7].

The most popular method of sulfur polymerisation is boiling with water, calcium hydroxide and precipitation with chloride acid. The secondary known method of polymerisation is hydrolysis of sulfur admixture by alcohol solutions. The resulting substances are widely used in the ceramic, electrical and chemical industries as an insulating material. The use of sulfur polymers in technical concrete also improves its mechanical properties, such as strength about 45-50 MPa by compression. Also known are bacterial deterioration properties of bacteria and sunlight of sulfur concretes [4-7.

2 Materials and methods

The sulfur concrete consists of the waste sulfur and Portland cement CEM I 42.5 R [8] (from Cement Ozarow SA). The granular sulfur (99.9%) was bought from refinery in Poland. The aggregates were natural sand and gravel with the granulation according to the ASTM D 3515 [9].

To obtain the sulfur cement, the different percentages of elemental sulfur have been mixed in a mechanical stirrer in sand bath. Then the mixture was heated to 150 °C and flurried for a 1 hour. Simultaneously the aggregates had been heated to 105 °C in laboratory dryer. Afterward the filler was added to the cement and then mixed. Next step was adding mixture of the aggregates and blending. The final product was concrete that was casted in the molds that were heated to 150 °C at the end vibration was done on a vibration table.

Table 1. Composition of the concrete mixes (weight per m^3 of concrete, kg).

Reference	Cement	Sulphur	Water	Sand	Gravel
0	352	0	176	676	1205
10	316.8	35.2	176	676	1205
20	281.6	70.4	176	676	1205
30	246.4	105.6	176	676	1205

The experiment includes four series of samples. According to ACI 548.2R-93 [10] for the maximum aggregate size of 12.7 mm, the percent by weight of the cement (sulfur) has been used as 0, 10, 20 and 30% (Table 1). From each batch three 10x10x10 cm cubic samples were casted. The samples were dried in the ambient temperature. After drying its bulk density (ρb), watertightness (w), absorptivity (a) and total porosity (Pt) was taken [11].

* Corresponding author: lukgrabo@agh.edu.pl

DOI:10.1051/e3sconf/20171801032

3 Results

The concrete with sulfur addition has higher bulk density than control probe. Mixes with 20 and 30 wt% of sulphur addition have similar values of density. The higher bulk density indicates on higher watertightness and lower values of absorptivity and total porosity. The concrete with 30 wt% of sulphur admixture proved the most watertight and least porous (Table 2).

Table 2. Average values of bulk density (ρ_b), watertightness (w), absorptivity (a), and total porosity (P_t) for all of the samples.

Reference	ρ_b	w	a	P_t
0	2.11	69.8	10.3	30.2
10	2.30	93.9	1.12	6.1
20	2.41	95.3	0.36	4.7
30	2.49	96.8	0.1	3.2

4 Summary

Obtained research was a successful experiment, which could develop the process of depth impregantion of concrete surfaces. The sulfur concrete surfaces were more resistant to water absorption than concrete made from Protland cement. Presented test results confirmed possibilities of sulfur concrete usage as a surface protector form for corrosive environment.

5 Conclusions

It was found that replacing part of the cement with waste sulfur in technical concrete changes soma of its physical properties. This change is hardly dependent on the concentration of sulfur. Based on the test result presented in the paper, the following general conclusions could be formulated:

- The obtained results showed that the physical properties of concrete are influenced with sulfur admixture. The watertightness of 10 wt% sulfur addition is similiar to Portland cement. For other mixtures the watertightness was higher than concrete without sulfur admixtures ;
- The absorptivity of concrete containing 20 and 30 wt% of sulfur were lower than a reference probe;
- The sulfur admixture decreased the total porosity.

References

1. E. Kwon, J. Ahn, B. Cho, D. Park., Constr. Build. Mater. **83**, 174-180 (2015)
2. E. Güneyisi, M. Gesoglu, N. Naji, S. İpek. Arch. Civ. Mech. Eng. **16**-1, 9-19 (2016)
3. M.Y. Cheng, M.T. Cao., J. Civ. Eng. Manag. **22**-5, 711-720 (2016)
4. M. Książek. Constr. Build. Mater., **68**, 558-564 (2014)
5. M. Książek. Engineering **62**(62C), 137-142 (2014)
6. M. Książek. Engineering, **58**(58C), 378-385 (2013)
7. M. Książek. Experimental research on the surface protection of concrete by polymer sulfur composite, Mag. Concrete Res. **64**(10), 945-955 (2012)
8. EN 197-1:2011. Cement. Composition, specifications and conformity criteria for common cements
9. ASTM D 3515. Standard Specification for Hot-Mixed, Hot-Laid Bituminous Paving Mixtures
10. ACI 548.2R-93. Guide for Mixing and Placing Sulfur Concrete in Construction
11. EN 1936:2010. Natural stone test methods. Determination of real density and apparent density, and of total and open porosity

E3S Web of Conferences **18**, 01033 (2017) DOI:10.1051/e3sconf/20171801033
MEC2017

Testing the possibility of leaching salt debris obtained from underground excavations

Piotr Pierzyna[1], *Marcin* Popczyk[1], and *Tomasz* Suponik[1*]

[1]Faculty of Mining and Geology, Silesian University of Technology, Gliwice, Poland

Abstract. For many years, the Salt Mine in Bochnia has conducted the extraction and restructuring of underground working to obtain new touristic routes. The works generate a large amount of rock salt material, colloquially called "salt debris", containing a large part of silty minerals. Due to the large concentration of NaCl which constitutes a threat to the environment, the salt debris may only be stored at hazardous material dumps. This is why the mine works towards the industrial use of the material. Currently, the debris is used to produce a stabilizing backfill mixture that is used for filling the liquidated "Moszczenica" shaft. The liquidation shall be completed in the beginning of 2018 and to continue the management of the salt debris, the mine plans to leach it in a newly constructed system and the saturated brine obtained in the process shall be introduced to underground workings. The paper presents the results of semi-industrial tests of leaching intensity of salt debris, depending on the graining and the time taken by the process. The tests were conducted for three grain classes of the salt debris (0-6, 0-10 and 0-20 mm), at a temperature of 15°C. The kinetic models of the process has been presented for the obtained results. The obtained results shall be used for the optimization of the operation of the future industrial system and shall be treated as input data for the automation control system.

1 Introduction

Within the process of revitalization of tourist routes and creating new underground heading elements, The Salt Mine in Bochnia obtains large amounts of waste material in the form of salt debris.

In case of active mines, the salt debris is successfully used to liquidate old and abandoned workings by backfilling [1]. In the inter-war period, in "Kałuszy" mine, wastes from the processing plant consisting of solid parts such as halite, anhydrite and silts (salt debris) as well as leach, served the production of a 1:2 mixture (solid parts:leach) that was used for the liquidation of underground workings [2]. As of today, the Salt Mine in Bochnia is not active and there are no possibilities to obtain new locations (abandoned workings) for depositing the debris in such a form. In special cases, crushed salt debris may also be used as a filling material, packing the underground waste dumps [3, 4]. German experience has exhibited that the material which constitutes insulating seals for radioactive and hazardous wastes dumps may also be based on the anhydrite salt debris. [5]. Due to the lack of underground waste dumps in Poland, this method of management of the salt debris may not be considered in the nearest time. An alternative for the above examples of management of salt debris, may be constituted by storing it on surface dumps. Due to the large concentration of the NaCl compound, however, such storage constitutes a hazard to the natural environment and should be stored at hazardous wastes dumps. Considering the above and the current possibilities of the "Bochnia" Salt Mine, the only possible method of management of the salt debris is its use in mining shaft backfilling technologies. For many years, this method was used in case of the completed liquidation of the "Siedlec" shaft and the pending liquidation of the "Moszczenica" shaft. The concept of the liquidation consists in filling the shaft tube with the backfilling mixture comprised of salt debris (Fig. 1), brine and hydraulic binding agent with proper characteristics and strength parameters [6, 7].

Fig. 1. Salt debris stored at a storage site by the "Moszczenica" shaft. Own source.

* Corresponding author: tomasz.suponik@polsl.pl

DOI:10.1051/e3sconf/20171801033

The works are conducted in line with the assumed shaft liquidation plan, approved by the Mining Authority and financed by the Fund for Environmental Protection.

The use of the salt debris for the liquidation of the "Moszczenica" shaft will continue until the end of 2017. After that term, the Mine will face a problem of how to manage the salt debris that will continue to be retrieved from underground headings in a manner that would be safe for people and the environment. One of the possibilities to solve the problem is the method of leaching the salt debris proposed by the authors of the paper. The method allows to obtain a saturated brine which could be introduced to the inactive salt chambers of the mine. The process of introducing the brine to the rock mass is currently conducted by the mine, but brine pumped out from underground workings that is subsequently transported with water carts to the place of application is used for that purpose. The technology proposed by the authors shall allow for:

- ensuring the maximum use of the salt debris obtained in the revitalization works of the underground headings,
- eliminating the storage of the salt debris at surface dumps and thus eliminating the adverse impact on the environment,
- management of the produced saturated brine using the inactive underground workings.

Such a solution has not yet been applied in Poland and the concept should be concerned innovative. Currently the mine is at the stage of applying for financial measures from the National Fund for Environmental Protection for the construction of the salt debris leaching system in line with the technology described above and the investment is to be completed by the end of 2017.

The aim of the study was to optimize the process of leaching salt debris in terms of grain size and process times. The brine from the installation of leaching will be directed through the existing well into the saline excavations which are isolated by natural rock layers. The condition and properties of the excavations and the surrounding rock allow for the safe functioning of the reservoir.

The paper presents the laboratory tests of possibilities of leaching salt debris, conducted on a semi-industrial scale at variable process parameters such as graining and leaching time. The saturation of the brine was determined by using two methods: indirectly, by measuring the density and by chemical analysis of the chlorides.

2 Methods and materials

2.1 Physical characteristics of the salt debris

The graining of the salt debris is highly variable. The grain-size analysis presented in Fig. 2. exhibits that approximately 50 % of the salt debris grains corresponds to the size of approximately 16 mm, while grains above 100 mm constitute approximately 10%. The petrographic composition of the salt debris exhibits that it is composed mostly of three types of rocks, namely: the NaCl salt compound (30-50 %), silty rock (40-50 %) and anhydrite (5-15 %).

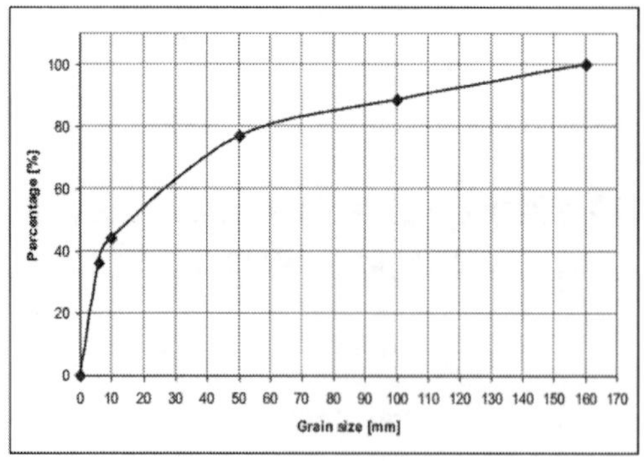

Fig. 2. Grain-size analysis of the salt debris.

2.2 Leaching test facility

The salt debris leaching tests facility is located at the technological hall owned by the Institute of Mining. The system is equipped with measurement and control devices such as: densimeter, flowmeters, temperature sensor and level sensors Aplisens S.A. The diagram of the installation is shown in Figure 3.

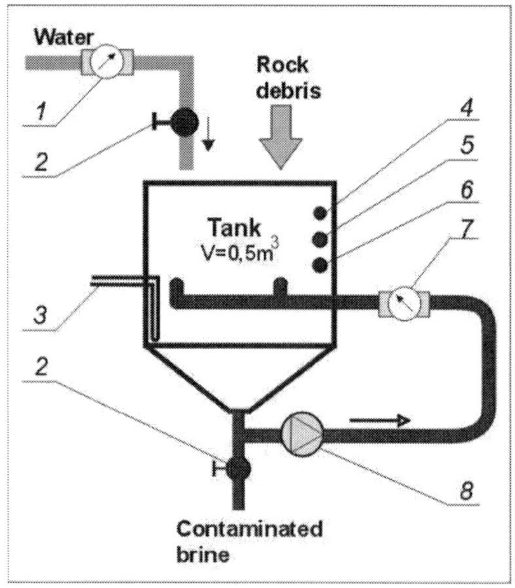

Fig. 3. The diagram of installation for leaching salt debris (1-Flowmeter water, 2-Valve, 3-Heater, 4-Densimeter mixture, 5-Level sensor, 6-Temperature sensor, 7-Flowmeter mixture, 8-Pump).

Additionally, the tank is equipped with a heater allowing for the operation in temperature higher than the temperature of the supplied water. The diameter of the pipe fitted in the system (100 mm) fully correspond to the actual diameters that may be fitted in the future industrial system. The system operates as a closed cycle and is equipped with a PH65 Powen's rotary pump, which serves the mixing, pumping and exchange of the mixture in the tank. The piping is equipped with a series of flap valves allowing for the control of the flow through the selected diameters of the pipes.

The buffer of the system is constituted by an approximately 0.5 m^3 tank, to which fresh water and salt debris are introduced. The view of the facility and the control system has been presented in Figs. 3 and 4.

Fig. 4. General view of the salt debris leaching system.

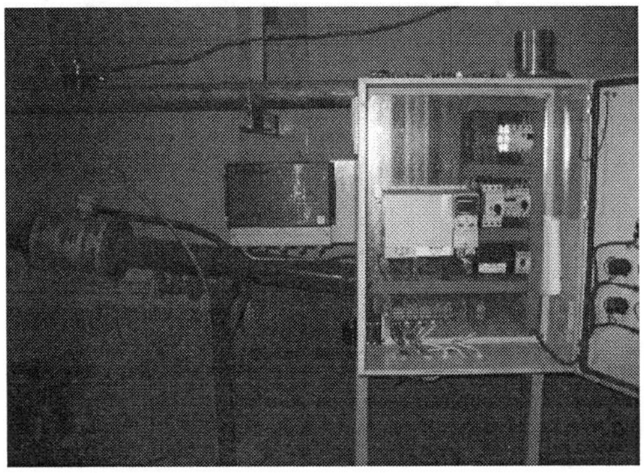

Fig. 5. The measurement and control device.

2.3 Research methods

The following research methods have been assumed in the semi-industrial scale tests. As the first step, municipal water at a temperature of approximately 12°C was introduced to the tank. The water was subsequently heated up to the temperature of 15°C±0,5°C that was assumed in the tests. The heating process (stabilization of the leaching temperature) was conducted during the operation of the pump which pumped the water in the closed-cycle of the system.

After stabilizing the temperature of the entire system, the salt debris was introduced to the tank in the mass ratio of debris to water 0.8: 1. This enabled the brine to reach saturation at a density of 1315 ± 2 g/dm^3, in laboratory conditions. This value has been reached in the initial laboratory tests and corresponds to the density of saturated brine at a temperature of approximately 5°C, contaminated by the post-leaching deposits consisting of mainly a clay.

The tests were conducted for three grain classes: 0-20mm, 0-10mm, and 0-6mm. The leaching temperature was set at 15°C. During the course of leaching, the brine density in the tank was measured at fixed time intervals and it was sampled for laboratory tests to determine the level of chlorides. The density was measured by a densimeter placed in the tank. Every other test was verified by a pycnometer measurement in accordance to PN-EN 1097-7:2001. The obtained pycnometric densities were correlated with the readings of densimeter placed inside the system, which has been previously calibrated. Chloride level analysis was based on PN-ISO-9297.

The temperature, pH and the electrolytic conductivity were determined using electrometric methods (with the following meters: PORTAMESS 913 pH with a SenTix 41 combination electrode; PORTAMESS 913 Cound). Also, the following determinations were made with regards to water:

- Concentrations of Cr^{6+}; SO_4^{2-}: UV-Vis Spectrophotometer DR5000 HachLange,
- Concentrations of Ba, Cr, Zn, Co, Mg, Mn, Cu, Ni, Ca, Fe: ICP-AES using the JY 2000 spectrometer,
- Carbonate and total hardness: the analyses were made using a titration with ethylenediamineteraacetic acid (EDTA).

2.4 Mathematical description

Determining the kinetics of the leaching of salt solution from salt debris is essential in the modeling and process design on industrial scale. The initial concentration of Cl - ions in the aqueous solution used for kinetics testing was 0.084 g / dm3. The tests were carried out at 15 ° C.

The changes in the concentration of chlorides as a result of the leaching process were calculated in accordance with the equation:

$$q_t = \frac{(C_0 - C_t)\cdot V}{m} \qquad (1)$$

where :

q_t – the amount of chlorides released from the mass of salt debris, mg/g,

C_0 – the concentration of chlorides in saturated solution (187.15 mg/dm^3), mg/dm^3,

C_t – The concentration of solution saturation by chlorides (difference in chloride concentrations in saturated solution (187.15 mg/dm^3) and in unsaturated solution after the certain time of leaching), mg/dm^3,

V – the volume of water used in the test, dm^3,

m – the mass of salt debris used in the test, kg.

In order to determine the mechanism of salt debris leaching and the factors limiting its speed, the two nonlinear kinetic models were used being the pseudo-1st-order (eq. 2) and pseudo-2nd-order (eq.3) models.

$$q_t = q_e(1 - exp(-k_1 t)) \qquad (2)$$

where:

q_t , q_e – The amount of chloride (mg) released from salt debris (kg) to solution (dm^3) in time t (min) and after reaching saturation, respectively, (mg/g)

k_1 - rate coefficient of the 1st order reaction (min^{-1})

 DOI:10.1051/e3sconf/20171801033

$$q_t = q_e \frac{q_e k_2 t}{1 + q_e k_2 t} \quad (3)$$

where:

k_2 - rate coefficient of the 2nd order reaction (g ·mg^{-1} ·min^{-1}).

3 Test results

3.1 Kinetics of the leaching process

On the basis of the conducted tests concerning the leaching of salt debris on the semi-industrial scale with variable process parameters, the following measurements were obtained (Table 1). In case of brine contaminated with clay, which has a significant effect on the density of the mixture, the measurement of its density may be treated as an indirect measurement indicating the density of the brine itself. Before that, it is necessary to carry out laboratory tests of the brine mixture and clay density and the corresponding density of brine, which allows to determine the density of the mixture in saturated state. In the case of systems operating on an industrial scale the measurement of brine saturation using a densimeter under the conditions of heavy clay content in the salt debris may be appropriate, but in the laboratory conditions where greater accuracy is required, it is necessary to measure the level of chlorides.

Table 1. Leaching status of salt debris of different graining at 15°C in relation to time.

Time debris leaching, min	Densities and level of Cl^- chlorides for different grain size					
	0-6mm		0-10mm		0-20mm	
	ρ, g/dm^3	Cl^-, g/dm^3	ρ, g/dm^3	Cl^-, g/dm^3	ρ, g/dm^3	Cl^-, g/dm^3
0	1002	0.084	1002	0.084	1002	0.084
1	1295	142.15	1288	121.80	1256	119.98
3	1309	185.77	1302	181.86	1277	182.11
5	1316	186.50	1313	185.84	1293	183.29
7	1314	186.45	1316	186.87	1304	185.08
9	1314	186.39	1315	186.57	1315	186.31
11	1315	186.35	1314	186.56	1316	186.95
13	1314	186.48	1314	186.53	1316	186.79
15	1316	187.15	1316	186.43	1317	186.74

The leaching process was carried out in a closed circuit at a site with a buffer tank with a capacity of approximately 0,5m^3 using a PH-65 pump with a capacity of approximately 30-35m^3/h, which allowed for a one-fold change in tank volume per minute. As can be seen in Table 1, leaching time for the brine from salt debris also decreases as the debris grain size decreases. The shortest leaching time of about 5min was observed in the sample with 0-6mm debris graining. The longest leaching time of about 9min was observed in the sample of 0-20mm debris graining. The leaching process using the centrifugal pump is characterised by a high intensity from the beginning. Already in the first minute we obtain the mixture density in the range from 1256 g/dm^3 for graining of 0-6 to 1295g/dm^3 for graining of 0-6mm. These densities correspond to approximately 95-98% of full brine saturation. After that the intensity of the process decreases and in the next few minutes the mixtures very slowly increase in saturation to its maximum level.

Based on the results obtained, it should be concluded that the optimal conditions for leaching salt debris and for reaching saturation of the solution occur at a grain size of 0-6mm and a process duration of 5 min.

As it may be noted in Table 1, the level of Cl^- chlorides determined in the laboratory environment increases along the increase of the mixture density. The concentration of chlorides in saturated brine was ca. 187 g/dm^3.

3.2 Mathematical description of kinetics

Figures 6, 7, 8 present the kinetics of the release of chlorides into the aqueous solution from the salt debris. The results show that the solutions were saturated after approximately 5, 9 and 11 minutes, for grain classes 0-6, 0-10 and 0-20 mm, respectively. The fastest leaching proceeded within first 60 seconds of the process. Later the aqueous solutions have reached slowly the saturation.

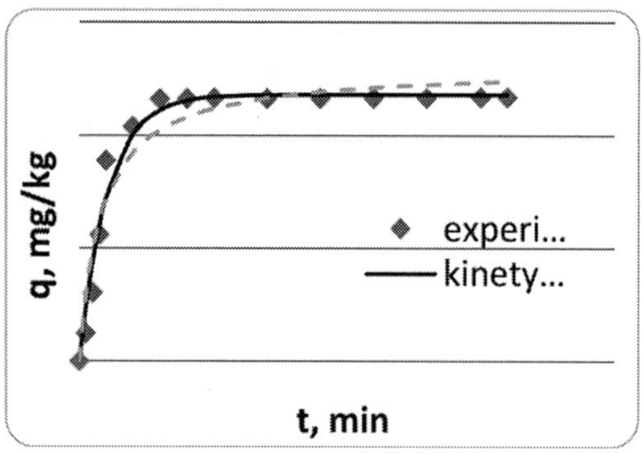

Fig. 6. The kinetics of the release of chlorides from the salt debris of grain size 0-6 mm to solution, T = 15°C, C_o (saturated solution) = 187.15 mg/dm^3.

The values of the rate coefficients (k) and other parameters of the kinetic models as well as the determination coefficient values are shown in Table 2. The highest value of R^2 was the pseudo-1st-order model. The rate coefficients of this model were 0.9484, 0.7484, 0.7198 1/min, for grain classes 0-6, 0-10 and 0-20 mm, respectively.

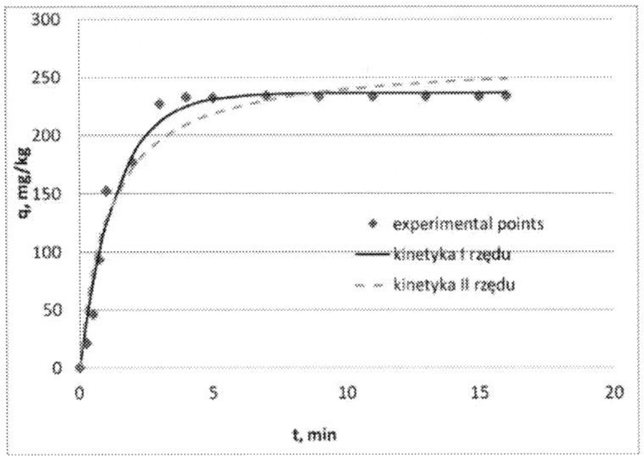

Fig. 7. The kinetics of the release of chlorides from the salt debris of grain size 0-10 mm to solution, T = 15°C, C_0 (saturated solution) = 187.15 mg / dm³.

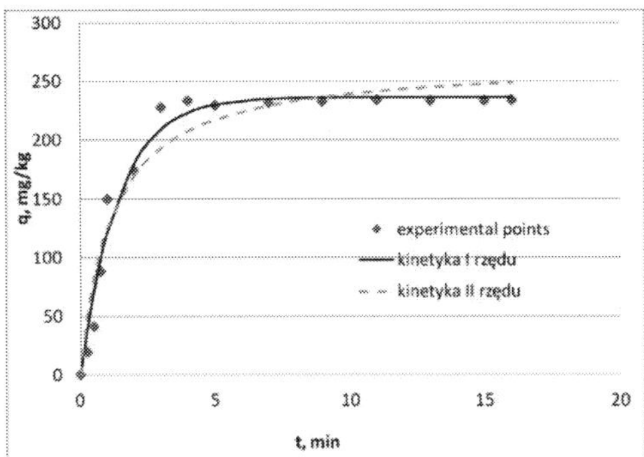

Fig. 8. The kinetics of the release of chlorides from the salt debris of grain size 0-20 mm to solution, T = 15°C, C_0 (saturated solution) = 187.15 mg / dm³.

Table 2. Parameters for kinetic models.

grain size, mm	The parameters of the kinetic model of:					
	pseudo-1st-order reaction			pseudo-2nd-order reaction		
	k_I, min^{-1}	q_{eI}, mg·kg^{-1}	R^2	k_{II}, kg·mg^{-1}·min^{-1}	q_{eII}, mg·kg^{-1}	R^2
0-6	0.9484	236.1	0.9950	0.0047	230.3	0.9885
0-10	0.7484	236.5	0.9955	0.0035	265.1	0.9893
0-20	0.7198	236.5	0.9947	0.0033	266.2	0.9883

3.3 Brine chemical tests

The main purpose of leaching of salt debris is to obtain saturated brine, which can be introduced to the underground workings through the hole from the surface. In order to employ this solution it is necessary to obtain – in the process of leaching of salt debris- the saturated brine. An additional requirement is that the brine obtained from the system and introduced into the rock is characterised by a similar chemical composition to the brine currently pumped from the workings in the other parts of the mine Bochnia.

The results of the studies presented in Table 3 were performed for brine obtained under optimum leaching conditions, i.e. for particle size and process time of 0-6 mm and 5 min, respectively. As can be seen in this table, pH of brine and post-leaching brine was 7.06 and 6.75 respectively, and the electrolytic conductivity and concentration with chlorides and sulphates were 180.4 mS/cm, 185.133 g/dm³, 3.750 g/dm³ and 178.3 mS/cm, 187.156 g/dm³, 1.750 g/dm³ respectively. Both types of brine are characterised by low carbonate hardness and very high concentration of calcium and magnesium. Low levels of barium, zinc, copper, manganese, and iron were also identified, with the concentration of iron in the brine probably being higher than presented in Table 3, as the tests were conducted a few days after obtaining the samples, and a brown hydroxide sediment was observed at the bottom of the tank. The concentration of other analysed metals was below the limit of quantification.

In summary, both analysed types of brine exhibit the same chemical composition and are characterised by very high concentration with chlorides and very high non-carbonate hardness represented by sulphates and chlorides, mainly of calcium and magnesium.

Table 3. Test results of brine from Salt Mine in Bochnia workings and brine obtained as a result of salt debris leaching.

Substance/parameter, unit	Brine from Salt Mine in Bochnia	Brine obtained from salt debris leaching
pH	7.06	6.75
Conductivity, mS/cm	180.4	178.3
Cl^-, g/dm³	185.133	187.156
SO_4^{2-}, g/dm³	3.750	1.750
Carbonate hardness, mmol (Ca+Mg)/dm³	1.41	0.90
Total hardness, mmol (Ca+Mg)/dm³	77.90	65.85
Cr^{6+}, mg/dm³	<0.01	<0.01
Temperature, °C	19.2	19.6
Ba, mg/dm³	32	22
Cr, mg/ dm³	<0.01	<0.01
Zn, mg/ dm³	0.94	0.72
Co, mg/ dm³	<0.006	<0.006
Mg, mg/ dm³	1021.44	792.83
Mn, mg/ dm³	2.684	1.678
Cu, mg/ dm³	0.25	0.34
Ni, mg/ dm³	<0.006	<0.006
Ca, mg/ dm³	1426.84	1322.64
Fe, mg/ dm³	0.08	1.08

4 Conclusion

The system used for testing operated in a hydraulically closed loop using a buffer tank of approximately 0.5m3, a set of connection pipes, and a PH-65 pump with a capacity of approximately 30-35m³/h, allowing for one-fold change in tank volume per minute (approximately). At fixed intervals during the testing the density of the mixture and the level of chlorides were measured, and on that basis the degree of brine saturation was determined. Based on the results of the tests, it can be concluded that:

1. the leaching system used is characterised by a high leaching efficiency, which allows to obtain approximately 95% saturation level only after one minute,
2. the maximum density achieved during the leaching of salt debris was within the range from 1314 to 1317g/dm^3. Conducted tests on the levels of Cl^- indicated their value at approximately 187.15 g/dm^3 at these densities. Comparing this value to the results of levels of Cl- in the amount of 185.13 g/dm3 in saturated brine supplied from the mine workings, it must be concluded that in the system the mixture achieved the state of full saturation,
3. with the increase of graining of salt debris the time required to saturate the brine increases,
4. the pseudo-1st-order model described best the kinetics of the release of chlorides from the salt debris to aqueous solution. The value of R2 ranged from 0.9947 to 0.9955. The rate coefficients of this model were 0.9484, 0.7484, 0.7198 1/min, for grain classes 0-6, 0-10 and 0-20 mm, respectively.
5. the optimal conditions for leaching salt debris and for reaching saturation of the solution occur at a grain size of 0-6mm and a process duration of 5 min. The brine obtained under such conditions had a similar chemical composition to the brine from the mine. The brine from the installation of leaching will be directed through the well into the saline excavations which are isolated by natural rock layers. The condition and properties of the excavations and the surrounding rock allow for the safe functioning of the reservoir.

In summary the results of leaching time tests, from the point of view of their applicability in the design of future industrial systems constructed by mines, are positive and fully acceptable by the mine. The debris leaching technology presented by the authors will allow to:

- ensure the full use of salt debris obtained from revitalisation works of underground workings,
- eliminate the storing needs of salt debris in landfill sites, therefore cancelling the negative environmental impact,
- manage the saturated brine produced in inactive underground workings.

This solution has as of yet never been applied and the idea should be seen as innovative. Currently the mine is in the process of applying for funding from the National Environmental Protection Fund for the construction of a salt debris leaching system according to the technology presented above, and the commencing of the investment is scheduled for the end of 2017.

References

1. K. Poborska-Młynarska, Przeg. Górn. **12**, 140-146 (2015)
2. K. Poborska-Młynarska, W. Andrusikiewicz, Geologia **36**, 3, 273–285 (2010)
3. C. Ludeling, W. Minkley, ARMA **2**, 1395–1402 (2014)
4. M. B. Dusseault, D. Mraz, J. Unrau, C. Fordham, ARMA **77**, 313–322 (1985)
5. H. Mischo, Glueckauf-Forschungshefte **63**, 126–138+118 (2002)
6. J. S. Mahlaba, E. P. Kearsley, R. A. Kruger, Min. Eng. **24**, 923–929 (2011)
7. F. Plewa, M. Popczyk, P. Pierzyna, T. Migdas, Polit. Energ. **14**, 317–323 (2011)

 DOI:10.1051/e3sconf/20171801034

The influence of ultrasounds on the process of uranium desorption

Eugenia Panturu[1], *Antoneta* Filcenco-Olteanu[1,*], *Aura Daniela* Radu[1], and *Marius* Zlagnean[1]

[1]E1Research & Development National Institute for Metals and Radioactive Resources, 70 Carol I Bvd., 020917, Bucharest, Romania

Abstract. In the present paper, we have studied the intensification of the desorption process of the complex anions of uranium U(VI), from ion exchange resins under the action of the ultrasonic field. The dynamics of uranium desorption from the anionic resins Purolite A600U/3472 and PM 611 was analyzed, both in standard conditions and with ultrasounds. The use of ultrasounds leads to the substantial growth of desorption kinetics and of the uranium recovery degree. The tests for stability of the resin grains size, mechanical resistance and the capacity of loading with uranium, after the repeated action of the ultrasounds field, emphasized the physical-chemical stability of resins used in the experiment (loading capacity of untreated resin 30.36 mgU/g resin and treated in continuous ultrasonic field 30.85 mgU/g resin).

1 Introduction

The industrial application of ionic exchange is widespread in areas starting from water softening, wastewater treatment and analytical chemistry to the purification of pharmaceutical products and of precious and radioactive metals [1-4]. The use of anions-exchange resins for uranium recovery from uranium-containing solutions is possible if the resin provides a high selectivity for uranium complexes, allows the desorption of the retained complex with simple agents, which give rich elutes and has the ability of multiple reuse [5-11].

As the ion desorption from the resin is the reverse process of adsorption, the physical and chemical factors that inhibit adsorption will favor desorption and, respectively, those who favor adsorption will inhibit desorption. Although desorption can be improved by increasing the temperature or by using concentrated extractants solutions, these measures may not be efficient from the costs' point of view. The use of ultrasounds as a means of enhancement of property transfer processes is reported in the specialty literature since the 1960s up to the present.

Thus, the Russian researchers [12] have shown that both the ultrasonic agitation and the mechanical agitation can accelerate the ionic exchange equilibrium, as the authors noticed the ultrasonic agitation is much more efficient than the mechanical one. Later on, it was noticed that the ionic-exchange equilibrium reached by applying an ultrasounds field lasts more than the one obtained in the case of mechanical agitation. Cheng and Wang [13] claimed that energy of the shockwave that occurs when the cavitation bubble collapses would interrupt film diffusion of the ions. Other studies [14] have shown that in the presence of an ultrasonic field, the capacity of ion-exchange resin increases with approx. 15-20%, and the rate of adsorption likewise increases. In other studies, [15, 16] it has been proved that the ultrasounds' effect is not limited only to the surface layer of the resin particle. In the presence of an ultrasound field, during cavitation a side effect that leads to an exposure of micro pores from the sorbent particles is produced. At the same time, the pressure pulsation induced by cavitation can facilitate the penetration of the solution in the micro capillaries relatively inaccessible from the sorbent particle.

Liu et al. [17], Mukherjee et al. [18] have demonstrated that ultrasonication can dislocate the organic matter adsorbed on a specific area of aquatic sediments and, consequently, to significantly increase the bioavailability of the sorbent.

Newman et al. [19] have conducted similar studies on granulated pieces of brick impregnated with copper oxide as a model for the decontamination of polluted soils. They noticed a decrease by 40% of the copper content through the washing of the impregnated pellets with a stream of water, in an ultrasonic tank at a frequency of 20 kHz.

Qin et al [20] investigated the effect of ultrasound on the desorption of phenol from impregnated resins and noticed. That ultrasound had ''spot energy effects'' which would enhance the process. Likewise, Rege et al. [21] showed that ultrasounds can lead to the increase of phenol adsorption rate from activated carbon and polymeric resins.

Küncek and Sener [22] used ultrasonication as a pre-treatment process to increase the adsorption capacity of sepiolite, a natural clay-based adsorbent, used to remove the methylene blue. The physical effect of the ultrasonication (20kHz, 5 hours) lead to a significant increase of the sepiolite specific surface area, which led to a higher retention capacity.

Recently [23] an experimental method has been developed for the efficient removal of methylene blue in

* Corresponding author: antonetafilcenco@yahoo.ro

the presence of ultrasounds (40 kHz), using the process of adsorption on activated coal impregnated with gold nanoparticles.

In recent years the research regarding the application of ultrasounds in hydrometallurgy, in dissolution operations [24-37], ionic exchange [38-47], and solvent extraction [48-53] has expanded.

Balasubrahmanyam et. al [54] have been performed experiments for leaching of uranium from MgF_2(byproduct from the uranium ore recovery process) in presence and in absence of ultrasound, using various conditions of leach acid concentration, different energy dissipation rates and different MgF_2 particle size distribution. It was obtained that the overall recovery has been increased by the application of ultrasonication and the main advantage of using ultrasound was the decreasing of the leaching operation time.

Kalsi et. al [55] have been studied the recovery of uranium form fluoride matrix under the influence of ultrasonication (22 ± 3 kHz and 150 W) and it was obtained a 76 – 91% removal of uranium form MgF_2 slag. The major advantages in this case are minimizing labor, remotisation of operations to reduce exposure, use of less nitric acid and the reduction in the reaction time.

Research on sono-chemical leaching of uranium [56] in nitric acid and sulfuric acid media has been carried out to understand the effect of ultrasonication on leaching process. Thus, it has been observed that the leaching rate under the influence of ultrasound is higher with low acid concentration of nitric acid, and it is high at a high leach acid concentration in case of sulfuric acid, compared to mechanical agitation, due to the faster rate in presence of ultrasonication of oxidative conversion mechanism of uranium tetravalent to hexavalent form.

Bikram et. al [57] conducted studies regarding the application of ultrasonication in density determination of uranium loaded organic phase in solvent extraction. Their method using dual frequency based ultrasound density measurement of uranium loaded organic samples in solvent extraction process can be used for critical density determination at extraction and reprocessing in the fuel cycle.

Hamadaoui and Naffrechoux [58] in their studies have been investigate the mechanism of ultrasonically enhanced desorption of 4-chlorophenol from granular activated carbon. Experiments of desorption were made in absence and in presence of ultrasonication at 516 kHz of different intensities and the conclusion was that the use of ultrasound field improves both the amount and the rate of desorption. Desorption increases with increasing temperature and ultrasound intensity, the mechanism of ultrasonically enhanced desorption in due both to the thermal and non-thermal effects of ultrasound field.

Korean researchers [59], studying the batch absorption of uranium from acidic solutions on ion-exchange resins, have discovered that when applying simultaneously the mechanical agitation and an ultrasonic field, the uranium adsorption speed on the ion-exchange resin increases because of both the increase of collision frequency, and the mixing intensification due to cavitation.

These studies suggest that ultrasonic radiation can be an effective way to improve the adsorption/desorption of ion species from the ion-exchange resins.

The aim of the research presented in this paper is to determine how ultrasounds help improve uranium desorption from ion-exchange resins type Purolite A 600U/3472 and PM 611.

2 Experimental

2.1 Reagents and materials

The reagents used were chemically or analytically pure (to perform the analysis) and used without additional treatment. The synthetic uranium solution was prepared from uranyl sulphate (UO_2SO_4) in distilled water, and for the pH correction was used anhydrous sodium carbonate (Na_2CO_3). The chemical composition of the synthetic solution is: U = 35.37 mg/L, Na_2CO_3 = 3.18 g/L, $NaHCO_3$ = 0.504 g/L and pH = 9.5.

Uranium desorption was made by using a chloro - sodic solution having the composition: 100 g/L sodium chloride (NaCl), and 10 g/L sodium carbonate (Na_2CO_3).

The resin Purolite A600U/3472, supplied by the company Purolite SA, Victoria, Romania (Purolite Co. Philadelphia, USA) is a strong basic anionite, with gel structure, specially designed for the recovery of soluble anionic complexes of uranium from acid or alkaline media. Structurally, this anionite is a polystyrene divinylbenzene polymer, with excellent mechanical integrity and resistance to stress.

Resin PM 611, provided by the company Suzhou Bojie Resin Technology Co. Ltd China,is a strong basic anionic exchanger, with a gel-type structure type 1, designed to separate the soluble anionic complexes of uranium in acidic or alkaline medium.

2.2. Experimental methodology

The resins were activated for the adsorption of uranium through sequential treatment with sodium hydroxide (NaOH) 4% and hydrochloric acid (HCl) 4%, using Berzelius glasses and magnetic stirrer type IKA RH-KT/C (manufactured by WERKE, Germany); the resin washing was made with distilled water (pH = 3 – 3.5) so that the resin remains in the Cl^- form.

The initial adsorption process was carried out under dynamic conditions, adsorption – desorption cycles conducted in a fixed bed, in four identical glass columns, with a length of 12 cm and an inner diameter of 1.5 cm where were placed 5mL (3.5 g) resin (Purolite A600U /3472 and PM 611). The columns were provided at the ends with frits in order to prevent losses of resin through intromission in effluent. The uranium solution flow rate through the ion exchange resin was 5 BEV/h (BEV = Bed Equivalent Volume of resin in the column). After uranium loading, the resin was washed with water and stored for the desorption tests. It is considered that the resins are fully loaded when the uranium content in the effluent equals the influent concentration. The loading capacity of the Purolite A600U/3472 resin was 30.36 mgU/g_{resin}, while that of the PM 611 resin was 25.64 mgU/g_{resin}.

Desorption was carried out both under standard conditions and in a continuous ultrasound system.

For carrying out the desorption tests in fixed bed in continuous ultrasound system, the column loaded with resin was immersed in the ultrasonic bath type Retsch UR1 (manufactured by Retsch GmbH, Germany), equipped with a high-frequency generator type PZT35 kHz and a power of 480W. In order to maintain the temperature of the ultrasonic bath to a value of approx. 20^0C, the bath water was changed continually with fresh tap water.

The supply with chloro-sodic solution necessary to the uranium desorption was carried out by means of a metering pump type 7521-47 (manufactured by Barington Bamant Co., USA). The chloro-sodic solution flow rate was 1 BEV/h (5 mL/h).

Uranium ion concentration in solutions was determined through the colorimetric solution, with the use of a UV-VIS CECIL 1011 spectrophotometer (manufactured by CECIL, Great Britain).

The mean momentary resin loading capacity with U (VI), q_i, was determined by difference between the amount of uranium in the initial and final solution, per volume of resin (equation 1) [60].

$$q = \frac{\sum_{j=1}^{i} G_v (c_{uin} - c_{ufj}) \tau_j}{M_r} \quad (1)$$

where c_{uin} is the uranium concentration of initial solution, g/L, c_{Ufj} is the momentary uranium concentration at bed exit solution, g/L, M_r is the mass of resin, g, and G_v represent the volumetric flow rate of solution which traverse the resin bed ($G_v = 25 * 10^{-3}$ L/h).

The stability of the resin grains size to the repeated action of ultrasounds was established after a granulometric comparative analysis carried out with a set of vibrating bolters type Retsch AS 200 (manufactured by Retsch GmbH, Germany) with a loop size of 0.2, 0.25; 0.5; 0.63; 0.8; 1 and 2mm, and a SEM analysis (Scanning Electron Microscopy, Quanta 3 D FEG, FEI Company, USA). Before and after the ultrasonication (6 and 18 hours), the resin samples were filtered and dried at room temperature for 24 hours.

The determination of the resin beads' mechanical resistance was carried out through grinding for 10 minutes in a mill with porcelain balls with the following features: mill diameter 98 mm, mill length 105 mm, 20 porcelain balls of 15 mm diameter, 30 balls with a 9.5 mm diameter, total ball load weight 130 g, mill speed 100 rpm. The swollen resin, in Cl^- form, was placed on a bolter with 0.8 mm loops, and class - 0.8mm was removed. 100 cm^3 of the class +0.8 mm was subjected to crushing. After the crushing, the resin is separated from the balls and placed on a bolter with a loops' diameter of 0.63 mm. Fractions A (+ 0.63mm) and B (-0.63 mm) are subjected to draining on a filtering cone with a G5 frit and is dried at room temperature for 24 hours. The fractions are weighed on an analytical balance type Precisa XT220A (manufactured by Precisa Gravimetrics AG Moosmattstrasse 32 CH 8953 Dietikon). The mechanical resistance is expressed by equation 2 [28]:

$$R_m = \frac{A}{A+B} \times 100 \quad (2)$$

3 Results and discussion

3.1 The dynamics of uranium desorption from the anion-exchange resins Purolite A-600U/3472 and PM 611 without/with ultrasounds

After loading the resin with uranium, it was subjected to the process of uranium desorption with chloro-sodic solution. Adsorption/desorption of uranium from carbonated solutions occurs according to equation 3:

$$4(R^+ - Cl^-) + 4Na^+ + [UO_2(CO_3)_3]^{4-} \Leftrightarrow 4R^+ - [UO_2(CO_3)_3]^{4-} + 4(Na^+ - Cl^-) \quad (3)$$

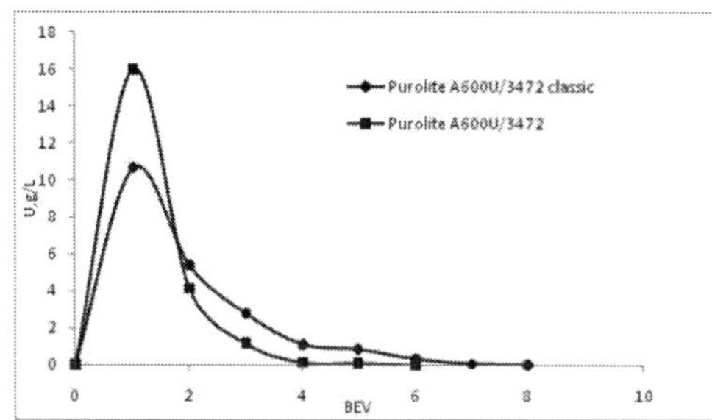

Fig. 1. Desorption isotherms of uranium fixed on Purolite A600U/3472 resin both in classic conditions and in the presence of ultrasonication.

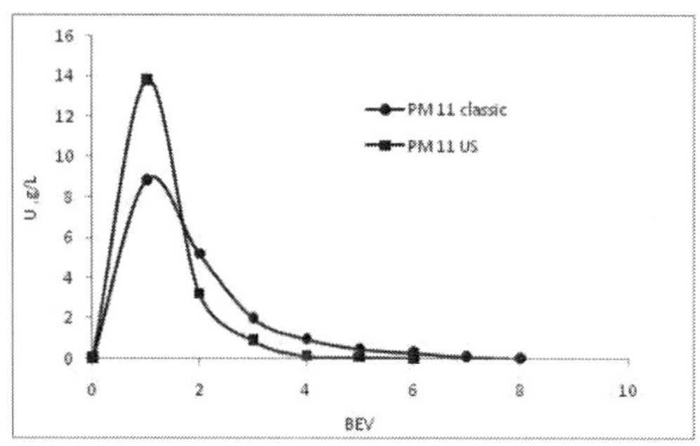

Fig. 2. The desorption isotherms of uranium fixed to the PM 611 resin under classic conditions and in the presence of ultrasonication.

For the comparison of the experimental results obtained and the highlight of the positive effects of ultrasounds on the desorption process of uranium complex anions from the ion-exchange resin, desorption isotherms have been plotted (figure 1), both in classic conditions and with ultrasounds for the resin Purolite A600U/3472.

In Figure 2 are comparatively presented the desorption isotherms of uranium complex anions from resin PM 611 in classic conditions and with ultrasounds.

As can be noticed in the figures shown, the use of ultrasounds considerably improves desorption. While in the absence of ultrasounds the complete desorption from the two resin types (Purolite A600U/3472 and PM 611, respectively) is carried out with 8 BEV of chloro-sodic solution, in their continuous presence throughout the process, the complete elution takes place with only 4 BEV chloro-sodic solutions. The use of a smaller volume may be due to desorption intensification towards the end of the process, when the removal of uranium from the resin on the classic operation is carried out very slowly (4 BEVs are needed to remove the last 5 – **6%** from uranium quantity.

The dynamics of uranium elution from Purolite A600U/3472 resin with and without ultrasounds is shown in Figure 3, and the dynamics of the uranium elution from PM 611 resin is shown in Figure 4.

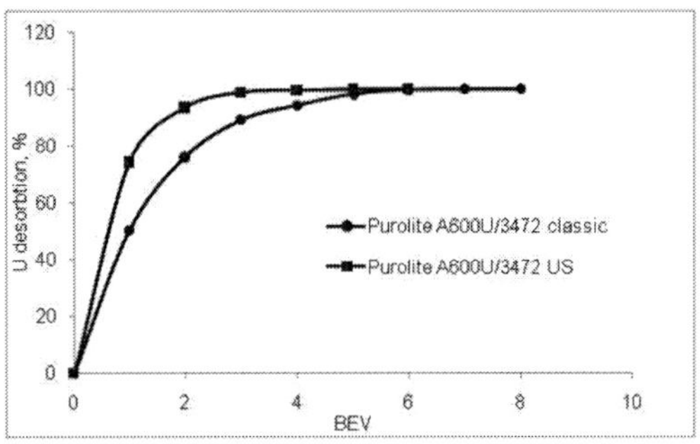

Fig. 3. Dynamics of uranium desorption from e Purolite A600U/3472 resin with and without ultrasonic field.

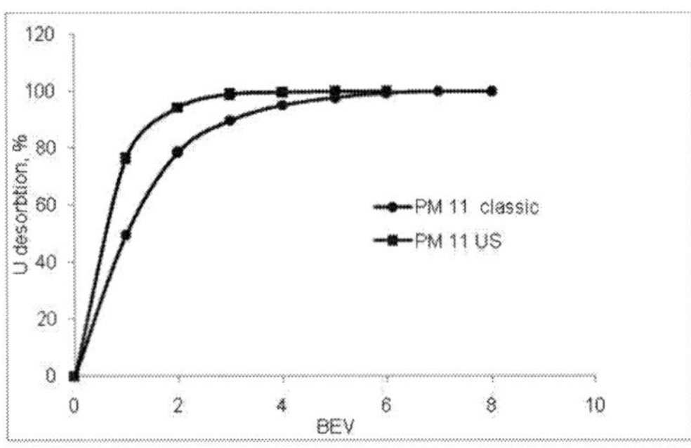

Fig. 4. Dynamics of uranium desorption from the PM611 resin with and without ultrasonic field.

The analysis of uranium desorption dynamics shows the fact that under classic conditions the uranium is desorbed from the two resin types in a percentage of 99% in a volume of 7 BEV chloro-sodic solution, whereas under the continuous ultrasounds' action this happens with a volume of 3 BEV. The difference between the two resins consists in the uranium content in the effluent and the recycled uranium percentage, which is in favor of the resin Purolite A6000 U/ 3472. In Table 1 are shown both the effluent composition that is obtained (eluate) and the percentage of recycled uranium from the rest of obtained eluate.

Table 1. Uranium content in effluent function of desorption conditions.

Resin type	**Desorption conditions**	**No. BEV**	**Effluent uranium content, g/L**	**Recycled uranium, %**
Purolite A600U	Classic	3	6.27	*10.88*
	US	3	7.09	*1.07*
PM611	Classic	3	5.36	*10.41*
	US	*3*	*5.95*	*1.03*

As a result of these experiments it was demonstrated the feasibility of using ultrasounds to enhance desorption of the uranium ions from the ion-exchange resins. In these experiments, particle diffusion and film diffusion are the rate determinant steps. Because ultrasounds produce acoustic vortices which cause micro turbulences both inside the particle pores and on the interface solid-liquid, these improve the desorption rate by increasing the rate of transport through diffusion in particle and in film, which leads to a high level of U in the first 2-3 volumes, which are precipitated. The rest of up to 8 volumes (with low uranium level) will be recycled.

3.2 Influence of ultrasounds on the physical features of resins Purolite A-600U/3472 and PM 611

To establish the influence of ultrasounds on the physical properties of the two strongly basic anionic resins, stability tests and SEM analysis were conducted regarding the grains' size and mechanic resistance.

3.2.1. Stability of the resin grain size

Friable solids such as resin grains existing in an aqueous ultrasonated solution might be broken or eroded due to high pressure shock waves resulting from cavitations implosions.

In order to assess the integrity of polymeric resins subjected to the ultrasounds field, swollen resin samples are placed in a glass beaker immersed in the ultrasounds bath and ultrasonated for 6 and 8 hours, respectively.

In Figures 5 and 6 is shown the content evolution of the granulometric classes of interest 0-0.2 mm; 0.8-1 mm and 1-2 mm for the two resins, depending on the length of the ultrasounds field.

DOI:10.1051/e3sconf/20171801034

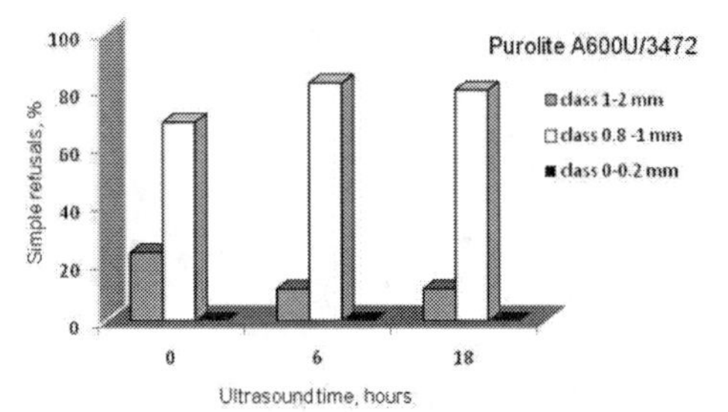

Fig. 5. Variation of particle size classes1-2 mm; 0.8-1 mm and 0-0.2 mm depending on the time in ultrasound field for Purolite A600U/3472 resin.

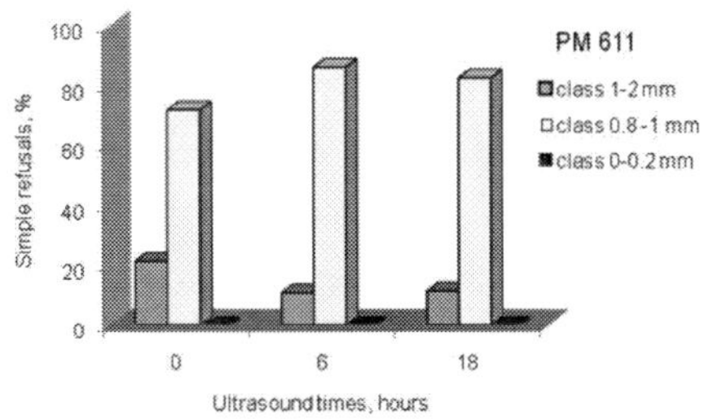

Fig. 6. The variation of particle size classes 1-2 mm; 0.8-1 mm and 0-0.2 mm depending on the time in ultrasound field of PM 611 resin.

Analyzing the results obtained, both for the Purolite A 600U/3472 resin and for the PM 611 resin is seen the fact that after the first 6 hours of ultrasonication, a decrease of content for the class 1-2 mm takes place, after which the granulometric characteristics remain constant.

3.2.2 The stability of mechanical resistance of the resin grains

Mechanical resistance, an essential characteristic (along with loading capacity of the resin with uranium) was established by the granulation process described in the experimental part. In the experiments carried out, resin fractions A (+0.63 mm) and B (-0.63 mm) were volumetrically determined right after sizing the milled resin 6 hours. The results obtained are shown in Table 2.

From the data presented are determined the following:

- in the case of untreated resins, the loss in passing sieve of 0.63 mm were almost insignificant, of only 0.25 for the resin Purolite A600U/3472, respectively of 0.50% for resin PM 611;
- in the case of afterwards ultrasonicated resins for 18 hours, the loss in passing sieve of 0.63 mm were slightly larger than in the case of untreated resins, namely 2.50 in the case of resin Purolite 600U/3472, respectively 4.00% for resin PM 611;
- the degradation of resin PuroliteA600U/3472 is due to the uniform grain use; the grains that reached in passing sieve kept their spherical shape but not the size;
- during the milling of resin PM 611 the grains don't keep completely their spherical shape, in the water appear resin detached fragments resembling some "fluff" or "flakes";
- the mechanical resistance obtained for the two types of resins, in the two cases was:

- for the non-ultrasonated resin - R_m = 99.75% (Purolite A600U/3472); R_m = 99.50% (PM611)
- for the ultrasonated resin for 18 hours -R_m = 97.50% (Purolite A600U/3472); R_m = 96.48% (PM611)

Table 2 The experimental results of the grinding test for 6 hours.

Resin	Resin volume, mL			Loss relative to baseline (cumulative passes, %)
	Fraction A +0.63 mm	Fraction B -0.63 mm	R_m, %	
Purolite SGA600U/3472 absence of ultrasound	99.75	0.25	99.75	*0.25*
Purolite SGA600U/3472 presence of ultrasound 18 hours	97.50	2.50	97.50	*2.50*
PM611 absence of ultrasound	99.50	0.50	99.50	*0.50*
PM 611 presence of ultrasound 18 hours	*96.00*	*3.50*	*96.48*	*4.00*

3.2.3. SEM analysis of resin granules

SEM analysis imagines of the resins evaluated in this study, magnified from 60,000X to 2,000,000X are shown in Figures7 -12.

SEM analysis of the resins, after sonication showed that the Purolite A600U/3472 resin maintains its spherical shape, without modification of the surface (Figures 8 - 9), while resin PM11 after 6 hours of ultrasound is not spherical, it flattens the shape. That fact finds after 18 hours of sonication also (Figures 11-12).

The analysis of micro - compositional quality of samples analyzed is the spectral lines main Kα of each sample and the analysis of micro - compositional quantity thereof, presents the evidence in the detection limit of the machine, expressed as a percentage by mass (Wt %) and the atomic percent (At %). Initial values of granules for each type of resin is not identical but they are in controlled limits so small variations in the values appear in micro-compositional analysis.

SEM analysis of the resin granules test results confirm the stability and mechanical strength of scale, being able to state that ultrasound does not affect the physicochemical properties of studied resins

DOI:10.1051/e3sconf/20171801034

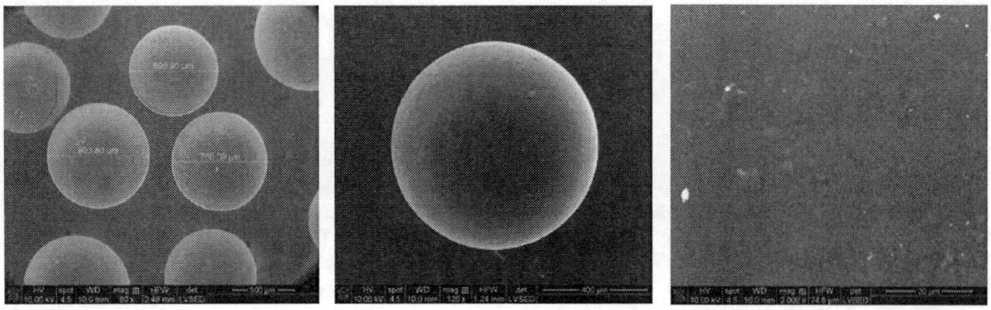

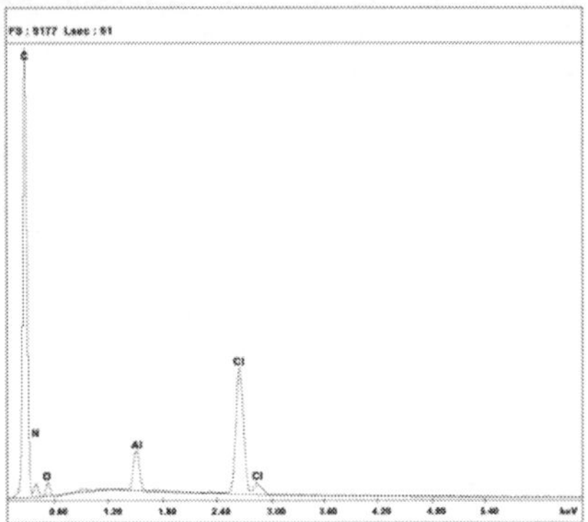

EDAX ZAF Quantification (Standardless)
Element Normalized
SEC Table : Default

Element	Wt %	At %
C K	66.88	81.71
N K	4.93	5.17
O K	2.01	1.84
AlK	3.45	1.88
ClK	22.72	9.41
Total	100.00	100.00

Element	Net Inte.	Bkgd Inte.	Inte. Error	P/B
C K	529.00	0.92	0.56	577.39
N K	18.65	1.57	3.20	11.88
O K	20.35	3.55	3.29	5.73
AlK	82.26	19.19	1.71	4.29
ClK	314.79	11.71	0.75	26.87

Fig. 7. SEM analysis of unultrasonated Purolite A600U/3472 resin – structure and composition.

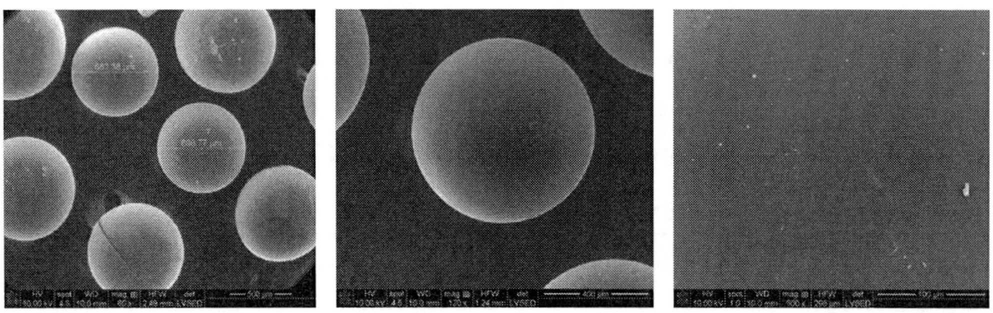

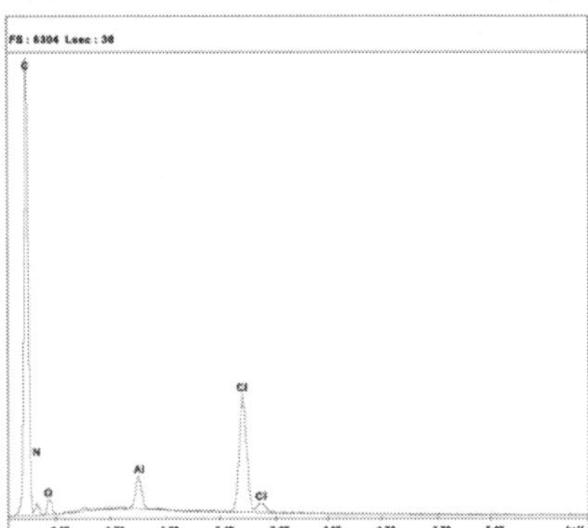

EDAX ZAF Quantification (Standardless)
Element Normalized
SEC Table : Default

Element	Wt %	At %
C K	69.12	83.17
N K	4.31	4.45
O K	2.41	2.18
AlK	2.79	1.50
ClK	21.37	8.71
Total	100.00	100.00

Element	Net Inte.	Bkgd Inte.	Inte. Error	P/B
C K	579.33	0.81	0.67	713.81
N K	16.23	1.41	4.35	11.48
O K	24.53	3.09	3.66	7.94
AlK	66.60	20.16	2.51	3.30
ClK	296.29	12.10	0.98	24.50

Fig. 8. SEM analysis of 6 hours ultrasonated Purolite A600U/3472 resin– structure and composition.

* Corresponding author: antonetafilcenco@yahoo.ro

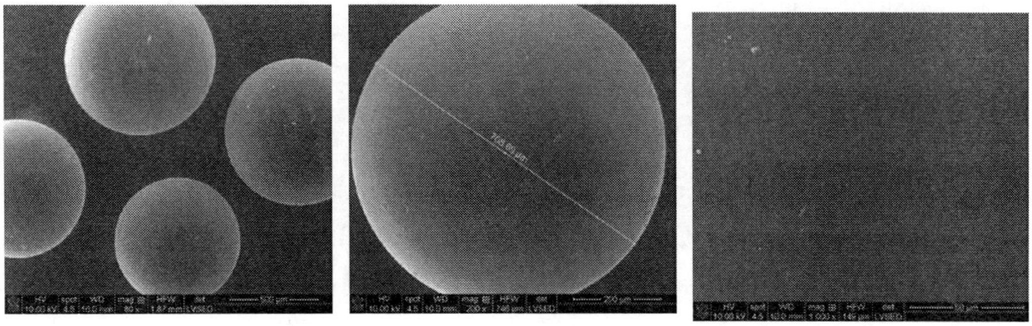

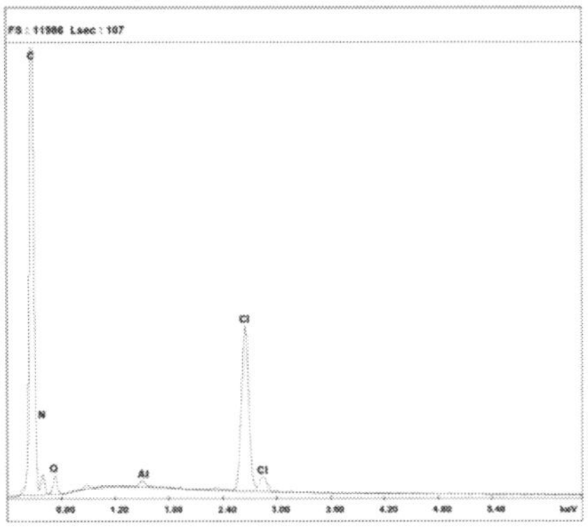

EDAX ZAF Quantification (Standardless)
Element Normalized
SEC Table : Default

Element	Wt %	At %
C K	68.05	82.12
N K	5.81	6.01
O K	2.27	2.05
AlK	0.41	0.22
ClK	23.46	9.59
Total	100.00	100.00

Element	Net Inte.	Bkgd Inte.	Inte. Error	P/B
C K	505.24	0.74	0.43	679.37
N K	20.08	1.27	2.28	15.77
O K	20.78	2.10	2.32	9.89
AlK	8.94	17.44	7.14	0.51
ClK	299.94	11.25	0.58	26.67

Fig. 9. SEM analysis of 18 hours ultrasonated Purolite A600U/3472 resin– structure and composition.

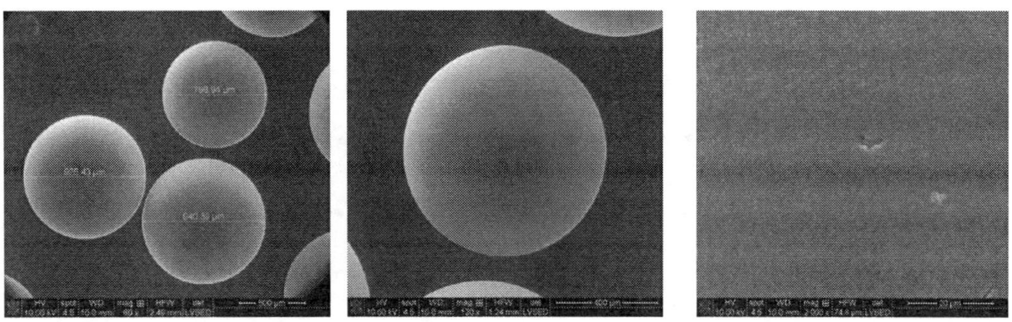

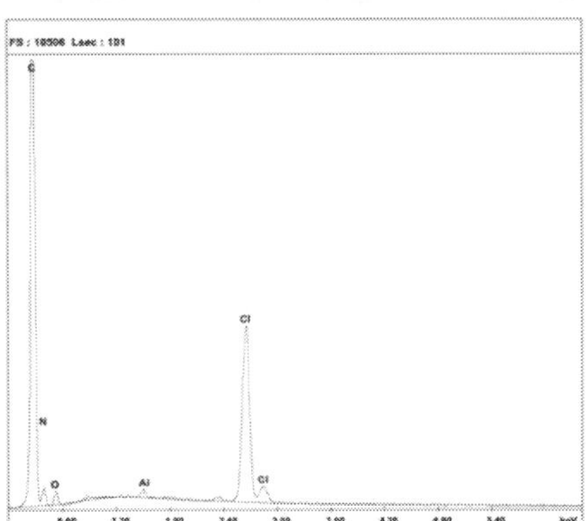

EDAX ZAF Quantification (Standardless)
Element Normalized
SEC Table : Default

Element	Wt %	At %
C K	68.52	83.14
N K	4.93	5.13
O K	1.53	1.40
AlK	0.40	0.22
ClK	24.62	10.12
Total	100.00	100.00

Element	Net Inte.	Bkgd Inte.	Inte. Error	P/B
C K	468.81	0.86	0.46	545.10
N K	16.00	1.46	2.70	10.94
O K	13.37	3.17	3.30	4.21
AlK	8.39	17.59	7.82	0.48
ClK	300.79	10.66	0.59	28.23

Fig. 10. SEM analysis of unultrasonated PM 11 resin - structure and composition.

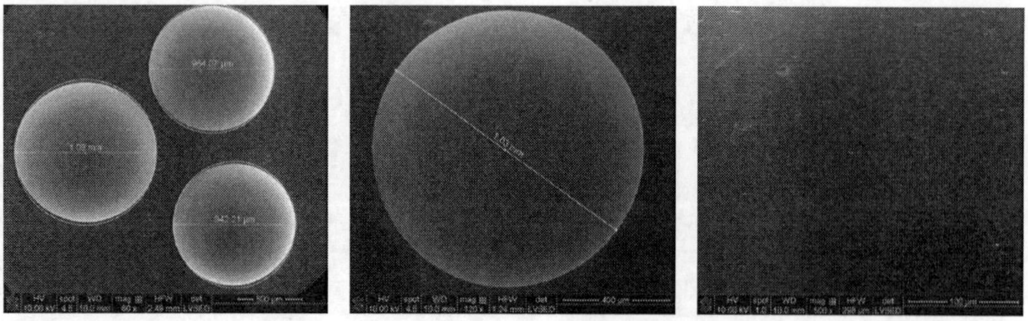

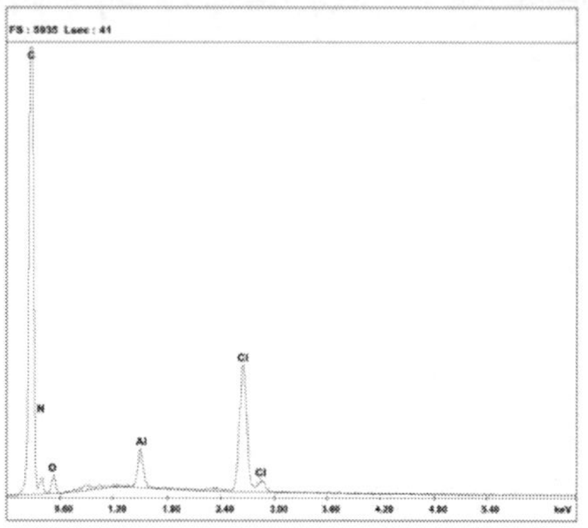

EDAX ZAF Quantification (Standardless)
Element Normalized
SEC Table : Default

Element	Wt %	At %
C K	69.29	82.69
N K	5.48	5.61
O K	2.34	2.09
AlK	2.77	1.47
ClK	20.12	8.13
Total	100.00	100.00

Element	Net Inte.	Bkgd Inte.	Inte. Error	P/B
C K	642.34	0.98	0.61	652.85
N K	22.10	1.63	3.53	13.54
O K	25.08	2.69	3.41	9.33
AlK	69.81	19.20	2.31	3.64
ClK	294.76	11.06	0.94	26.64

Fig. 11. SEM analysis of 6 hours ultrasonated PM 11 resin – structure and composition.

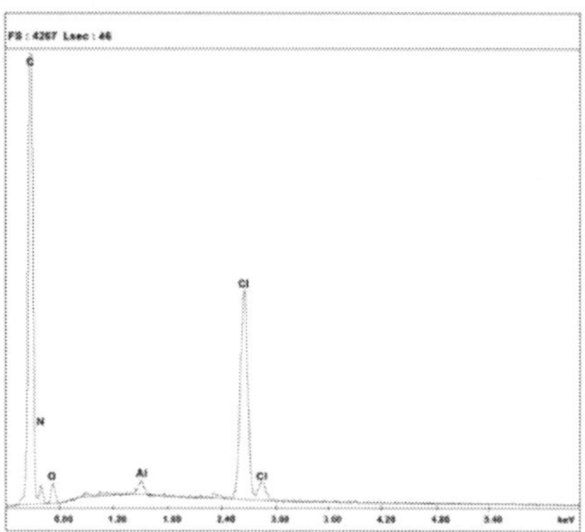

EDAX ZAF Quantification (Standardless)
Element Normalized
SEC Table : Default

Element	Wt %	At %
C K	66.14	81.53
N K	4.83	5.17
O K	2.14	1.98
AlK	0.84	0.46
ClK	25.99	10.85
Total	100.00	100.00

Element	Net Inte.	Bkgd Inte.	Inte. Error	P/B
C K	422.34	0.74	0.72	572.26
N K	15.72	1.26	4.00	12.48
O K	18.47	2.11	3.80	8.77
AlK	17.30	17.89	6.20	0.97
ClK	312.18	11.81	0.86	26.44

Fig. 12. SEM analysis of 18 hours ultrasonated PM 11 resin – structure and composition.

3.3 The influence of ultrasounds on the uranium loading capacity of the resins

3.3.1 The dynamics of uranium desorption from the anion-exchange resins Purolite A-600U/3472 and PM 611 without/with ultrasounds

Since the adsorption/desorption operations are cyclical, it is important to know if applied ultrasound only on desorption operation influences the loading capacity of the ultrasounded resin compared with non-ultrasonated in the adsorption process.

Adsorption experiments performed in dynamic conditions, on the two types of ultrasounded resin for 18 hours in similar conditions with the initial adsorption experiment (non-ultrasounded resin), led to the results presented comparatively in Figures 13 and 14.

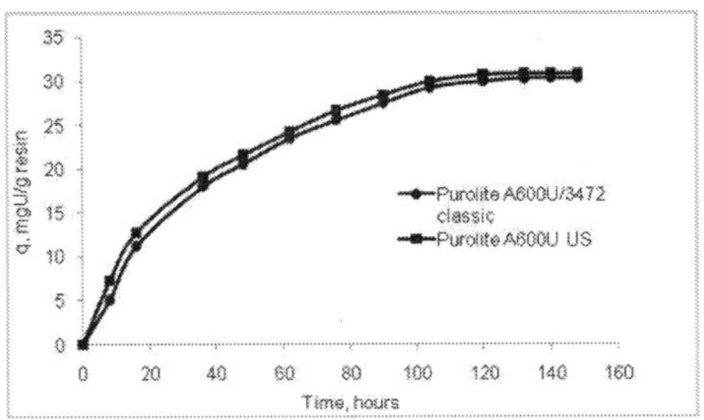

Fig. 13. Comparative evolution of uranium (VI) adsorption dynamics on Purolite A 600U/3472 resin previously untreated with ultrasound field and in the field of ultrasound for 18 hours (from solution with the following characteristics: U = 35.37 mg/L Na_2CO_3 = 3.18 g/L, $NaHCO_3$ = 0.504g/L and pH = 9.5).

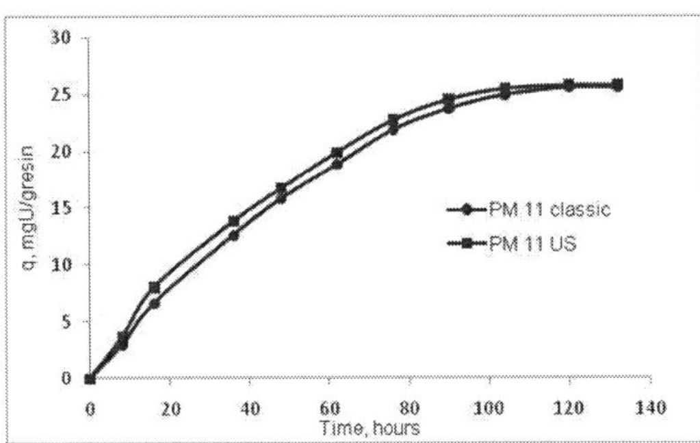

Fig. 14. Comparative evolution of uranium (VI) adsorption dynamics on PM 611 resin previously untreated with ultrasound field and in the field of ultrasound for 18 hours (from solution with the following characteristics: U = 35.37 mg/L, Na_2CO_3 = 3.18 g/L, $NaHCO_3$ = 0.504 g/L and pH = 9.5.

The experimental results shown in Figures 13 and 14, emphasize the fact that ultrasounds do not affect the adsorptive properties of the resins, as the uranium loading capacity is basically similar.

The value of effective adsorption capacity of ultrasounded Purolite A- 600U/3472 resin was 30.85 mgU/g resin, compared with non-ultrasounded of 30.36 mgU/g resin, and in case of PM 611 resin the values are comparable also, respectively 25.79 mgU/g ultrasounded resin and 25.64 mgU/g non-ultrasounded resin.

SEM analysis of the resin granules, reveals that their surface did not change significantly after ultrasonication, thus making the transfer area (grain size) to maintain the same initial limits. This would be the reason for the similar values of loading capacity of both situation studied.

It can be affirmed that ultrasounds do not affect the adsorptive properties of the resins studied and they can be used consecutively without special treatments in the adsorption – desorption process.

4 Conclusions

The desorption of the uranium complex ions from two ion-exchange resins Purolite A600U/3472 (Romania) and PM 611 (China), both in classic conditions and under the action of the ultrasonic field has been studied.

Uranium desorption in presence of continuous ultrasonic field is 99% with 3 BEV of chloro-sodic solution, for two types of studied resins, while 7 BEV in classical conditions. The difference between the two resins can be observed in the uranium content of effluent and in the recycled uranium, which is in favor of Purolite A600 U/3472 resin.

The ultrasound influence study on physical properties of the resins, on uranium loading capacities and SEM analysis demonstrated that it does not affect the adsorptive properties of resin.

The mass transfer intensification in desorption process is due to acoustics vortexes, which causes micro turbulences in the interiors of pores of the granule and the solid-liquid interface.

Application of ultrasound in the uranium desorption process has led to a decrease in the process's duration and the decrease of desorption reagent volumes, which leads to an increase of the uranium content by 16% in the first two or three volumes of effluent that are going to precipitation and also to reduction of reagents costs by over 50%.

References

1. *Uranium extraction Technology Technical Reports*, Series No. 359 IAEA (Vienna,1993)
2. W. Holl, *Fundamentals of Ion Exchange* (Institute for Technical Chemistry, Karlsruhe,1997)
3. F. Dardel, T. V. Arden, Rohm and Haas Brochure, **A(14)** (1989)
4. C.E. Harland, Ion Exchange: Theory and Practice, (Royal Society of Chemistry, 1994)
5. T. Ionescu, *Schimbul ionic în tehnică*, (Editura Tehnică, Bucuresti, 1969)
6. A. Botez, T. Dobre, E. Panturu, A. Filcenco-Olteanu, Cent. Eur. J. Chem. **12 (7)**, 769–773 (2014)

* Corresponding author: antonetafilcenco@yahoo.ro

7. F.T. Shostak, M. V. Vittikh, G. A. Savel'eva, G. S. Kozlov, L. S. Malinovskii, Akad. Nauk Kaz. SSR, Tr. Resp. Soveshch., 43–45 (1963)
8. K.L. Cheng, Zenpu Wang., Microchim. **78**, 399-406(1982)
9. A.S. Medvedev, V.V. Puchkov, N.N. Khavskii, R.G. Surukhanov, I.I. Tarasova, G.A. Evgrafova, *Izv.Vyssh. Uchebn. Zaved. Tsvetn. Metall.* (1986)
10. B.A. Agranat, M.A. Dubrovin, N.N Khavskii, G. Eskin, *Basics of Physics and Technique of Ultrasound* (Vysshaya Shkola, Moscow, 1987)
11. O. Tetsuo, J. Chromatogr. A **793**, 365-369 (1998)
12. D. Liu, I. Aoyama, H. Okamura, B.I. Dutka, Toxicol. Water Qual., **11**, 195–200 (1996),
13. S. Mukherjee, R. Matthew, H.N. Hsieh. Geotech. Spec. Publ., ASCE, **65**, 122 (1997)
14. A.P. Newman, J.P. Lormer, T.J. Mason, K.R. Hutt, Ultrason. Sonochem, **4**, 153-156 (1997)
15. W. Qin, Y. Yuan, Y. Dai, Chem. Ind. Eng. Soc., **II**, 600 (1997)
16. S.U. Rege, R.T. Yang, C.A. Cain, **44**, 1519–1528 (1998)
17. I. Küncek, S. Şener, **17**, 250-257 (2010)
18. M. Roosta, M. Ghaedi, A. Daneshfar, R. Sahraei, A. Asghari, Ultrason. Sonochem., **21**, 242-252 (2014)
19. A. Vildan Beşe, Ultrason. Sonochem.,**10**, 790-796 (2007)
20. F. Xie,L. Haiying ,M. Yang,L. Chuncheng, C. Tingting, H.Zhiyuan, Y.Gaoqing, J. Hazard. Mater. **170**, 430-435 (2009)
21. L. Chuncheng, X.Fengchun,M. Yang,C. Tingting,L. Haiying, H. Zhiyuan,Y. Gaoqing, J. Hazard. Mater. **178**, 823-833 (2010)
22. J.M. Timothy, S.S. Phull, M. Sternagel, J.P. Lorimer,B. Pollet, Ultras. in Environm. Eng., **25**, 181 (1999)
23. B.G. Pollet, R. Viennet, J.Y. Hihn, J.Y. Doche, J.P. Lorimer, T.J. Mason, *Hydrodynamic and Mass Transfer behaviour near the Ultrasonic Horn at Low Frequencies,* Applications of Power Ultrasound in Physical and Chemical Processing Conference (Paris, 2001)
24. B. Pollet,S.S. Phull, J.P. Lorimer,T.J. Mason, *The Use and Applications of Ultrasound in Environmental Protection – Environmental Sonochemistry,*S.E.T. '99 at the House of Commons – Young Research Scientists, Engineers and Technologists,(1999)
25. J.M. Timothy, F. Doust, L. Paniwnyk,B. Pollet, *Mass transfer measurements at small electrode-ultrasonic horn separations,* 10th Meeting of the European Society of Sonochemistry (Hamburg, 2006)
26. T.J. Mason, J.P. Lorimer, L. Paniwnyk,B. Pollet, Mater World, **7(2)** (1999)
27. A.D., Radu, A. Woinaroschy, E. Panturu, Rev. Chim., **65(4)** 470-474 (2014)
28. E. Panturu, Ghe. Jinescu, R. Radulescu, A. Filcenco Olteanu, C. Jinescu, Rev. Chim. **59(9)** 1036-1040 (2008)
29. W. Xue Jiang,J. Chen,Yl. Xiangbo,Xl. Wang,J. Zhang,J. Huang,J. Zhao, J. Ind. Eng. Chem. (2015)
30. E. Panturu, Method acid extraction of uranium from ore in ultrasonic field-Patent No. 122642 (2009)
31. W.C. Kyeong, J.K. Chul, H.S. Yoo, Highly efficient uranium leaching method using ultrasound, US Patent N0.: 8,470,269 B2 (2013)
32. Process intensification of chemical decontamination of radioactively contaminated soils using ultrasound, Patent No. 123554/08 (2013)
33. D. Feng, C. Aldrich, Hydrom., **55**, 201–212 (2000)
34. E. Panturu, Ghe. Jinescu, F. Aurelian, R. Rădulescu, A. Filcenco-Olteanu, Rev.Chim., **58(6)** 551-556 (2007)
35. K. Aneel, J. Takemura, Int. J. Env. Sci. Devel. 1, **3**, 260-263 (2010)
36. D. Feng, H. Tan, J.S.J. van Deventer, Miner. Eng., **16(3)**, 257-264 (2003)
37. J.K. Chul, W.C. Kyeong, H.S. Yoon, *Sonochemical ion exchange in RIP system, ALTA,* Uranium Conference 97 (2011)
38. D. Chen, S.K. Sharma, A. Mudhoo, *Handbook of application of ultrasound: sonochemistry for sustainability,* CRC Press (2011)
39. T.J. Mason, D. Peters, *Practical Sonochemistry. Power Ultrasound Uses and Applications,* Ellis Horwood Publishing Limited (2002)
40. M. Breitbach,D. Bathen,H. Schmidt-Traub, Ind. Eng. Chem. Res., **42(22)**, 5635–5646 (2003)
41. W.B.Zhang,M. Deng,C.X. Sun,S.B. Wang,. Eng. Chem. Res., **53(1)**, 333–339 (2014)
42. E. Panturu, Method of optimizing the uranium desorption from ion exchange resins under the effect of ultrasound, No.brevet 122 767 (2010)
43. J. K. Youngs, Solvent extraction process, US 5463177 A (1995)
44. B.Pesic, Technical Report: Application of ultrasound in solvent extraction of nickel and gallium, Contract DE-AC07-941D13223, Interior Department's Bureau of Mines and US Department of Energy, Idaho, **78** (1996)
45. X.P. Chen, W.X. Wang, S.B. Li, J.L. Xue, L.J. Fan, Z.J. Sheng, Y.G. Chen,. Polym. **80**, 944–948 (2010)
46. M. Sillanpää, T.D. Pham, R.A. Shrestha, *Ultrasound Technology in Green Chemistry,* (Springer, 2011)
47. M. Bimakr, R.A. Rahman, F.S. Taip, N.M. Adzahan, Z.I. Sarker, A. Ganjloo, Molecules, **17**, 11748–1176 (2012)
48. T. Li, X.Y. Qu, Q.A. Zhang, Z.Z. Wang, Ind. Crop. Prod. **35**, 98–104 (2012)

DOI:10.1051/e3sconf/20171801034

49. A. Balasubrahmanyam, S.B. Roy,S. Chowdhury, K.N. Hareendran, A.B. Pandit, , Ind. Eng. Chem. Res **45(22)**, 7639–7648 (2006)
50. P.K. Kalsi, B.S. Tomar,K.L. Ramakumar,V. Venugopal, Journal of Radioanal. Nucl. Chem., **293**, 863-867 (2012)
51. A.Balasubrahmanyam, S.B. Roy,Y. Ladola, S. Chowdhury, K.N. Hareendran, A. B. Pandit, Chem. Eng. Proces.: Process Intens., **47**, 2107–2113 (2008)
52. R. Bikram, S. Bhattacharya, S.B. Roy, Int. J. Nucl. Energy Sci. Techn., **8**, 171-178 (2014)
53. O. Hamdaoui, E. Naffrechoux, AIChE Journal **53**, 363–373 (2007)54
54. K.W. Chung, C.J. Kim, H.S. Yoon, Uranium ion exchange adsorption method using ultrasound, S Patent 8,475,746 B2 (2013)
55. EPA (U.S.Environmental Protection Agency), *Understading variation in partition coefficient, values, Vol I, The Kd model, methods of measurement, and application of chemical reaction codes*, EPA 402-R-99-004A, U.S. Environmental Protection Agency, Washington D.C (1999)
56. A. Balasubrahmanyam, S.B. Roy,S. Chowdhury, K.N. Hareendran, A.B. Pandit, Ind. Eng. Chem. Res., **45(22)**, 7639–7648 (2006)
57. P.K. Kalsi,B.S. Tomar,K.L. Ramakumar, V. Venugopal, Journal of Radioanal. Nucl. Chem., **293**, 863-867 (2012)

E3S Web of Conferences **18**, 01035 (2017) DOI:10.1051/e3sconf/20171801035
MEC2017

Impact of spiral separator geometrical parameters on the density separation of various fine-grained materials

Jan Szpyrka[1,*], *Tomasz* Suponik[2], and *Marcin* Lutyński[1]

[1] Silesian University of Technology, Faculty of Mining and Geology,44-100 Gliwice ul. Akademicka 2, Poland

Abstract. The study aims at the assessment of the impact of geometrical parameters of spiral separators on the efficiency of density separation of fine-grained materials. Experiments were carried out on three spiral separators: Krebs 2.85, Reichert LD-4 and Reichert LG-7. Three materials were used for the tests: raw coal, coal waste and mix of sand and magnetite as the model material. Results of raw coal and coal waste upgrading showed that density separation was most efficient in Reichert LD-4 spiral . This is due to the fact that this device had the highest amount of coils, height of sluice as well as was equipped with additional dense product collector and additional water sluice for transport water. The lower slope of sluice and larger height made separation even more efficient. Analysis of separation of model material, that is the mix of sand and magnetite, showed that in this case the existence of additional water sluice does not have an impact on product separation and best results were obtained in the Reichert LG-7 spiral separator. The shape and width of sluices did not have a significant impact on the separation process.

1 Introduction

Spiral separators are commonly used for the separation of fine-grained materials with the particle size not exceeding 3 mm. The lower limit of particle size is 0,1 mm for the low density materials and 0,05 mm for heavy minerals [1,2]. The arrangement of material in accordance with its density in the sluice is determined by the liquid pressure which moves particles in the water stream [3]. As larger and heavier particles sink to the bottom of the sluice faster and experience more drag from the bottom, they travel slower, and so move towards the center of the spiral. Conversely, light particles stay towards the outside of the spiral, with the water, and quickly reach the bottom. Other forces also act on the particles and these are: gravity where direction is always constant for each particle and the force depends on the particle density; centrifugal force which results from the spiral movement of the water stream; friction which is the result of the slide of the particle on the surface of the sluice [4,5]. These forces act on the particles and lighter particles move with a higher speed whereas centrifugal force pushes particles to the outer side of the spiral [6]. The friction increases towards the inner side of the sluice whereas the centrifugal force decreases towards the inner side. Hence, dense particles tend to concentrate on the inner side of the sluice [7].

The efficiency of separation in the spiral separator depends on various factors such as feed parameters i.e. particle size distribution, density fraction and slurry density or geometrical parameters of spiral separator i.e. number of coils, angle and width of spiral as well as location of separating slots (fans) and number of heavy product collectors. Geometrical parameters have an impact on the precision of separation, yield of products and water stream movement in the sluice. Water stream in the sluice has varying height which increase towards the outer side of the sluice. The speed of the water stream changes similarly and the highest speed is at the outer edge where light particles are concentrated.

The purpose of this study was to assess the impact of geometrical parameters of three types of spiral separators (Krebs 2.85, Reichert LD-54, Reichert LG-7) on the efficiency of separation of various fine-grained materials. The practical application of this study was to select highly efficient spiral separator for the concentration of coal from coal waste slurries deposited in impoundments.

2 Materials and methods

Three types of materials with different density fractions were used in the study, these were: raw coal, coal waste from a dump and model material consisting of sand and magnetite mix. All three materials were separated in the spiral concentrator with the same slurry density of 300 g/dm^3.

Each separator is characterized by different geometrical parameters mainly number of spirals, slope of the sluice, width and shape of the sluice, diameter of the spiral, location of separating fans (slots) and number of heavy product collectors. These parameters are as follows:

- Krebs 2.85 separator: height 1750 mm; spiral diameter 900 mm; number of coils 2,85; sluice slope 15°; sluice width 280 mm – including additional sluice for

* Corresponding author: jan.szpyrka@polsl.pl

DOI:10.1051/e3sconf/20171801035

transporting water of 70 mm; shape of sluice – circular sector with flattened bottom, the outer part of the sluice is of 8 cm in height and bend 90°. Additional sluice for transporting water is the u-shape sluice 6,5 cm in depth; distance between separating cuts for heavy product, semi-product and light product is 100, 80, 85 mm respectively; heavy product collectors – underflow at the bottom of spiral separator; spiral is made of plastic and covered with liners. There are two sluices with three coils each.

- Reichert LD-4 spiral: height 2500 mm, spiral diameter 900 mm; number of coils 6; sluice slope 15°; sluice width 280 mm - including additional sluice for transporting water of 60 mm; shape of sluice – circular sector with flattened bottom; the outer part of the sluice is of 8,5 cm in height and bend 90°; additional sluice for transporting water is a flat shaped sluice 6 cm in depth; distance between separating cuts for heavy product, semi-product and light product is 100, 80, 80 mm respectively; heavy product collectors – underflow at the bottom of spiral separator and one additional collector in the middle of the spiral (third coil); spiral is made of plastic without liners. There are two sluices with six coils each.
- Reichert LG-7 spiral: height 2500 mm, spiral diameter 580 mm; number of coils 6; sluice slope 20°; sluice width 200 mm; shape of sluice – circular sector with flattened bottom; the outer part of the sluice is of 6,5 cm in height and bend 90°; no additional sluice for transporting water; distance between separating cuts for heavy product, semi-product and light product is 40, 60, 70 mm respectively; heavy product collectors – underflow at the bottom of spiral separator. This spiral separator, in comparison to the other two, is much narrower and the sluice slope is more inclined hence the flow speed of material and water is higher.

Table 1. Ash content in particle size fractions of raw coal and coal waste.

δ, g/cm^3	Raw coal		Coal waste	
	γ, %	λ_{ash}, %	γ, %	λ_{ash}, %
< 1.4	79.48	3.08	53.56	7.46
1.4 – 1.6	5.28	22.04	5.28	27.91
1.6 – 1.8	3.03	37.27	11.58	70.29
> 1.8	12.21	80.19	29.58	70.68

δ - density fraction; γ - yield; λ_{ash} - ash content

Table 2. Ash content in density fractions of raw coal and coal waste.

Particle size, mm	Raw coal		Coal waste	
	γ, %	λ_{ash}, %	γ, %	λ_{ash}, %
>3	12.80	19.89	9.30	32.86
3 - 2	14.69	24.01	18.96	35.83
2 - 1.5	11.03	22.36	14.87	33.02
1.5 - 1	12.95	19.64	21.84	31.36
1 - 0.5	16.43	20.21	28.26	33.16
< 0.5	32.10	21.77	6.78	49.13
Average		21.39		34.31

γ - yield; λ_{ash} - ash content

Both particle size and denisty analysis were carried out in accordance with national standards. Ash content was measured according to the polish standard PN-G-04512:19809/A/1:2002. Magnetic separation was carried out on the Satmagan separator.

Ash content in each particle size fraction of raw coal and coal waste is presented in Table 1 whereas the ash content in denisty fraction of raw coal and coal waste is shown in Table 2.

Model material was created by mixing sand and magnetite in the mass ratio of 1:4. The particle size range of feed was 0-2 mm. Average magnetite content in the feed of the model material was 24,28%.

3 Results and discussion

Results of ash content analysis λ_{ash} in separation products of raw coal, coal waste and model material (magnetite and sand mix) are shown in Tables 3, 4 and 5 respectively. In Table 3 the product of density separation of LG-7 separator is not shown since it was not possible to separate fine-grained raw coal in this device. High slope of sluice in LG-7 separator causes high flow speed of the slurry which results in slurry overflow from sluice and inefficient separation.

Initial assessment of density separation presented in Tables 3, 4 and 5 indicates that Krebs 2.85 and Reichert LD-4 separators are more efficient in upgrading raw coal and coal waste, whereas Reichert LD-7 is more efficient in upgrading sands and ores.

Detailed analysis of density separation results of raw coal shown in Table 3 revealed that this feed was upgraded more efficiently with the use of Reichert LD-4 separator. Ash content in concentrate of Krebs 2.85 separator was slightly lower, yet the yield of concentrate in Reichert LD-4 was much higher. Comparison of the two separators showed that tailings of the Reichert LD-4 have high concentration of ash and lower yield.

When comparing results of coal waste upgrading, the Reichert LG-7 spiral was the least efficient as well. Although yields of products were satisfactory (see Table 4) but the overall ash content in tailings and middlings was considerably high. Among other two separators Reichert LD-4 was more efficient in density separation of coal waste. The ash content in Krebs 2.85 was only slightly lower than in Reichert LD-4 yet the concentrate yield for the latter one was considerably higher. Overall, the efficiency of separation for the Reichert LD-4 was higher when considering the ash content and yields in middlings and tailings.

Particle size and ash content analysis of separation products of raw coal (Table 6) indicate that the low efficiency of Krebs 2.85 separator might be caused by the low number of coils. Because of that the residual time of particles feed in the separator is low and as a result particles do not separate resulting in high ash content in tailings. The fine particle size fraction of middlings from this separator had a low ash content indicating the process of agglomeration of fine particles with coarse ones and as a consequence – decrease in separation efficiency. Relatively high content of coal in tailings confirmed low separation efficiency which was caused be the short (height) of the spiral separator despite the fact that it is equipped with additional water sluice.

Table 3. Ash content (λ_{ash}) in products of raw coal upgrading in Krebs 2.85 and Reichert LD-4.

Products	Krebs 2.85		Reichert LD– 4	
	γ, %	λ_{ash}, %	γ, %	λ_{ash}, %
Concentrate	29.88	7.93	70.66	8.37
Middlings	32.88	13.65	9.69	20.66
Tailings	37.24	32.98	19.65	55.02

Table 4. Ash content (λ_{ash}) in products of coal waste upgrading in Krebs 2.85, Reichert LD-4 and Reichert LG–7.

Products	Krebs 2.85		Reichert LD–4		Reichert LG–7	
	γ, %	λ_{ash}, %	γ, %	λ_{ash}, %	γ, %	λ_{ash}, %
Concentrate	24.90	13.88	48.86	14.91	81.20	30.56
Middlings	29.10	24.27	21.65	31.58	16.06	50.35
Tailings	46.00	47.89	29.49	53.61	2.74	64.21

Table 5. Magnetite content (λ_m) in products of model material (magnetite+sand) in Krebs 2.85, Reichert LD–4 and Reichert LG–7 separators.

Products	Krebs 2.85		Reichert LD–4		Reichert LG–7	
	γ, %	λ_m, %	γ, %	λ_m, %	γ, %	λ_m, %
Tailings (light product)	10.37	52.70	2.50	86.07	48.45	8.33
Middlings	11.96	9.70	1.11	60.70	41.87	31.00
Concentrate (heavy product)	77.67	20.93	96.39	19.63	9.39	81.07

More detailed analysis of products from the Reichert LD-4 separator (Table 6) indicate that high ash content in the concentrate was caused by the inflow of light products above 3 mm in size which should have been separated in the additional sluice. These particles were "thrown out" to the outer side of the sluice. Middlings and tailings had slightly better properties than the ones obtained in the Krebs 2.85 separator which could be attributed to the higher number of coils in the latter one.

High ash content in the concentrate of Krebs 2.85 separator (table 7) was also caused by the passing of coarse grains (>2 mm) with high ash content into this product. These particles should have stayed in the additional water sluice where separation should have taken place. Low ash content in dense particle size fractions showed that separation was insufficient due to the low number of coils. Additional transport water (sluice) was not able to separate coal from the dense fraction (tailings).

Relatively high ash content in the concentrate from the coal waste of the Reichert LD-4 spiral (Table 7) can be attributed to the content both of fine and coarse particles. The latter ones were pushed out to the outer side of the sluice. When comparing particle sizes and density fraction of dense products from this separator with Krebs 2.85 it is evident that better separation efficiency is obtained in spiral separators with higher number of coils.

Table 6. Ash content (λ_{ash}) in particle size fractions of raw coal upgrading in Krebs 2.85 and Reichert LD-4 separators.

Particle size, mm	Ash content λ_{ash} (%) in particle size fractions					
	Krebs 2.85			Reichert LD–4		
	Light product (concentrate)	Semi-product (middlings)	Heavy product (tailings)	Light product (concentrate)	Semi-product (middlings)	Heavy product (tailings)
> 3	6.08	25.46	39.39	17.03	32.96	73.66
3 – 2	7.01	20.72	44.86	8.78	40.93	74.51
2 – 1.5	4.26	14.80	41.82	9.84	43.07	73.63
1.5 – 1	5.33	9.14	36.00	6.49	31.61	68.41
1 – 0.5	3.58	14.96	31.48	4.55	11.69	52.99
< 0.5	14.3 9	6.52	25.77	9.03	9.88	40.33

Table 7. Ash content (λ_{ash}) in particle size fractions of coal waste upgrading in Krebs 2.85 and Reichert LD-4 separators.

Particle size, mm	Ash content λ_{ash} (%) in particle size fractions					
	Krebs 2.85			Reichert LD–4		
	Light product (concentrate)	Semi-product (middlings)	Heavy product (tailings)	Light product (concentrate)	Semi-product (middlings)	Heavy product (tailings)
> 3	19.53	36.85	43.06	20.96	42.25	52.15
3 – 2	18.15	35.21	55.52	19.16	49.33	58.86
2 – 1.5	11.46	29.81	50.70	13.98	42.91	58.19
1.5 – 1	7.45	20.68	46.89	10.32	32.6	55.84
1 – 0.5	6.54	13.52	44.81	8.42	18.64	49.05
< 0.5	18.30	21.30	54.25	34.99	28.39	59.65

Results of separation of sand and magnetite mixture (Table 5) indicate that the most efficient separation was in Reichert LG-7. Highest yield of light product with low content of magnetite and lowest yield of the heavy one with high content of magnetite were obtained in this separator. In the other two separators (Krebs 2.85 and Reichert LD-4) results were not as expected, in both cases the highest content of magnetite was observed in the light product (Table 5). Also the yields of products showed inefficient separation process.

4 Results and conclusions

For the purpose of coal upgrading Reichert LD-4 spiral is the best choice. This is due to the following reasons:

- higher number of coils and height causes longer residual time of particles in the sluice and as a consequence more accurate density partition,
- existence of additional dense product collector which is located in the middle of the separator height and receives tailings separated in the first phase of separation,
- existence of additional water sluice which increases efficiency of separation of a "clean" heavy product (with high ash content) minimizing at the same time its yield. Lack of this sluice in Reichert LG-7 spiral causes contamination of concentrate grains with light product

grains which as a consequence increases ash content in concentrate (table 4),
- Lower sluice slope inclination which causes lower flow rates hence more efficient separation,
- Larger sluice width – the LG-7 separator sluice width is 200 mm and free separation of material is reduced which results in the mixing of products.

One significant observation was made during the course of experiments – the location of "cuts" does not have a major impact on the efficiency of separation. Location of cuts (slots) was almost the same in case of Krebs 2.85 and Reichert LD-4 i.e. the largest gap was for the tailings – 100 mm, smaller for the middlings and concentrate – 80 mm and 85 mm, for Krebs 2.85 the gaps were 80 and 80 mm respectively. Despite of it, analysis of products showed some large differences.

The shape of sluice did not differ significantly in the separators and did not have a substantial impact on the process of separation.

Analysis of separation of model material, that is the mix of sand and magnetite, showed that in this case the existence of additional water sluice does not have an impact on product separation. Best results were obtained in the Reichert LG-7 spiral separator which is not equipped with the additional sluice. In case of Krebs 2.85 and Reichert LD-4 a small amount of magnetite was found in the additional sluice. The number of coils did not have an impact on the efficiency of separation which was proven by the fact that the least favorable results were obtained in the Reichert LD-4 spiral with the same number of coils as the Reichert LG-7. The shape and width of sluices did not have a significant impact on the separation process.

References

1. A. B. Holland–Batt, P.N. Holtham, Miner. Eng., **4**, (1991)
2. R. G. Richards, D. M. MacHunter, P. J. Gates, M. K. Palmer, Miner. Eng., **13**, (2000)
3. Y. M. Stokes, Computing Flow in a Spiral Particle Separator (14th Australasian Fluid Mechanics Conference Adelaide University, Adelaide, 2001)
4. W. Blaschke, S. Blaschke. *Technika wzbogacania grawitacyjnego. Wzbogacalniki strumieniowe* (Wyd. Inst. Gosp. Sur. Min. i Energią PAN, 1999)
5. J. Drzymała. *Podstawy mineralurgii,* (Oficyna Wydawnicza Politechniki Wrocławskiej, Wrocław, 2009)
6. Z. Blaschke, Inżynieria Mineralna, **1(2)**, 33-36, (2000)
7. D. K. Henderson, D. M. MacHunter, Mineral Technologies, (2003)

AUTHOR INDEX

Alakangas Lena 161
Anger Ildiko 114
Bakalarz Alicja 31, 131
Baranek Wit 109
Bodénan Françoise 78
Charan Thonangi Gouri 70
Chattopadhyay U.S. 70
Cichy Krystian 37
Dirlich Stefan 78
Drzazga Michal 109
Drzymala Jan 43
Duchnowska Magdalena 31, 131
Eslamian Atiyeh 121
Fatyga Miroslaw 143
Filcenco-Olteanu Antoneta 161, 177
Fiuza Antoio 161
Foszcz Dariusz 64
Gawenda Tomasz 64
Gherghe Sandica Liliana 104, 114
Gliniak Maciej 169
Grabowski Lukasz 169
Graczykowski Cezary 57
Grotowski Andrzej 15, 78
Guezennec Anne-Gwénaëlle 78
Jha G.S. 70
Jura Jakub 152
Kolacz Jacek 1
Konowrocki Robert 47
Kopyto Dorota 109
Krawczykowska Aldona 143
Krawczykowski Damian 64
Krzeminska Malgorzata 5
Kubacz Norbert 8
Leszczynska-Sejda Katarzyna 109
Luszczkiewicz Andrzej 31, 131
Lutynski Marcin 73, 188
Manka Arkadiusz 37, 143
Marijnissen Michal J. 57
Matusewicz Maciej 109
Matusewicz Tadeusz 109
Matusewicz Wojciech 109
MehdiJebraeeli Seyed 121
MichajLow Maciej 47
Mlynarczykowska Anna 27
Moise Georgiana 114
Myczkowski Zbigniew 109
Niedoba Tomasz 21
Oleksik Konrad 27
Pajor Grazyna 37, 143
Palimaka Piotr 143
Panturu Eugenia 177
Pawlos Witold 5
Pawlowska Agnieszka 78
Pierzyna Piotr 73, 99, 171
Pieta Paulina 21
Pietrzyk Stanislaw 143
Pilsniak-Rabiega Magdalena 94
Pochanke Andrzej 47
Polek Daria 139, 169
Pomykala Radoslaw 125
Popczyk Marcin 99, 171
Radu Aura Daniela 177
Radzyminska-Lenarcik Elzbieta 84, 89
Ratajczak Tomasz 149
ReguLa Cezary 37
Rezaee Bahram 121
Rojek Jerzy 57
Sadowski Zygmunt 78
Saramak Daniel 27, 64
Sarvi Atefe 121
Sen Kalyan 70
Sinha K.M.K. 70
Skorupska Bozena 8
Stepien Michal 143
Suponik Tomasz 99, 171, 188
Surowiak Agnieszka 21, 155
Szolc Tomasz 47
Szpyrka Jan 188
Szubert Agnieszka 78
Tobiczyk Kornel 37
Tomus Nicolae 161
Tora Barbara 125, 143
Trusca Roxana 114
Ulewicz Malgorzata 152
Wejman Katarzyna 94
Wieniewski Andrzej 8
Witecki Kajetan 15, 78
Witt Katarzyna 84, 89
Wolosiewicz-Glab Marta 64
Zabihi Abolfazl 121
Zlagnean Marius 161, 177